KT-483-959

Catalogue of the Universe

Catalogue of the Universe

by Paul Murdin
and David Allen

with original
photographs by
David Malin

Cambridge University Press
Cambridge London Melbourne

Published by the Syndics of the
Cambridge University Press
The Pitt Building, Trumpington Street,
Cambridge CB2 1RP
Bentley House, 200 Euston Road,
London NW1 2DB
296 Beaconsfield Parade, Middle Park,
Melbourne 3206, Australia

Created, designed and produced by Reference International Publishers Limited.

First published 1979.

Printed in the United States.

ISBN 0 521 22859 X

Contents

Paul Murdin
Paul Murdin is an astronomer at the Royal Greenwich Observatory, formerly on leave at the Anglo-Australian Observatory near Sydney, Australia. His special interests in astronomy have been relating the discoveries of x-ray astronomy to the optical universe. He has played a key role in identifying Cygnus X-1 as a black hole and the Vela pulsar as a flashing star, one of the faintest stars known. He has written dozens of technical papers, and four popular books which bring astronomy into perspective for the layman.

David Allen
Born 1946 in Cheshire, David Allen has two degrees from the University of Cambridge. He was a Carnegie Fellow at the Hale Observatories, Pasadena, Ca. After three years at the Royal Greenwich Observatory he joined the Anglo-Australian Observatory in 1975. He is the author of *Infrared—the new astronomy*, written for a wide audience, as well as dozens of technical papers on many fields of astronomy.

David Malin
David Malin began his career in chemical analysis using optical and electron microscopy and x-ray diffraction techniques, which he was later to put to use in astronomy. He joined the Anglo-Australian Observatory as its research photographer in 1975. In astronomical photography the principles are: get the photographs in as much detail as possible, as fast as possible; and get as much information off the photo as there is. Making original contributions in all these areas, David Malin has particularly exploited the "unsharp masking technique" to bring out the hidden detail in photos of nebulae and galaxies. These pictures appear in this book for the first time outside the specialist journals.

Preface

Most books on astronomy, including those claimed to be Atlases, introduce basic principles and illustrate them with examples. In this book we are trying to do what an Atlas should do, which is the reverse approach: to illustrate a comprehensive selection of objects in the Universe and to write about the examples chosen, bringing the basic principles out of the objects themselves. This is the way research is done: studying particular objects teaches astronomers astronomy, and we hope to show how.

We have chosen a wide range of those celestial objects which can be pictured—galaxies, stars, nebulae, the planets—and in the captions we give some of the background history of the object and say why we find it interesting. We are courageous (or foolhardy) enough to put forward what we think is a "best-buy" theory for the objects which we illustrate. We hope that an astronomer browsing in the sections which lie outside his special field thinks that the best-buy choices are justified, even if he happens to disagree with us in the area he knows about.

We have tried to approach completeness in the range of cosmic phenomena we write about, and this means that we have used some diagrams to show concepts which can't be shown photographically. But the photographs are the core of the book, because they are closest to the objects themselves. Even a jaded professional astronomer in the prime focus cage of a big telescope, with his back to the stars, and facing a reflection of the Universe in a mirror, feels a thrill from knowing that real photons from distant galaxies are exposing the photograph he is taking.

About a third of the photographs have been specially prepared for this book. We have drawn on the plate files of two new observatories in Australia because the project started while we were all three on the staff of the Anglo-Australian Observatory. This explains the emphasis in our section on southern celestial objects, though we think we are correcting the previous northern hemisphere bias in popular astronomy books, rather than contributing to hemispheric chauvinism ourselves. The photographs we prepared ourselves have been made from originals taken by many colleagues at the two observatories, the U.K. Schmidt Telescope Unit of the Royal Observatory, Edinburgh, and the Anglo-Australian Observatory. Individual thanks are given in the Acknowledgements, which identify the initials and full names of the Observatories to which each picture is credited (in the lower right hand corner). The reader will notice that the AAT and UKSTU black and white pictures in this book are not starkly black and white, as pictures usually appear in astronomy books. In many of our pictures the tonal range is compressed by the soft or unsharp masking technique explained, for the photographically inclined, at the end of this book. In others, the tonal range is deliberately restricted so that the sky appears gray. Both methods make visible significantly more detail than the hard contrast of stars against a jet-black sky. In fact, the sky is not black and even on the darkest nights it shines with the feeble light of distant stars and the soft glow of faint aurorae. In addition to revealing more, the pictures you will see here are thus closer to the real appearance of the sky.

We are grateful to Robin Scagell and Martin Self, our two editors, for all the effort which they have put into this book.

How objects are catalogued

The sky contains a large number of nebulae—objects which at first sight in a telescope appear as vague misty patches. (The word *nebula*—plural nebulae, pronounced "nebulee"—is Latin for "cloud.") Quite often what in a small telescope looks like a nebula turns out on closer inspection to be a closely packed star cluster. Many catalogues of nebulae, therefore, contain all types of object mixed together as they appear in the sky.

Two catalogues of nebulae, galaxies and clusters are in widespread use. The first was compiled by the French comet hunter Charles Messier between 1771 and 1784 and in its present form comprises 109 entries. The catalogue contains the brightest nebulae, galaxies and star clusters visible from France and is especially widely used by amateur astronomers. Objects in it are referred to by Messier's initial and a serial number. Thus in this catalogue the well-known Crab Nebula is called M 1, and the Pleiades star cluster (the Seven Sisters) is M 45.

The second catalogue is a compilation of all nebulae known to the Danish astronomer Johan Dreyer in 1888. The majority of the objects were discovered earlier by Sir William Herschel and his son Sir John in the course of their systematic surveys of the whole sky, including the parts of the southern hemisphere hidden from northern latitudes. Dreyer's compilation is called the *New General Catalogue*, or NGC, and lists 7840 objects. Thus the Crab Nebula is also known as NGC 1952.

A further 5086 objects were added to the NGC in two supplements called the *Index Catalogue*, or IC.

Astronomers have since compiled many other specialist catalogues. Usually the objects are referred to by the compiler's name or initial and a serial number. Another way to describe objects is by reference to the constellation, or large-scale star pattern, in which they reside. The naked-eye constellation patterns are often of great antiquity, yet they are still used as a convenient guide to parts of the sky. Thus we get Centaurus A, the first radio source to be discovered in the southern hemisphere constellation of Centaurus.

Finally, large numbers of especially famous galaxies, nebulae and clusters are known by individual names, often derived from their appearance and their more or less fanciful resemblance to terrestrial objects.

Units of measurement

The following measurements are used in this book.

Number
1 million = 1,000,000
1 billion = 1,000 million = 1,000,000,000
1 trillion = 1 million million = 1,000,000,000,000

Distance
1 meter = 39 inches (approx)
1 millimeter = 1/1000 meter = 1/25 inch (approx)
1 micron = 1/1000 millimeter = 1/25 of one thousandth of an inch (approx)
1 kilometer (km) = 1000 meters = 3300 feet (approx) = 5/8 mile (approx)
1 light-year (l.y.) = distance traveled by light in a year = 9.4 trillion km (approx)

Speed
1 kilometer per second (km/sec) = 2 250 miles per hour (approx)
speed of light = 300,000 km/sec (approx) = 186,000 miles/sec (approx)

Temperature
Freezing point of water = 0°C = 32° Fahrenheit = 273 Kelvin
Boiling point of water = 100°C = 212° F = 373 K
Absolute zero = 0 K = −273°C

1. The Universe of Galaxies

This is a photograph of a small patch of sky. If you were to stretch out your hand at arm's length, a finger would just about cover the actual area of sky shown.

In this patch of sky, the unaided human eye would see nothing, for none of the stars is bright enough to be seen. But by using a telescope as a giant camera, and keeping the shutter open for a long time to record faint images, many stars and astronomical objects become visible.

Although this is only a tiny part of the sky, it is fairly representative of the sky as a whole. It is, however, the only photograph in this book which has been chosen to show nothing in particular. So what does it reveal? Even a glance shows that there are two types of object. The circular, compact dots are stars while the fuzzier blobs which are not always circular are the much more distant galaxies.

The star images vary in size, some being as much as a few millimeters across. But these images do not represent the true sizes of the stars. The spreading of light in the photographic emulsion and in the telescope, and the turbulence of the Earth's atmosphere through which the photograph was taken, combine to enlarge the star images. If these factors could be bypassed, the diameters of the star images would be 10,000 times less.

The slender spikes on the brighter star images are also an artefact of the telescope. They are caused by the spreading of the starlight round the struts for the assembly which holds the photographic plate, and are called diffraction spikes. There are diffraction spikes around the fainter star images, too, but these are too faint to have been recorded.

Despite their faintness and tiny apparent size, the stars on this photograph are all objects very like the Sun. Indeed, many are actually larger and brighter than our Sun. It is only because the Sun is so close to us that it is so prominent in our lives. The stars on this photograph are tens of millions of times farther away than the Sun.

The galaxies are yet more remote by a similar factor. Knowing that they are much more distant, yet that they appear only a little fainter than most of the stars, indicates that galaxies are intrinsically much brighter objects than stars. In general, galaxies achieve this enormous luminosity not by being larger versions of stars, but by containing hundreds of millions of stars. The individual stars in a galaxy are widely separated, so even at their extraordinary remoteness from us, the galaxies have appreciable size. They can therefore be distinguished from nearby stars.

A part of the sky UKSTU

To write the distances of these galaxies in miles or kilometers would need 23 digits. It is more convenient to use light years, which are based on the travel time of light. One light year is the distance light travels in one year, about ten million million kilometers. Galaxies can be photographed several billion light years away. By comparison the Sun is only eight light minutes away, and the Moon a little over a light second away.

The light travel time bears another, related, piece of information—how long ago the light we are recording began its journey. As we examine objects at great distances, we are obliged to view them in the past. Many of the galaxies in this photograph are seen as they were a billion years ago; if we want to know how they look now, we must wait a billion years for their light to reach us!

Stars exist only in galaxies. Although galaxies appear numerous on this photograph, they only occupy about a hundred-millionth of the vast volume of space of the known Universe. The view of the sky from a typical point in space is empty of individual stars, and usually only distant galaxies can be seen. But our own sky is full of stars, as shown on this photograph, which indicates that we live in a privileged place in the Universe. Far from being in a random part of empty space, our environment is atypical of the Universe.

This is not surprising, since without the Sun there would be no life and the Sun, being a star, lies in a galaxy. From our special position not only are we able to look out on the multitude of galaxies, many like our own, and learn of their nature and variety, but at the same time we can look into our Galaxy and discover an even greater variety of objects. The pages of this book are a gallery of what we find.

Clusters of Galaxies

Signposts to the expanding Universe

Galaxies rarely exist in isolation. They are mostly found in groups which may number from two or three to 1000 or more. These photographs illustrate the different types of large clusters of galaxies, from sparse (few members) to rich (many members), and from near (galaxies appear large) to far (galaxies appear small).

As in all photographs of galaxies in this book, foreground stars of our own Galaxy confuse these pictures. The brighter stars have diffraction spikes, and faint stars produce circular dots. The cluster galaxies are all fuzzy, and most are elongated.

In all clusters, the member galaxies are distributed with no more pattern than a tendency to congregate towards the center. What holds them together? The answer is gravity—each galaxy is a captive of the gravitational pull of all its neighbors. Just as the Earth's gravity holds the Moon in orbit, so the cluster members' joint gravity holds the cluster members together. The tug of gravity acts as if all the cluster's mass were concentrated at the very center, even if there happens to be no object there. Each cluster galaxy will travel around and through the cluster, probably for its entire life, locked in the gravitational embrace of its fellows, and at the same time adding its weight to the force which holds them in place. This is just one example which demonstrates the overwhelming importance of the force of gravity in astronomy.

Large clusters of galaxies measure many tens of millions of light years across. Yet even these may not

Cluster of galaxies POSS

Abell 1146 AAT

Abell 1060 AAT

be the largest organized structures in the Universe. The sketchy data which reaches us from the most distant parts of the Universe suggests that clusters of galaxies themselves group into giant superclusters, whose scale is greater by yet another factor of 100 or more.

The name of the American astronomer Edwin P. Hubble is inescapably associated with the early study of galaxies. In the 1920's Hubble measured the velocities of galaxies in a number of clusters and found that the faint clusters are receding from us at great speed. Taking the brightest galaxy in a cluster, the fainter it is, the faster it is moving away. Because the brightest galaxy in a cluster is always around the same true brightness, its faintness indicates how far away it is. So the more distant a cluster of galaxies, it seems, the greater its speed of recession from us. This relationship is called *Hubble's Law*.

Thus Hubble showed in 1929 that the Universe is expanding. The most distant galaxies have the greatest velocities and have therefore receded farthest from us. Put another way, if we could run time backwards very fast, we would see the galaxies converging on a small region, and all arriving there at much the same time. But we cannot tell when precisely they would converge, or where the convergent point would be.

Today, the majority of astronomers accept the Big

Centaurus cluster AAT

Fornax cluster UKSTU

Bang Theory of the origin of the Universe. This suggests that the Universe we now see arose from the explosion of some indescribably massive but relatively small object. The energy of that explosion is inconceivable, for observations show that as we probe deeper and deeper into space, with improved techniques, objects still continue to recede increasingly fast, at velocities which are appreciable fractions of the Universe's speed limit, that of light (300,000 km/second). Furthermore, the galaxies do not seem to be thinning out, even at vast distances.

Yet every time Hubble's expansion law is confirmed to continue to twice the previously known distance, the energy required in the Big Bang is increased thirty-two fold—and by an even larger factor as the speed of light is approached.

Hubble's Law is the only tool astronomers can use to estimate distances in the farthest reaches of the Universe. A measurement of the recession velocity (the so-called redshift) implies the distance to an object. Nearby galaxies act as standards, since their distances can be worked out in other ways, but this is a difficult business, and the calibration may be considerably in error. The currently accepted value for the expansion rate—it is known as the Hubble Constant—is 15 km/sec for every million light years

Virgo cluster UKSTU

distance, and this is used in this book. The most distant cluster of galaxies on these photographs is receding from us at about 60,000 km/sec, implying a distance of four billion light years.

Why does material collect into clusters of galaxies, stars and planets? Once again, gravity is the important force. The expanding Universe would not have been entirely uniform. Patches denser than the rest must have existed from the beginning. Having a greater concentration of mass, the gravitational pull of these denser patches attracted gas from less dense regions. Each became the seed of a growing concentration. The collapsing clouds themselves contained smaller denser patches which became the nuclei of smaller collapses, and so on. Today we view this process at a stage when much of the gas in the Universe has collapsed to form stars and planets, lying within galaxies, clusters and maybe even superclusters of galaxies.

The Coma Cluster

An x-ray cluster

Much of the constellation of Coma Berenices is covered by a large cluster of galaxies, 350 million light years distant and 16 million light years across. This photograph reproduces the central regions of that cluster. Here you can see more galaxies than foreground stars of our own Galaxy, and virtually every galaxy is a member of the cluster. The Coma cluster is of particular interest to astronomers because it emits x-rays.

Clusters of galaxies are not static phenomena; their member galaxies must be in continual motion to counteract the gravitational force which is trying to pull them together. Astronomers have measured the speeds of the Coma cluster galaxies and find them to be surprisingly high. Rapid motion means that the galaxies are being acted upon by a strong gravitational pull, and gravity implies mass. The mass of the cluster is calculated to be seven times the expected total mass of all the visible galaxies. Some of the missing mass must be present in the form of invisible gas between the galaxies. Until recently there were few clues to the presence of this gas, but in 1971 the discovery of x-rays from the Coma cluster provided evidence of its existence.

As the galaxies move through the cluster gas, they heat it by friction. The stars of each galaxy produce no frictional drag — they pass easily through the gas. But each galaxy also contains its own quota of gas. Gas is sticky stuff; the galaxy's gas and the cluster gas rub together, and so are heated. Because the galaxies move very rapidly — at velocities up to 1000 km every *second* — the frictional heating is great. The gas is raised to a temperature of a few million degrees.

A law of physics is that objects of different temperature appear to have different colors. Cool objects are red and hot objects blue. The different colors are light of different wavelengths: red light has longer wavelengths than blue. Later in this book we will describe objects so cool that they radiate their energy beyond the red, in the infrared to which the eye is insensitive. Now we are concerned with very hot gas. So hot, indeed, that its energy is released beyond the blue, beyond

Coma cluster KPNO

even the ultraviolet—as x-rays.

X-radiation does not penetrate the Earth's atmosphere. If it did, life as we know it would not have developed. X-ray astronomers therefore ply their trade from satellites orbiting above the atmosphere. The Coma cluster is the brightest of a score of clusters already discovered to emit x-rays.

Shakhbazyan 1

The most compact cluster known

In 1957 the Armenian astronomer Shakhbazyan discovered a group of compact galaxies clustered tightly in a portion of the constellation Ursa Major. It was 16 years before Lick Observatory astronomers Lloyd Robinson and Joe Wampler decided to examine more closely Shakhbazyan's group. They confirmed that all the galaxies lie at the same distance from us, and they found two surprising facts.

The first surprise was the distance. Shakhbazyan 1 is more than two billion light years from us and is therefore one of the most distant clusters studied. From its distance, Robinson and Wampler were able to determine the size of the cluster as 500,000 by one million light years, the average member galaxy being 60,000 light years in diameter. The brightness of the galaxies was also unexpectedly high: many emit more than 100 billion times as much light as our Sun.

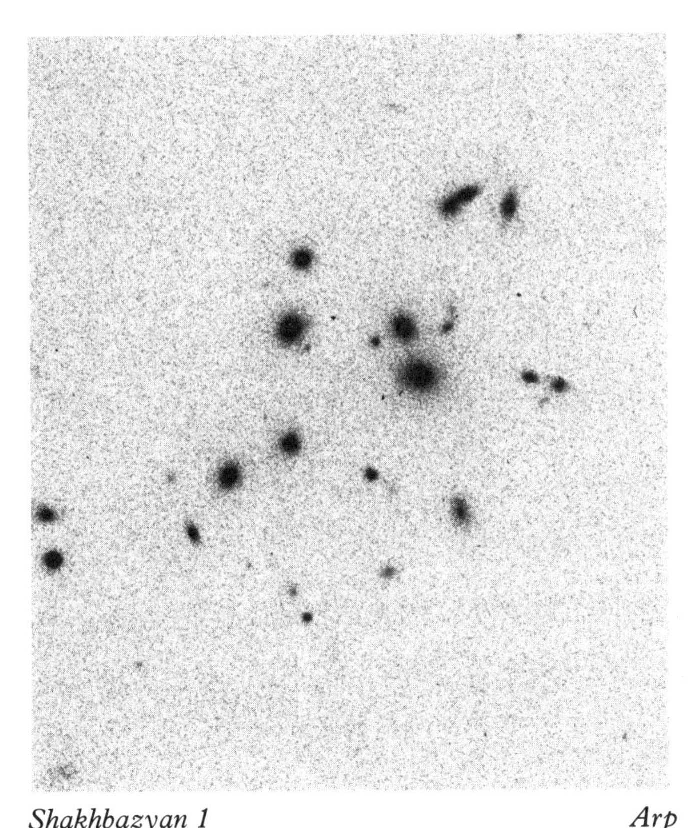

Shakhbazyan 1 *Arp*

As a second surprise, Robinson and Wampler found that the galaxies are all moving very slowly within the cluster. This indicates that the gravitational pull of each galaxy on its neighbors is small, and hence that the mass of the cluster is small. If the galaxies were more massive, they would have to move faster in order to avoid falling into one another. The total mass of the cluster came out at under 1000 billion times that of the Sun. This sounds huge, yet there are single galaxies closer to us in the Universe which weigh more than that. Because the individual velocities of the galaxies are small, the cluster has not spread out, hence its compact nature.

The galaxies of Shakhbazyan 1 are, for their mass, very efficient at giving out light. They must contain almost exclusively stars, and this means that they are old. Were they young, there would still be considerable amounts of gas within then, which would contribute to their mass but not to their brightness. Bearing in mind that we are seeing the cluster as it was two billion years ago, and that the galaxies were then old, it appears that Shakhbazyan 1 was one of the first clusters of galaxies to have formed in the Universe.

Normal Galaxies

Collections of stars

On a photograph of a rich cluster of galaxies we notice the collection of galaxies rather than the galaxies themselves. The scale of the photographs of clusters shown so far is too small to reveal the intricate nature of the individual galaxies. Now, however, we can look at some closer galaxies which show more detail.

Individual galaxies, whether isolated examples or members of clusters or groups, occur in a variety of forms. Like dogs, galaxies are immediately recognizable, but may be large or small, long and thin or short and fat, smooth or hairy. The common feature is their nature: all galaxies are giant agglomerates of stars relatively isolated in space from their neighbors. The average galaxy, if such a thing can be defined any better than the average dog, contains about a billion stars, together with a small mass of gas which may eventually form more stars. The ranges are extreme, however, for galaxies are known which are more than 1000 times larger and smaller than this figure.

What determines the size and nature of a galaxy? Current theories maintain that the Universe began as a cloud of gas expanding from some giant explosion. Gas in motion, like water in a stream, is turbulent and tends to fragment into eddies. In so gigantic a stream as the Universe, these eddies are vast.

Individual eddies themselves break up into smaller

NGC 4552 – elliptical galaxy  UKSTU

NGC 2997 – spiral galaxy  AAT

NGC 3109 – irregular galaxy ESO

whorls under the influence of gravity. Locally dense blobs of gas attract the atoms and molecules from more tenuous regions, and so form the nuclei of collapsing clouds of gas. The laws of physics prohibit a single cloud the size of a galaxy from collapsing indefinitely, but instead limit it to a few tens of thousands of light years diameter. This is indeed the size of most galaxies. Very young galaxies are believed to be clouds of gas—mostly hydrogen—of about this dimension. Such clouds would be very hard to detect, and there is disagreement about whether examples have been found.

Having attained the size of a galaxy, the cloud is not restricted from collapsing further under the strengthening influence of its own gravity. To do so, however, it must subdivide into smaller units. These units we know as stars. Stars are, in fact, the most fundamental units of the astronomers' Universe as they are the building blocks from which galaxies are made, and are easy to observe in their own right. It is the presence of large numbers of stars which makes galaxies visible. While the shape and size of galaxies is dictated by the collapse of a gas cloud, what we see is the light from millions of stars, unresolved from one another in all but the nearest galaxies. These stars merely map where gas once lay.

Elliptical Galaxies

Celestial footballs

Of all the varieties of galaxies, those described as elliptical are the least attractive to the eye. As the five photographs (p. 15-16) show, they are rounded objects, centrally condensed, and ranging in apparent shape from as round as an English soccer football to as elliptical—that is, flattened—as its American equivalent. Like footballs, they are not flat, but have depth. Unlike a football, however, a galaxy cannot be picked

NGC 4374 UKSTU

NGC 4459 UKSTU

up and examined, but must be viewed from a great distance and from only one side. Therefore astronomers cannot be sure whether a galaxy like NGC 4564 (page 16) is cigar-shaped, or is a disk of stars seen almost edge-on. The measurements needed to distinguish between these two possibilities are difficult to make and few have been made. However, astronomers expect cigar-shaped galaxies to be rare. Equally, galaxies which appear circular may be spherical, or cigar-shaped objects seen end-on, or disk galaxies seen from above. The proportion of galaxies which appear circular is sufficiently large to convince astronomers that many genuinely spherical galaxies do exist. But in very few cases can astronomers state whether a particular galaxy is spherical or not.

Elliptical galaxies range greatly in size. The largest galaxies in clusters are usually elliptical, and so are most of the smallest. There seems no tendency for small galaxies to be any more or less circular than large ones, but the very largest cluster galaxies, called cD galaxies, are usually of intermediate ellipticity, with a noticeable, but not pronounced, flattening. An example of a cD galaxy lies near the center of Abell 1146, shown on page 10.

Elliptical galaxies contain no young stars, since these are born only where there is gas enough to build

NGC 4406 UKSTU

NGC 4473 UKSTU

NGC 4564 UKSTU

Messier 104 KPNO

them and elliptical galaxies contain little or no gas. So most of the stars in elliptical galaxies are old and red, and the galaxies themselves appear red. Among galaxies redness is a sign of senility. It used to be thought that elliptical galaxies had used up all their gas making stars, and this may be true of the isolated examples that travel alone through the vastness of the Universe. However, it has recently been realized that a premature aging machanism exists in clusters. As described in the case of the COMA CLUSTER (p. 12), the gas content of galaxies can be swept out by frequent encounters with other cluster members. The effect of this stripping of gas is not only the generation of x-rays; but the removal of gas from a galaxy prevents it from rejuvenating its population of stars. Each pass through a cluster takes eons off a galaxy's life expectancy. This explains the observation that elliptical galaxies, with only old stars, are more likely to be members of groups or clusters, than isolated galaxies.

Spiral Galaxies

Celestial frisbees

Spiral galaxies are dashing young beauties beside their more commonplace elliptical neighbors. They make popular photographers' models, and certainly merit several pages in this gallery. The collection on this page (above) and opposite serves as an introduction to their style and variety.

The distinctive aspect of spiral galaxies, which gives them their title, is the set of arms, usually two in number, which wrap around their centers. These normally start outside a central, circular region called the bulge or nucleus, and wind out to several times its diameter. The arms lie in a flat disk, so that a galaxy would appear very different viewed face on and in profile. The sky is full of examples of spiral galaxies seen from all angles.

The spiral arms contain young, hot stars, glowing clouds of gas, and dark streaks of dust. When a spiral galaxy is viewed edge-on the dust clouds obscure the stars which lie behind, thus producing a series of dark blobs or, in some cases, a continuous dark line crossing the galaxy.

Hot stars are blue. The arms of spiral galaxies therefore appear blue, in contrast to the predominantly orange hue of the nuclei. During the 1940's, this clear distinction led the German-American astronomer Walter Baade to identify two families of stars which he called Population I and Population II. Stars in the arms of spiral galaxies, and also in the young, irregular galaxies to be pictured later, are Population I. Population II stars, however, are found in elliptical galaxies, and in the nuclear bulges of spiral galaxies. This hints that spiral galaxies are simply elliptical galaxies wrapped up in blue arms.

Indeed, inside most spiral galaxies lurk elliptical cores trying to take control. As long as gas remains in the spiral arms, young blue stars will continue to form there. But as that gas becomes exhausted, particularly by passage through galaxy clusters, spiral galaxies lose their distinctive shape and may evolve to resemble elliptical galaxies.

IC 5332 UKSTU

NGC 7793 AAT

NGC 4945 AAT

Each spiral galaxy rotates about an axis which is perpendicular to the galaxy's disk. For many years, astronomers could not be sure whether spiral arms rotate in advance of the rotation of the nucleus, or behind it. But measurements show that the inner portions are rotating faster than the outer and so coil more around the nucleus. Therefore the arms trail the rotation of the galaxy.

Spiral galaxies display a range of styles, from those with very prominent nuclei and weakly developed arms to examples in which the nucleus is scarcely seen and the arms are highly developed. The array of types can be seen in this and subsequent pages.

Our own Galaxy, which we see from within as the Milky Way, is a spiral galaxy. Although no one can truly comprehend the scale of our Galaxy, or any other, it is worth bearing in mind its size. It is 100,000 light years in diameter, contains as much mass as a trillion Sun-sized stars, and shines as brightly as 100 billion suns.

Messier 81

A radio view

Plate 1 shows a map of the spiral galaxy Messier 81. It was made by a radio telescope tuned to the wavelength

Messier 81 KPNO

emitted by hydrogen gas, 21 cm. Thus the view shows where the gas of the galaxy lies, and it can be compared to the optical photograph above, which shows where the stars lie. At the center of the galaxy (marked in Plate 1 by a cross) all the gas has been used up in making stars. The radio picture is dark there, whereas the optical picture is bright. The reverse is true in the spiral arms where there remains a great quantity of gas, and the radio picture is bright while the optical arms are thin and delicate.

It is much easier to measure the velocity of gas by radio techniques than to measure the velocities of the stars optically. The map on Plate 1 has been color-coded according the the velocity of the gas. Areas shown red are receding from us faster than the average; areas in blue are receding from us more slowly. We see at a glance that the upper edge of the galaxy is receding faster than the lower edge. This does not mean that the galaxy is tearing apart—rather, it is rotating. The right-hand side lies nearer to us than the left. This can be seen from the photograph because dark clouds in the spiral arms are silhouetted against the nucleus on the right, but lie behind the nucleus on the left and do not show as boldly. Therefore the galaxy rotates counter-clockwise as we view it, with the spiral arms trailing.

Barred Spiral Galaxies

Siamese-twin galaxies

The galaxies illustrated on the right are obviously spiral, but they are distinguished from other members

NGC 2442 AAT

of their kind by the structure of their nuclei. Instead of having rounded, ellipsoidal nuclei, the central portions are long and narrow, and are described as bars. The

NGC 6384 AAT

spiral arms generally start at the ends of the bar.

One feature of a barred galaxy is that there is comparatively little mass at its center. If that were the case, then the bar would wind up just as the spiral arms do. Instead, the mass must tend to reside at the ends of the bar, somewhat like a dumbbell.

It is by no means abnormal for collapsing gas clouds, such as those which form galaxies, to develop two centers which coexist in celestial harmony. We see obvious examples of this in the case of double stars, and even double clusters such as h AND CHI PERSEI (p. 122). Barred spiral galaxies are probably another manifestation of the same phenomenon.

If we think of spiral galaxies as being ellipticals which have wound up into spiral arms, barred spirals would be an indication that cigar-shaped elliptical galaxies can exist. When spiral galaxies are seen nearly edge-on, it is difficult to be sure whether their nuclei are barred or circular. Of the nearly face-on examples which have been reliably classified, almost half are barred spirals. Thus barred spiral galaxies are as common as ordinary spiral galaxies. This is no more surprising than the fact that double star systems are about as common as single stars (see p. 81).

Messier 83

Face-on spiral

The color version of this photograph (Plate 1) clearly shows the difference between the stars in the nucleus (red-orange) and the stars of the spiral arms (blue and patchy). M 83 is a spiral galaxy, also known as NGC 5236. It lies about 27 million light years from us and is receding from us at a little over 500 km/sec. The visible galaxy appears flat, and we view it from nearly overhead. Radio maps of M 83 show that its gas extends beyond the visible spiral arms, and that it is not quite flat: one side bends up towards us and the other side curls away from us like the brim of an Australian soldier's hat.

M 83 weighs nearly one trillion times as much as our Sun, and its stars combine to shine 100 billion times as brightly—which is about the same as our own Milky Way galaxy. A large proportion of the stars in M 83 are of the young, hot Population I variety that lie in the spiral arms. The brighter ones among them die by exploding as supernovae. In our own Galaxy such an explosion occurs about once in 30 to 50 years but, in M83, supernovae occur every 10 or 15 years, more frequently than in any other galaxy.

NGC 4548 UKSTU

Messier 83 AAT

Messier 101

A giant spiral galaxy

Messier 101 is a typical, large spiral galaxy. Since it is nearby astronomically speaking, we see it in great detail; we view it almost face-on, so we are able to study its form and structure particularly well. M 101 has a small nuclear bulge and particularly well-developed spiral arms. Its popular name is the Pinwheel Galaxy.

Most descriptions of M 101 quote a distance of 15 million light years, but this may be considerably in error. Although astronomers can determine quite reliably the relative distances of galaxies, the numerical values of these distances are still uncertain to within a factor of two. Allan Sandage of the Hale Observatories has been particularly active in the attempt to determine more precisely the scale of the Universe, and he has derived a distance of 24 million light years for M 101.

The galaxy has a diameter of nearly 0.5°—the apparent size of the Moon. Using Sandage's distance, this converts to a linear diameter of almost 200,000 light years, which means that M 101 is one of the largest spiral galaxies known. As might be expected, so large a galaxy is also very weighty, with an estimated mass of $3x10^{11}$ times that of the Sun.

About 7% of the galaxy's mass is accounted for by hydrogen gas which has not yet been used in the manufacture of stars. The hydrogen lies mostly in the spiral arms, and congregates into clumps weighing a few tens of millions times as much as the Sun. Within each clump is a bright nebula, where hot, young stars have formed and now cause the gas to glow. The nebulae, some of which are bright enough to be seen in amateur telescopes, help to give rise to the blotchy appearance of the spiral arms seen clearly on this photograph.

M 101 is the most prominent member of a rather sparsely populated group of galaxies whose members span many degrees of sky.

Irregular Galaxies

Misfits of the world of galaxies

When Man views the immensity of the Universe, he is awed. In an attempt to ease the nature of the problem, he endeavors to define a small number of labels which he can attach to a large number of objects. The Universe contains more galaxies than a man could count in his lifetime, but it is reduced to manageable proportions if he can make such (untrue) statements as "30% of galaxies are elliptical and 70% are spiral."

Messier 101 KPNO

NGC 1313 AAT

Alas, Nature is not so simple. Try as he might to attach labels to the natural world, Man is forever confronted by the infinite diversity of his environment, and is doomed to fail. Throughout this book, many of the distinctive and unusual celestial objects will find the limelight, for by a study of the abnormal we learn better the nature of normality.

The simple statement made in the first paragraph is not valid, for spiral and elliptical galaxies do not account for 100% of all galaxies. After picking out the spirals and ellipticals, we are left with a varied assortment of individuals. The photographs shown here illustrate some of them. All are different. Galaxies they may be, for they have the right dimensions and comprise stars and gas, but the only word which properly describes them is *irregular*.

Since they are individuals, it is difficult to make valid generalizations about irregular galaxies. However, guarded statements can be made. Irregular galaxies usually contain very many hot, blue stars—Population I objects. They also contain proportionately more gas than ever spiral galaxies. These facts indicate that irregulars are young, for Population I stars can have formed only recently, and gas still remains because stars have yet to form from it. Irregularity of form might also be interpreted as evidence of gangling youth: only in middle age might a galaxy develop a regular structure.

Large irregular galaxies are rather rare. Some of the galaxies on this page are small, nearby objects in which the brightest stars may be seen as individuals on photographs. These are often further distinguished by a second label—dwarf irregular galaxies.

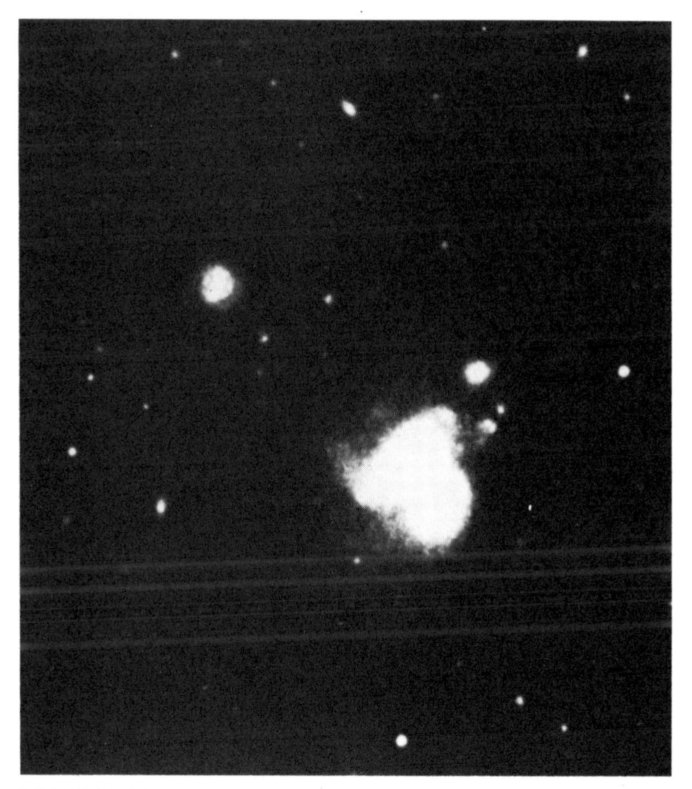

NGC 3690 POSS

NGC 520 POSS

The Sculptor Dwarf Irregular Galaxy

Smallest member of the Sculptor Group

The very name of the Sculptor dwarf irregular galaxy betrays its nature. It is an unpretentious conglomeration of stars lying in the southern hemisphere constellation Sculptor, and was discovered only in 1976. Its discovery resulted from the photographic survey of the southern sky being undertaken with new, large Schmidt-type survey telescopes at the European Southern Observatory in Chile and the U.K. Schmidt Telescope Unit in Australia. Astronomers working at both telescopes found it independently.

Our illustration is a negative picture in which the stars show as black dots rather than as white. Astronomers have found that it is easier to see faint stars on negative prints than on positive prints and we will use this method to show especially dim objects.

As the word dwarf implies, the galaxy is small, and it contains few stars. It might even be thought possible to count the stars in the galaxy, but this photograph reveals only the very brightest stars. There may be many millions of fainter stars which are unrecorded. The dimensions of the Sculptor dwarf galaxy are 2800 by 3700 light years. Its brightest stars outshine our Sun by a factor of about 40,000 and the whole galaxy emits nearly one million times as much light as the Sun.

This is a young galaxy which has formed, possibly less than 100 million years ago, from a relatively small cloud of gas. Much of the gas remains, having so far

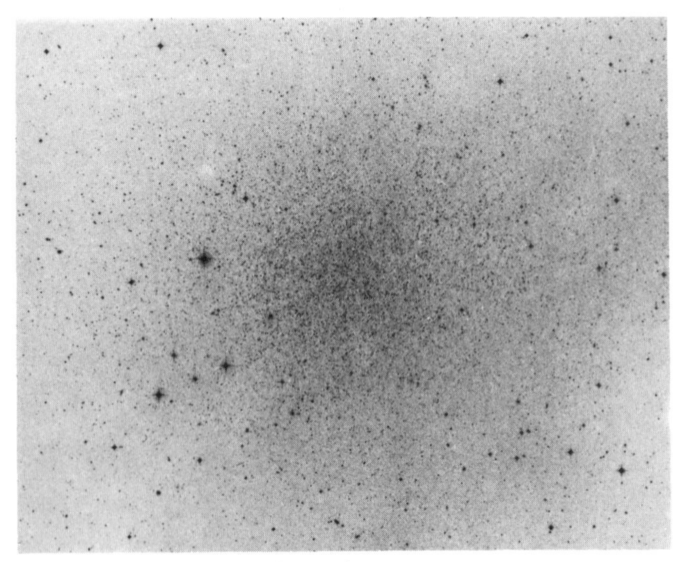

Sculptor dwarf irregular galaxy UKSTU

failed to condense to form stars. Radio astronomers can detect the hydrogen gas by its characteristic emission at a wavelength of 21 cm. They have determined that the mass of unused gas is some 10 million times the mass of the Sun, which accounts for one-third the total mass of the galaxy.

The constellation of Sculptor contains some large, bright galaxies, of which the most impressive is NGC 253, illustrated on p. 38. These bright galaxies are traveling together as a sparse cluster, and indeed form the nearest group of galaxies to our own. The newly-discovered Sculptor dwarf irregular galaxy also appears to be a member of the Sculptor Group.

NGC 1533

A galaxy larger than it looks

NGC 1533 is a lenticular galaxy. At first sight one might think that NGC 1533 is elliptical, and a few decades ago it was classified as such. The concept of lenticular galaxies was introduced when it became apparent that a very small number of galaxies show features intermediate between spirals and ellipticals.

The term lenticular means lens-shaped. The lens referred to is not the thin, curved glass of a pair of spectacles, but the rather fatter version found in telescopes. Lenticular galaxies, therefore, are disks thickest in the middle, and in this respect they may resemble many elliptical galaxies. The difference between elliptical and lenticular galaxies is most apparent at the edges of the disk: elliptical galaxies have rounded edges to their disks while lenticular galaxies taper towards their outermost parts. It is almost as though the lenticular galaxies are trying to extend farther in the planes of their disks and become spirals, but have run out of material. This effect is heightened by the occasional presence of a very faint lane of dark dust along the periphery of the lens. This looks like a weak version of the dark band crossing spiral galaxies seen edge-on.

Detailed studies of these objects show that lenticular galaxies bear greater similarities to spirals than to ellipticals. Astronomers classify lenticular galaxies as S0, meaning spiral but without arms.

In the mid 1970's, radio astronomers mapping the distribution of hydrogen gas in galaxies turned their attention to lenticular galaxies. Already, it was known that spiral galaxies contain considerable quantities of gaseous hydrogen—sometimes up to one-tenth of their mass—but that elliptical galaxies are very deficient in gas. The similarity of lenticular galaxies to spirals was greatly heightened by the discovery that NGC 1533 and other lenticulars are very rich in gas. So rich, in fact, that it is difficult to understand where the hydrogen lies within the visible galaxy. Where are the new stars and the nebulae which mark the gas-laden areas of spiral galaxies?

Very long-exposure photographs have since been taken of several lenticular galaxies. In most cases, extremely faint blobs can be seen beyond the edges of the lens. Sometimes these faint blobs sketch out tenuous spiral arms. Astronomers conclude that lenticular galaxies are spirals in which only a minuscule proportion of the gas of the arms has formed stars. Whereas a normal spiral galaxy may have one billion stars within its spiral arms, lenticular galaxies may support fewer than one million.

Computer simulations of the formation of galaxies predict that stars are easily formed in spiral arms. The absence of stars in the arms of lenticular galaxies, in which the stars of the nuclear bulge are old, is somewhat unexpected. It may be that lenticular galaxies are all fairly young spiral galaxies, and that if we were to

NGC 1533 UKSTU

watch for a further few million years, they would turn into conventional spirals. However, there is no evidence that the stars of the lens are sufficiently youthful for this explanation to be satisfactory—lenticulars seem to have been around for a long enough time already. Instead, astronomers must seek some other mechanism which in some cases inhibits the conversion of gas to stars. No such mechanism has yet been proposed.

NGC 1533 lies at a distance of 36 million light years. The visible lens is about 5000 light years in diameter, but the entire galaxy is probably three times as large. The mass of the visible stars is estimated at ten billion times that of the Sun, and a further three billion solar masses of hydrogen lies in the outer parts of the galaxy.

Low Surface Brightness Galaxies

Smudges of gas

Low surface brightness galaxies are newcomers to our understanding of astronomy, but they appear to be quite common. They are numerous, and there are many that can be found, but they are faint and so have easily been overlooked up to the present day.

Two things contributed to their discovery. First of these is the manufacture of photographic emulsions capable of recording them. In the 1950's, when the deepest photographic survey of the northern sky was made, the best photographic emulsions could not record faint features of low contrast. The late 1970's witnessed a survey of the southern sky by astronomers at the U.K. Schmidt Telescope Unit, using a new Kodak emulsion, called IIIa-J, which has a much greater ability to detect faint features. The second factor was the determination of astronomers using IIIa-J emulsions to produce perfect results. Faint smudges all too easily appear on large photographic plates. One by one, the causes of such plate faults were eliminated. Eventually, the only faint smudges left on the plates came in the same place on successive photographs, and therefore had to be real features in the sky.

Low surface brightness (LSB) galaxies are small and very faint. Their faintness is easily attributable to the small number of stars which they contain. Observations by radio astronomers however, show that LSB galaxies are extremely rich in hydrogen gas. Enough gas is present within them to make at least a small galaxy of stars, but the gas has shown little inclination to condense. LSB galaxies are the most immature of all star systems. Whether their immaturity is due to extreme youth or to slow development remains to be

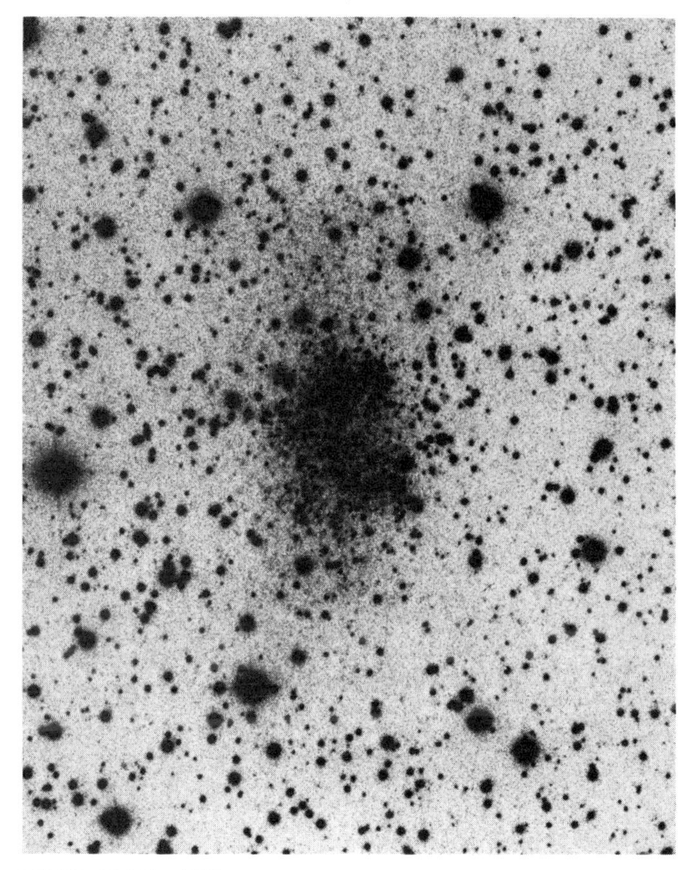

UKS 1927 –177 AAT

Low surface brightness galaxy UKSTU

UKS 1334–277 UKSTU

I Zwicky 1 POSS

determined. It does seem likely, however, that given time most LSB galaxies would form dwarf irregular galaxies.

We cannot readily discover LSB galaxies at great distances from us. Nonetheless, a good many examples have been found in the local area. Present indications are that LSB galaxies may be as numerous as all other types of galaxy known.

I Zwicky 1

Compact galaxy

I Zwicky 1 is a compact galaxy, but at first glance it looks like a star. Four decades ago, the Swiss-American astronomer Fritz Zwicky predicted that many of the faint blue stars found on general sky photographs, such as the first in this book, were not really stars at all, but galaxies. This seemed a preposterous suggestion: as the previous photographs in this book show, galaxies are large, structured, fuzzy objects that could not possibly be mistaken for stars. Zwicky maintained, however, that galaxies need not be large, flamboyant objects whose stars were widely separated. Some, he argued, might be more condensed, and these he called compact galaxies.

Zwicky searched for compact galaxies on photographs of the sky, and eventually discovered and listed several thousand of them. They look like stars, but on closer examination they can be seen to have slightly fuzzy edges. Like ordinary galaxies they are composed of millions of stars, but clustered much more tightly together. In many cases the constituent stars are hotter and bluer than those in conventional galaxies,

UKS 0242–724 UKSTU

resembling the stars in the arms of spiral galaxies.

Compact galaxies range from nearby, sparsely populated, dwarf systems which must be only a few hundred light years in diameter, to extremely bright and very distant galaxies which may be among the most massive in the Universe.

I Zwicky 1 is the first in Zwicky's catalogue and is particularly interesting. It contains a large quantity of hot gas, mostly hydrogen and helium, but in which the evidence for iron is unusually strong.

The other compact galaxy shown here was recently found on a plate taken by the new U.K. Schmidt Telescope, and appears even more compact than most of Zwicky's examples. It lies at a distance of two billion light years.

Why are compact galaxies compact? In some cases this is an illusion caused by the presence of unusually bright stars throughout the galaxy. Such a galaxy, at a very large distance from us, would appear quite bright although being extremely small. In other examples the stars do lie much closer together. This could result if conditions were favorable in the initial gas cloud from which the galaxy collapsed. In particular, if the gas cloud rotated very slowly, it could shrink unusually small.

Messier 64

The Black Eye Galaxy

M 64, a spiral galaxy discovered by the German astronomer Johann Bode in 1779, is unusual for two reasons. First, its spiral arms are extremely smooth, lacking the structure normally present in the arms of spiral galaxies. It seems that there are no large clouds of gas in the arms, either to produce local concentrations of stars or to glow as individual nebulae. Spiral galaxies with this attribute are called anemic. Second, an extremely dense cloud of dust encircles the nucleus about one-quarter of the distance out from the galaxy's center. This cloud produces a prominent dark patch in the galaxy by absorbing all the background starlight, and gives rise to the galaxy's name. As with M 81 and THE ANDROMEDA GALAXY M 31 we can tell which side of M 64 is the nearer. M 64 must contain a mass of gas mixed with the dust, and this makes the lack of gas in the outer spiral arms more bizarre.

M 64 is about 65,000 light years in diameter. It lies in the direction of the Virgo cluster of galaxies, but appears brighter than all members of that cluster. Measurements have shown that M 64 is not a member of the cluster and lies in front of it.

Messier 64 *Hale*

Groups of Galaxies

Clusters in miniature

Not all clusters of galaxies are so large and densely populated as the Coma cluster and the other examples just illustrated. A great many galaxies are found in

Longmore's Group AAT

Group of galaxies UKSTU

NGC 3623,7,8 POSS

small groups, numbering typically half a dozen major members. The group called Longmore's Group is an example of a very compact small collection of galaxies. The photographs on this page illustrate a selection of groups of galaxies of various sizes and complexity.

All the galaxies in a group are traveling together through the Universe, and probably will continue to do so for the lifetimes of the galaxies. In some cases the groups appear not to be stable; that is, the combined mass of the members is not enough to bind the group together. Most people these days are familiar with the example that in order to escape from the Earth, you need a high speed. Jumping is not enough—you are pulled back quickly—but if you can reach the speed of a space rocket, known as escape velocity, you will not fall back to Earth.

Similarly, when one of the individual galaxies is traveling faster than the escape velocity from the group, it may fly off into space, leaving the group behind. However, in order to determine the escape velocity of a group, astronomers must know its mass. As in the Coma cluster, there is probably gas within these small groups contributing to the mass. The gas, however, is not easily detected and we cannot therefore be certain whether any group of galaxies really is unstable.

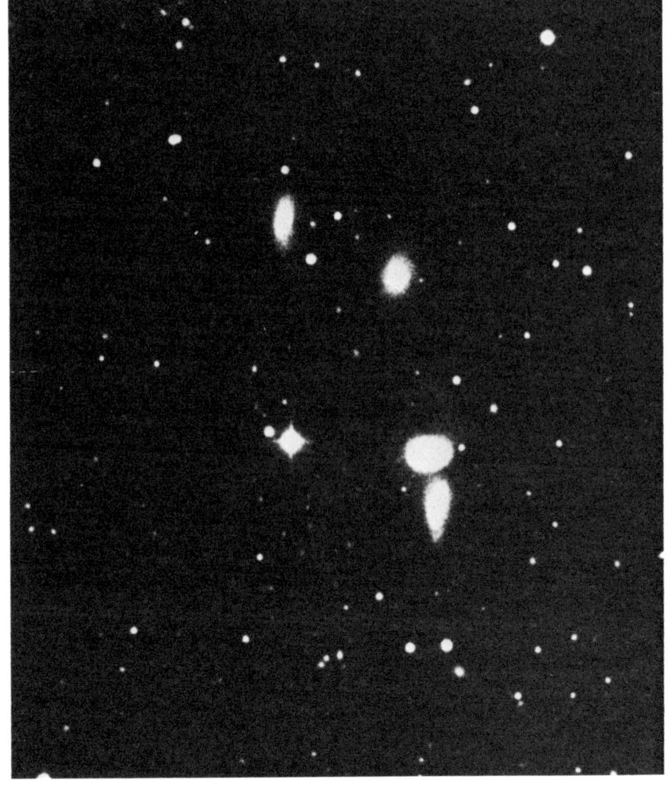

NGC 833,5,8,9 POSS

In some of the groups the galaxies are widely separated. In others they lie close together. In Longmore's Group there is enough material, probably stars, to bind all the members together. In a few of the groups illustrated, galaxies have been distorted by one another's gravitational fields and are now festooned with long streamers. These streamers comprise millions of stars which have been pulled out of one galaxy by the passage of another, a process known as tidal interaction.

Stephan's Quartet

Four, then five, back to four again

In the late 19th century, galaxies were still being discovered by visual inspection of the sky. Many famous astronomers of that time spent long, cold nights peering through eyepieces looking for new nebulous objects. One such observer, working from Marseilles, was named Stephan. His lists of discoveries are not very lengthy, and it is unlikely that his name would now be remembered but for his discovery in 1877 of four galaxies very close together in the sky. These galaxies now bear the catalogue numbers NGC 7317, 7318, 7319 and 7320.

As photographic plates became more widely used, astronomers recognized that this collection represented an unusually compact grouping of large galaxies. Moreover, NGC 7318 was seen to be two galaxies closely interacting. The group of four faint, nebulous objects discovered in 1877 therefore became known as Stephan's quintet. For many years Stephan's quintet was quoted as the best example of a small group of galaxies.

Doubts began to be raised when the velocities of the five galaxies were measured in the 1960's. While four of them had almost identical velocities, one—NGC 7320—was discrepant. Thus, while the five galaxies had the appearance of a group, Hubble's red-shift law indicated that one galaxy did not belong with the others. For a while, some wondered whether Stephan's quintet might be the exception that tested the rule, and that the problem really lay with our assumption that what we see as a red-shift of the galaxy's light is due to its velocity—an assumption on which much of our understanding of the Universe depends. Is there some deeper, unknown cause of the red-shift? But finally, almost a century after the discovery of the group, astronomers convinced themselves that NGC 7320 is not a member of Stephan's group, but lies in front. Curiously, NGC 7320 is an outlying member of another group of galaxies, a group much more scattered than Stephan's. So once again Stephan's collection is a quartet of galaxies.

Recent observations indicate that Stephan's quartet is merely the compact center of a small, scattered cluster of galaxies at a distance of about 250 million light years.

Stephan's quartet KPNO

Interacting Galaxies

A gallery of rogue galaxies

In the photographs on p.28, two or more galaxies have been distorted by mutual interaction. Such interactions are described as tidal, or gravitational.

Gravity is the overriding large-scale force in the Universe. By its effect we have weight which prevents us flying off the spinning Earth. By its effect the Moon orbits the Earth, and the Earth orbits the Sun. The gravitational pull of the Sun and Moon raises a bulge in the oceans and gives rise to the tides. Without gravity there would be no galaxies, for each star remains within a galaxy only because the other stars attract it. Indeed, without gravity there would be no stars, for they condense out of the gas under the influence of gravity.

Gravity is a property possessed by all objects in proportion to their mass. This book is gravitationally attracting every other object in the room, your neighbor, the Earth, the Sun, and even the galaxies pictured here, though the objects near to it are attracted most. As you move the book around, or even turn its pages, you change the book's gravitational effect, and hence influence the motion of every star and galaxy in the Universe. The influence is, however, utterly negligible because the book is small, and its effects on the rest of the Universe are quite imperceptible. The most sensitive laboratory experiment could not detect the gravitational pull of this book. If the book were placed in orbit around the Earth, it

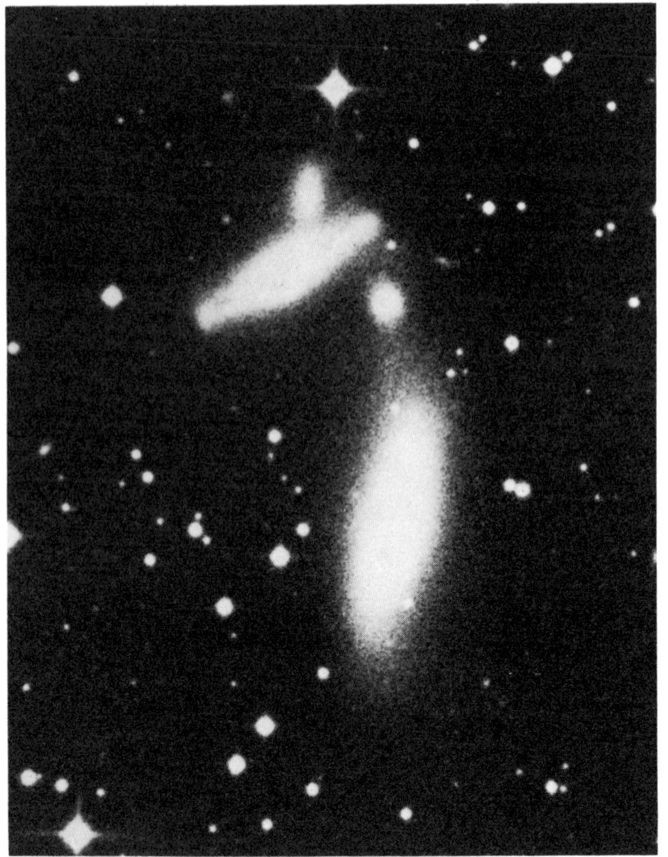

Interacting galaxies UKSTU

Interacting galaxies UKSTU

would cause our planet to wobble by only one ten million billionth of a millimeter.

Galaxies, however, are rather bigger than this book. A typical galaxy has a mass 10^{42} times greater, a number which defies comprehension. 10^{42} is larger by a factor of a million million than the number of grains of the finest sand which would be needed to construct the Earth. Galaxies, therefore, exert an impressive gravitational pull, extending over a very wide volume of space.

We cannot move galaxies to study the effects of their changing gravitational pull; we have no need to, for Nature has arranged many examples. When galaxies make close passes, they cause disruptions to one another's patterns of stars. A star in one galaxy may suddenly find itself being pulled hard in another direction by a second, passing galaxy. Then, instead of traveling contentedly round its parent galaxy in a simple orbit, as the Earth moves round the Sun, the star changes its motion to accommodate the new forces. It, and its neighbors, may move right out of their parent galaxy, a process often referred to as the gravitational slingshot. They may travel away with the interloping galaxy, or find themselves lost somewhere in the space between. Or they may fall back into their parent galaxy and greatly modify its appearance.

NGC 4435/8 UKSTU

NGC 2207 AAT

Fly's Wing UKSTU

These photographs illustrate some of the infinite variety of shapes which can be produced by gravitational encounters between galaxies. The luminous streaks between and around galaxies on these photographs are gas and stars torn out of one galaxy by an interaction. Astronomers believe that most of these shapes could be explained in detail by computer calculations which simulate the interactions. But time on a computer is expensive, so the majority will never be studied. Astronomers confine their attention to only the most distinctive examples, and some of these are described on the next page.

Messier 51

The Whirlpool Galaxy

Messier 51 is a pair of galaxies undergoing a relatively simple gravitational interaction. The larger galaxy, to the top of the picture (p. 30), bears the catalogue designation NGC 5194, while the companion is NGC 5195.

The pair of galaxies is not far off: astronomers estimate 13 million light years away. Of the two, NGC 5195 is slightly nearer. It is pursuing a curved orbit around the larger galaxy, and having passed in front will soon begin to move away behind. "Soon" in this context means within the next few hundred million years. NGC 5195 is an irregular galaxy, rather blue and probably rather younger than NGC 5194. It shows slight indications of developing an elongated center, and may eventually evolve into a barred spiral galaxy. The formation of the bar is probably being induced by the interaction with NGC 5194.

NGC 5194 is a large galaxy weighing roughly 100 billion times as much as the Sun, or about three times as much as its companion. It is clearly a spiral galaxy, and indeed was the first galaxy to be recognized to have spiral characteristics, by Lord Rosse in 1845. We view it from about 60° above the rim, and the left-hand side is the nearer. The nucleus of NGC 5194 contains a small core about 140 light years across which contains hot gas. Within that nucleus gas is being driven by some unknown mechanism at speeds up to 200 km/sec. Radio waves are emitted from a region about five times larger centered on the core: the origin of the radio waves is probably linked to the fast-moving gas. There are many young, hot stars in the nucleus, and these too contribute to exciting the gas which is heated to about 30,000°C. Dust is mixed in with the gas in the nucleus.

Computer simulations of the interaction of the two galaxies have been performed by the U.S. astronomers

Messier 51 KPNO

NGC 4038–9 (inset: their center, ESO) UKSTU

Alar and Juri Toomre. They are able in their simulations to reproduce the current appearance, positions and motions of the two galaxies with remarkable accuracy.

NGC 4038 and 4039

The Antennae

The pair of galaxies NGC 4038 and 4039 are seen immediately after a very close encounter. In the smaller picture, details of the central parts of the two can be seen. Both lack the simple nucleus of a normal galaxy, and show instead confused regions of blobs and gaps. Measurements of the velocities of the bright blobs reveal complex motions which were not at first understood. The galaxies are both rotating about their own axes and orbiting around one another, and when these motions are taken into consideration, the velocities can be explained.

The larger photograph shows that these galaxies have, in the course of their interaction, thrown off long curved tails—which give rise to their popular name, the Antennae. These tails contain stars which were ejected from the parent galaxies by the gravitational slingshot. The detailed appearance of the tails of these galaxies can be predicted by computer calculations. Such computations show that NGC 4038 and 4039 were quite normal galaxies before their interaction, and that some of their outlying stars were stripped off to form the tails.

Galaxies often contain gas as well as stars, and the gas is normally confined to the outer portions. Thus one might expect there to be gas in the tails too. Radio astronomers have confirmed this suspicion by detecting enough hydrogen in the tails to form 1.5 billion stars like the Sun.

The distance of the Antennae is 48 million light years, which means that the tails are about 100,000 light years long. This gives some indication of the time involved for the interaction of the two galaxies and the formation of their antennae, since the stars of the tails cannot travel at speeds remotely approaching that of light. In fact the galaxies have probably taken several hundred million years to make their mutual pass.

NGC 5291

Interacting with the Seashell

The two galaxies in the center of the photograph (p. 31) are interacting with each other. The smaller, the Seashell, is passing close to the larger galaxy, and suffering for the trespass. The gravitational pull of NGC 5291 is sufficient to disrupt the Seashell and has dragged out some of its stars to form a tail. This distortion has made the smaller galaxy resemble a whelk, hence its name. By comparison, NGC 5291 seems relatively unruffled by the interaction, and this suggests that it is the more massive of the two.

The small nebulae which lie above and below the pair of galaxies are part of the same system. Because of their great distance from us—over 250 million light years—they appear very faint; but they are galaxies of quite respectable size. They contain many young, hot stars which illuminate and excite large quantities of gas. From the observed velocities of these little galaxies, astronomers can deduce the mass of NGC 5291 and hence confirm that it is very large, about two million million times that of the Sun. Radio observations have shown that the gas of NGC 5291 extends as far as the small galaxies, making it one of the largest objects known, with a linear diameter of 600,000 light years. The visible galaxy is only the nucleus of this giant system, and the small nebulae are but tiny portions of the gas wherein young stars have formed.

Why have stars recently formed in these particular spots and not throughout the galaxy? Something must have caused this localized star formation. A clue to the trigger is afforded by velocity measurements which indicate that the small galaxies lie only around half of the rim of the gaseous disk which comprises the outer parts of NGC 5291.

NGC 5291 is an outlying member of a small cluster of galaxies, and is falling towards the center of the group. Its stars pass freely through the tenuous gas of the cluster, but its gas, being a sticky material, is slowed and compressed. The compression of the gas is the trigger we seek to produce star formation, and it acts first around the leading edge of the galaxy.

In this case it is possible to make that difficult step from a flat photograph to a three-dimensional picture. The stars on this photograph are all foreground members of our galaxy, while most of the galaxies lie at about 250 million light years distance, and are part of a cluster whose center lies off the picture to the right.

NGC 5291 is a thin disk of gas and can be visualized as a wheel seen nearly edge-on with its apparently longest dimension stretching most of the vertical height of the photograph. The bright, visible galaxy is the hub of the wheel; invisible gas occupies the remainder. The wheel is currently at the farthest edge of the cluster and is moving towards us and to the right as it falls towards the center. The leading rim of the wheel therefore forms a curved line, convex to the right, and along this line a series of small galaxies has formed. The nebulous knots are produced as the rim collides with the unseen gas of the cluster. Because the stars are unhindered in their interaction with cluster gas, the hub has moved farther to the right and appears projected against the rim itself. Finally, a random galaxy, the Seashell, has recently been attracted by the gravitational pull of NGC 5291. It has passed in front of the larger galaxy and is now moving away to the upper left.

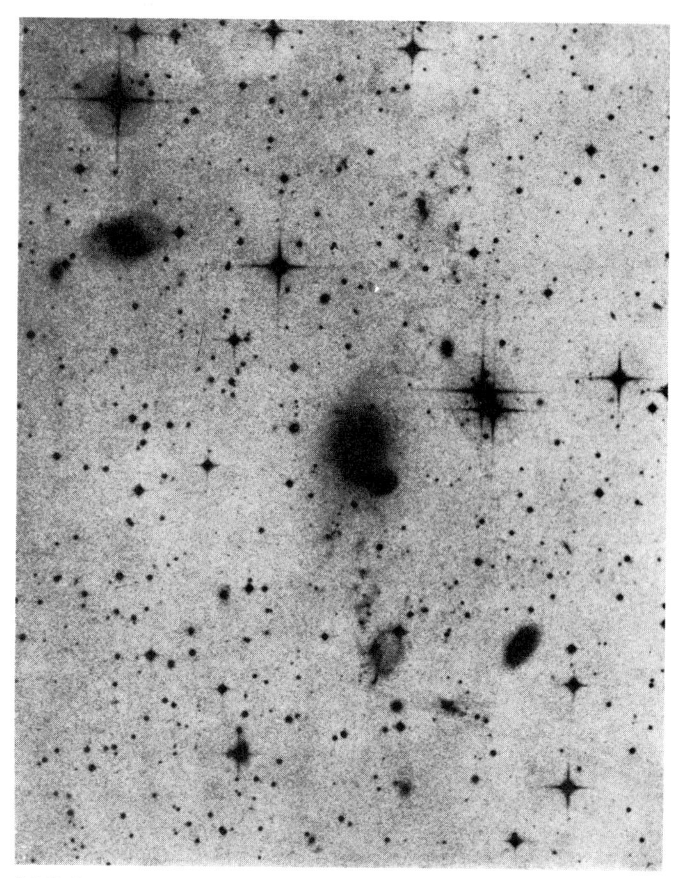

NGC 5291 UKSTU

NGC 1510 and NGC 1512

The case of the rejuvenating galaxy

The pair of galaxies illustrated in the photos on p. 32 received little attention until very recently, yet they have proved to be a particularly distinctive case of interaction, in which one galaxy has become rejuvenated at the expense of the other.

The photographs show, respectively, the inner detail and the faint outer extensions of the pair. At left on both photographs is NGC 1512, a barred spiral galaxy. At its center is a striking miniature spiral, a feature not known in other galaxies but probably fairly common in barred spirals. Linking the ends of the bar is an elliptical ring, a feature seen weakly in a small percentage of galaxies and particularly well developed here. Only a few years ago, astronomers were unaware that much material lay outside this ring, but as the

NGC 1510–12 AAT

long-exposure photograph shows there are faint spiral arms extending a considerable distance. Some of these show evidence of disturbance by the interaction with NGC 1510, on the right.

The total diameter of NGC 1512 on this photograph is over 8 arc minutes. What makes the galaxy remarkable is the presence of a vast, associated cloud of hydrogen fully four times as large as the visible galaxy. The hydrogen cloud appears bigger than the Moon, though it can only be "seen" by radio astronomical techniques. When account is taken of the distance of NGC 1512, which is nearly 50 million light years, the linear extent of the hydrogen cloud is found to be 400,000 light years, making it one of the biggest galaxies known. The hydrogen occupies a flattened disk, containing more than ten billion solar masses of gas. This is about 6% of the total mass of the galaxy.

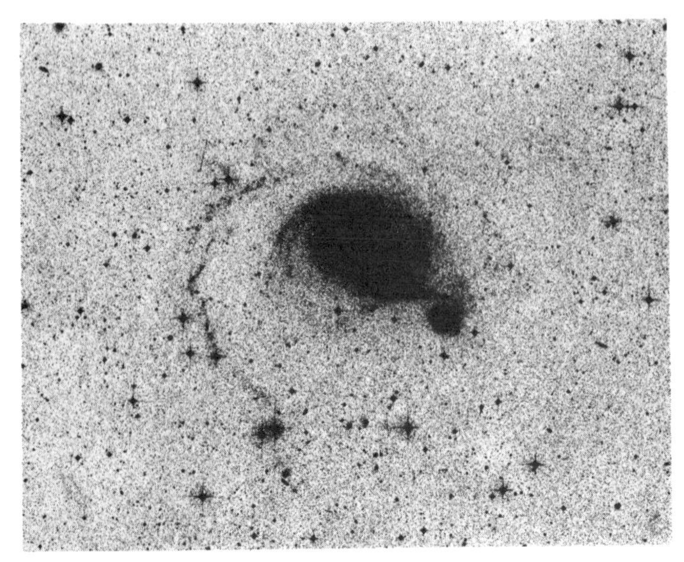

NGC 1510–12 UKSTU

By a remarkable coincidence, we happen to view this enormous galaxy at a time when it is being visited by the small elliptical galaxy, NGC 1510. Even more coincidental is the fact that the orbit of NGC 1510 lies almost within the plane of the larger galaxy. This means that instead of passing quickly through the hydrogen, NGC 1510 is plowing a long furrow through the densest part of the gas.

A study of the two galaxies has been made by a team of astronomers at the Anglo-Australian Telescope led by Tim Hawarden. They find that the effect on NGC 1510 of its encounter with so much hydrogen is spectacular. Elliptical galaxies normally comprise only cool, old, red stars, and contain no gas. At the center of NGC 1510, however are two compact nuclei, one of which is a gigantic cluster of young, hot, blue stars embedded in a cloud of gas. In its passage through the hydrogen outskirts of NGC 1512, the elliptical galaxy has scooped up vast quantities of gas. Hawarden and his collaborators estimate that a mass of gas between 100 million and a billion times the mass of our Sun has been stolen by NGC 1510 from the disk of NGC 1512. Gravity pulled the gas into the center of the elliptical galaxy, where it became compressed and therefore formed several million young, hot stars.

NGC 1510 is probably half way around its encounter with the larger NGC 1512. When the encounter ends it will have acquired a sizeable fraction of its mass as hydrogen, and will have built up a population of young stars totally atypical of normal elliptical galaxies. In doing so, the life expectancy of this elderly galaxy has been prolonged.

The Cartwheel Galaxy

A celestial splash

A glance at this galaxy shows why astronomers christened it the Cartwheel. It was distorted into the unusual shape by a head-on collision between two galaxies. The Cartwheel has a "rim" 170,000 light years in diameter made predominantly of glowing gas. Within this lie a "hub" and "spokes" made up of old, red stars. The distance of the Cartwheel from us is 500 million light years. It is believed once to have been an ordinary but very large galaxy, with a bright nucleus and a large disk of gas which had not then condensed to form stars.

A few hundred million years ago, a smaller galaxy passed directly through the center of the large galaxy. We can identify the culprit: it is the leftmost of the pair to upper left in this photograph. Computer simulations of such a collision between galaxies indicate that the stars of the nucleus are little affected by the events, but that a ripple passes out through the gas like the splash pattern surrounding a stone

Centaurus A

Messier 81 velocity map

Messier 83

PLATE 1

Messier 31 and its two companion galaxies

PLATE 2

Small Magellanic Cloud

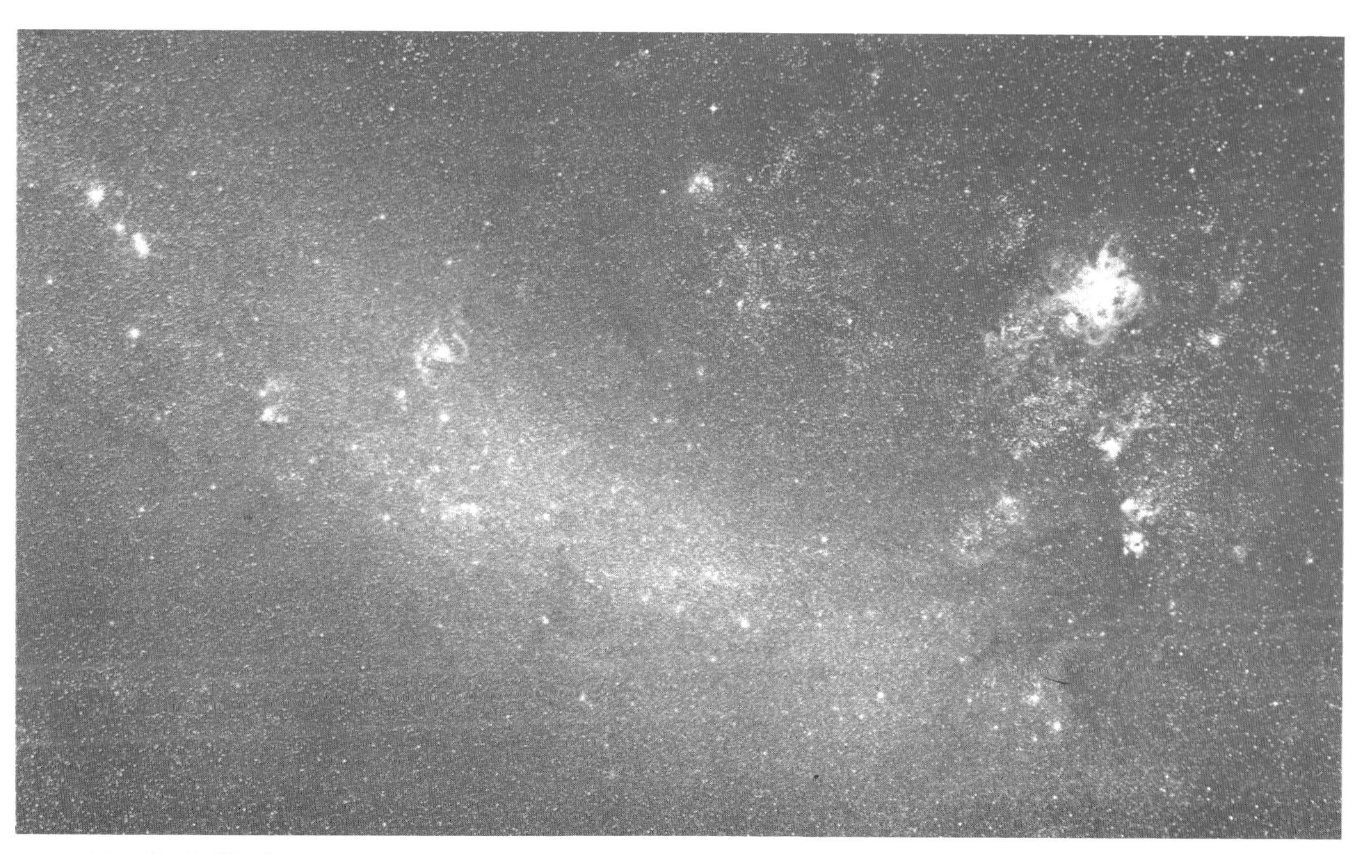

Large Magellanic Cloud

PLATE 3

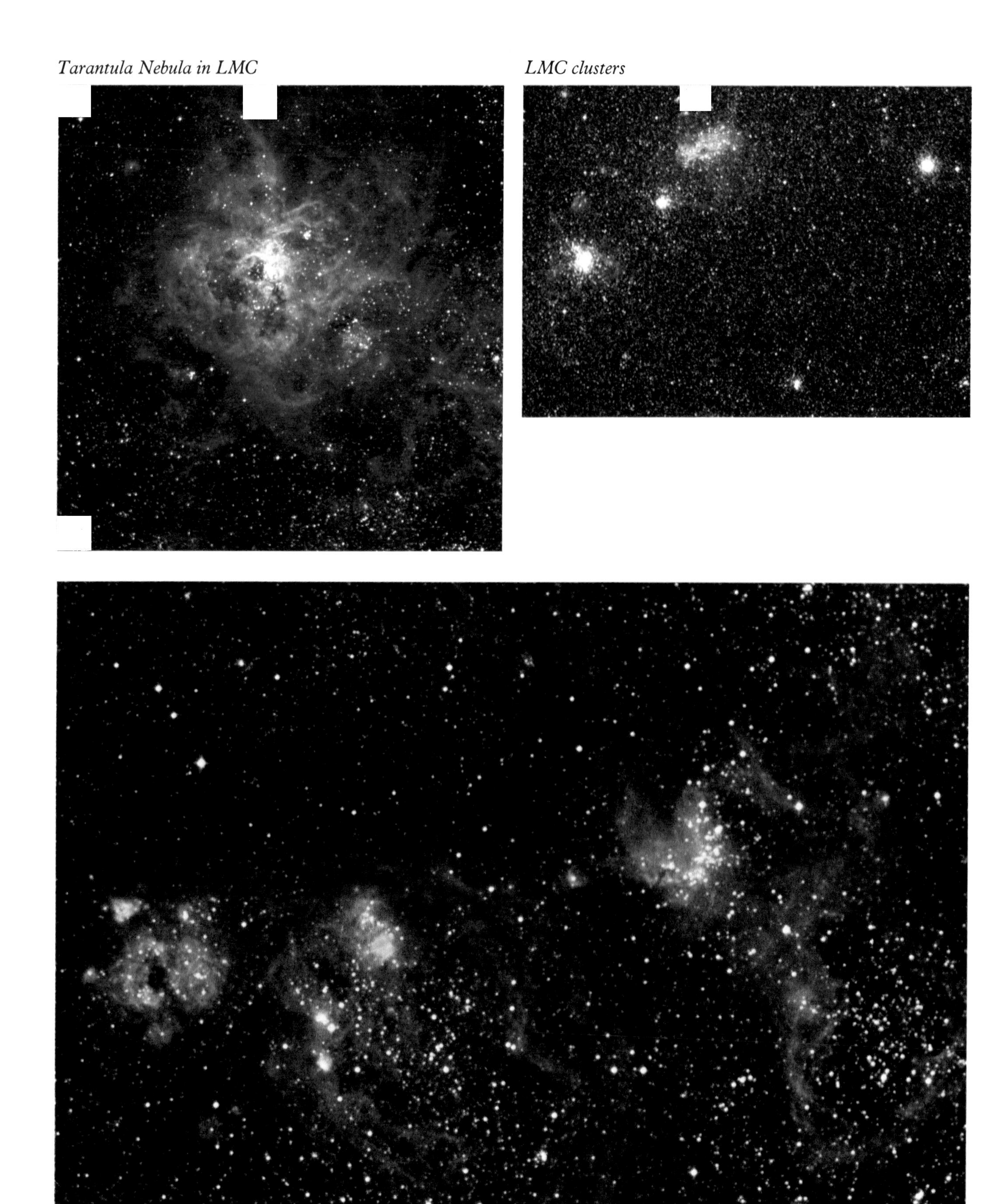
Tarantula Nebula in LMC

LMC clusters

LMC Nebulae

PLATE 4

dropped into water. The pileup of gas in the ripple sparked off a burst of star formation since the gas there suddenly became dense enough to contract into stars. Because this took place suddenly, there are in the rim exceptional numbers of young, bright stars. Being bright and hot, these stars are causing the remaining gas of the rim to glow.

In contrast with the effect of these catastrophic events on the gas, the original red stars of the galaxy's bulge survive as the hub and spokes. The computations predict that these stars will form weak spokes, a small inner ring and a tiny nucleus within this ring, which is exactly what is seen in the Cartwheel.

The intruder galaxy now lies some 250,000 light years along the Cartwheel's axis. Because of the attractive force of gravity, the hub has been pulled by the intruder galaxy somewhat out of its original position at the center of the Cartwheel's rim. The hub followed the intruder in its onward orbit, and is therefore seen offcenter in the wheel.

Because large numbers of stars were formed simultaneously in the rim, there will be correspondingly large numbers of stellar deaths among the short-lived stars. Such stars explode violently, producing what are called supernovae. Astronomers expect to see on average up to one bright star in the rim of the Cartwheel explode every year, about 100 times the normal rate of supernovae in a normal galaxy.

The Carafe Group

Two unusual galaxies

The three bright galaxies in the Carafe Group travel

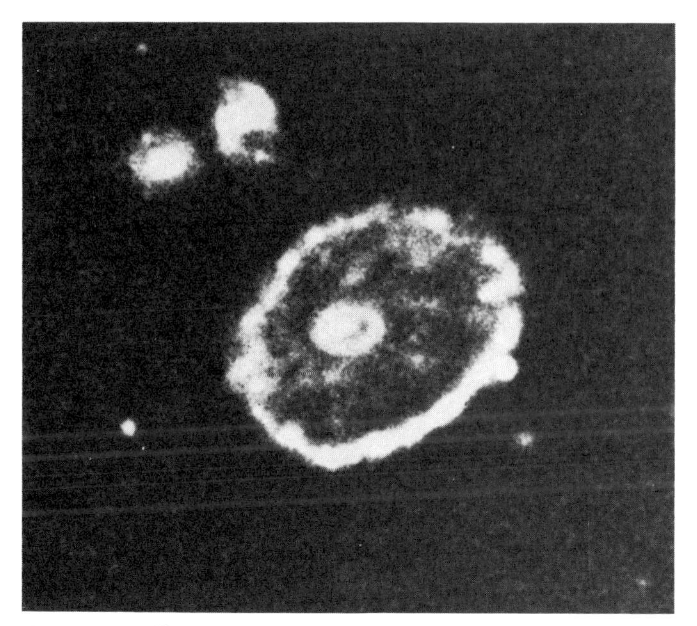

Cartwheel Galaxy UKSTU

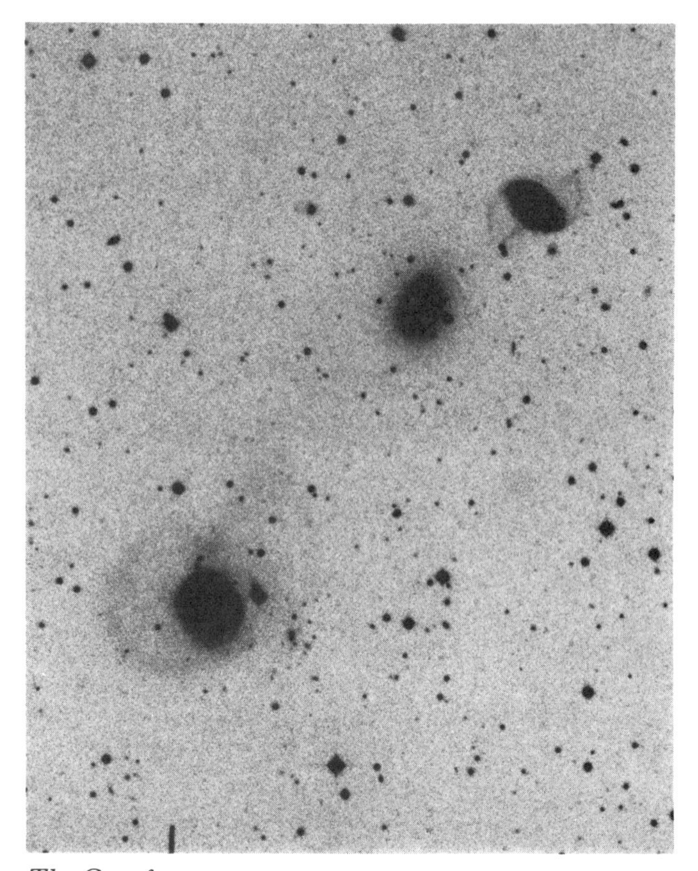

The Carafe UKSTU

together in space and form an unusual trio. NGC 1595, in the middle, is a conventional elliptical galaxy. To its upper right is NGC 1598, which at first sight appears to be a normal spiral galaxy. When examined more closely, a faint luminous line can be seen crossing its shorter dimension and apparently linking the very ends of the spiral arms. Whether this feature comprises stars or gas is not known. In either case, it is difficult to envisage how it could maintain its straightness in the plane of a rotating galaxy. Astronomers therefore believe that it represents two jets projecting out of the plane, probably along the axis on which the galaxy rotates. The apparent alignment with the ends of the spiral arms is but chance. Since stars or gas in this position would fall back into the galaxy relatively quickly, the jets must have been thrown out by some mild explosion in the nucleus of NGC 1598. The length of the jets is about 100,000 light years, comparable with the length of THE ANTENNAE (p. 30).

At lower left is the Carafe Galaxy. The peculiar shape of this object has not been explained. Some interaction with NGC 1595 has occurred, for there is material, believed to be stars, in a long, faint extension towards that galaxy. Whether the interaction can explain the faint, offcenter ring around the Carafe has astronomers puzzled. This ring, which is also thought to be composed principally of stars, is one of the largest known in any galaxy, with a diameter of 260,000 light years.

The nucleus of the Carafe is a small, bright object buried in the overexposed central image on this photograph. Gas in the nucleus is being driven, by some unknown force, at velocities up to several hundred km/sec. There is a pall of dust and gas immediately in front of the nucleus. If this were removed, the Carafe would outshine the other members of the trio.

NGC 1097

A galaxy with rays

The galaxy NGC 1097 is the focus of faint, luminous rays. These are shown on the deep exposure (larger) photograph; two project towards the upper left, and two to lower right. The rays make the galaxy resemble NGC 1598 (p. 33).

The short-exposure photograph (inset) reveals an important difference, however: NGC 1097 is an interacting pair of galaxies. The smaller, elliptical galaxy lies to the upper right, and is visible through the disturbed arms of NGC 1097. As we have seen, jets and tails can be thrown off by galaxies as they affect each other gravitationally. It seems quite possible to ascribe these rays to the gravitational interaction between the spiral and elliptical galaxies. However, an explosion in the galaxy's nucleus certainly cannot be ruled out, and is favored by many astronomers.

Observations made through different colored filters reveal that the rays which project to the upper left are blue, with very faint, fuzzy blue patches at their tips. The rays to lower right are red. One of these lies diametrically opposite one of the blue rays, hinting at an explosive origin. But the other, which is too faint to be seen clearly on this reproduction, is inclined at 11° to its corresponding blue ray. This faint red ray is also the longest; if thrown out of the galaxy by an explosion, its faintness may have resulted from the material being spread out over a longer path. The colors of the rays, and the fuzzy ends to the blue rays, are not understood by astronomers. NGC 1097 lies at a distance of 40 million light years, so the longest ray has a length of at least 200,000 light years—more if it projects somewhat towards or away from us.

Messier 82 and Bode's Nebula M 81

The exploding galaxy and its companion

The two bright galaxies on this photograph lie at a distance of about ten million light years from us and form a pair in space. Discovered by Bode in 1774, they are usually known by their numbers (81 and 82) in Messier's catalogue of nebulous objects, but only M 81 is known as Bode's Nebula.

M 81 is the larger of the pair and is obviously a spiral galaxy. We view it at an angle of 32° above the rim, so its graceful arms are wound into elongated ellipses. It has a large nuclear bulge containing many cool, red stars. At the very center of this bulge is a small nucleus containing gas and hotter stars, and this nucleus emits radio waves, as does our own GALACTIC CENTER. The spiral arms are thronged with hot stars and gas, but the unseen gas extends beyond the stars. Radio astronomers have mapped this gas, and find that it continues the same spiral pattern out to nearly twice the diameter of the visible galaxy. In addition, gas lies in a great curve stretching all the way from M 81 to its

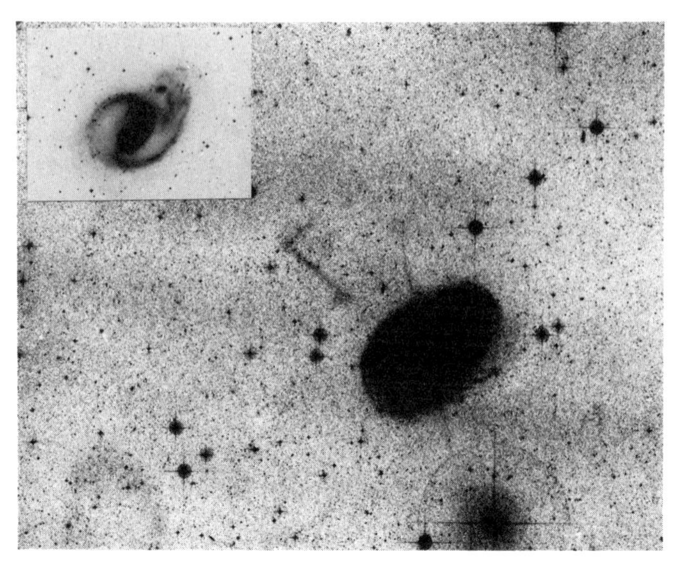

NGC 1097 UKSTU

Messier 81 and 82 POSS

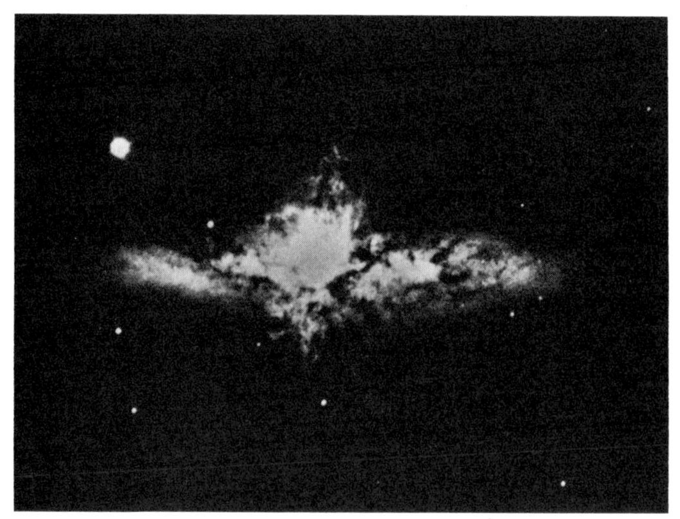

Messier 82 *Hale*

companion M 82. This bridge of gas was generated by a close pass of the two galaxies about 100 million years ago.

M 82 is much the more interesting and unusual galaxy and we show it in a separate photograph. Most astronomers have classified it as irregular, though there is some reason to believe that it is a disk galaxy seen edge on. If we could view the disk from above or below, we might find that M 82 has some spiral attributes.

Because dust lies in the plane of the disk, the center of M 82 is dimmed by about a factor of 50, and astronomers have therefore had some difficulty in studying it. By piecing together evidence secured in a variety of ways, they have learned something of the nucleus of M 82, and hence of the galaxy's unusual structure.

It seems that a few million years ago the center of M 82 exploded, for reasons which are still a mystery. The explosion ripped through the abundant gas of the galaxy, so that clouds of gas weighing many millions times as much as the Sun were thrown off at velocities of up to 500 km/sec. The motion of this gas produced the irregular structure in the galaxy's central regions.

Not all the gas was thrown out. The explosion left some portions of the nuclear gas intact. However, the shock waves emanating from the explosion caused this residual gas to be compressed, which resulted in the formation of a vast number of very hot stars. The hot stars now lie in gigantic clusters, each containing up to 10,000 stars, which have heated up the remaining gas and dust. The dust, in particular, is at temperatures around 100 K (−170°C), which is warm compared with the depths of space. It radiates in the unseen infrared. Astronomers find that the energy radiated by this warm dust alone is 50 million times as great as that output by our Sun. As seen in the telescopes of infrared astronomers, M 82 is the brightest galaxy in the sky after our own.

Amid this jumble of hot stars, gas and dust at the center of M 82 lies the remains of whatever caused the explosion. Radio astronomers have found a strong source of radio emission emanating from a region about 3000 light years across. However, this radio source appears similar to those in many other, more normal galaxies, such as its companion M 81. Thus we cannot necessarily relate it to the explosion although some astronomers have been tempted to do so.

NGC 3256

Two galaxies in collision?

Over the last few decades, many peculiar celestial objects have been described as colliding galaxies. In nearly all cases, subsequent research has shown this not to be the case. NGC 3256 is an object regarded by some astronomers as a prime example of a collision between galaxies, but by others as a single, distorted galaxy. At the present time no one can decide with confidence.

NGC 3256 is a very luminous galaxy. The long streamers seen on this photograph are certainly unusual, reminding one of THE ANTENNAE pictured earlier. Its distance is about 185 million light years, so the streamers which subtend 8 arc minutes on the sky (about a quarter of the Moon's diameter) have a total length of 440,000 light years. This is an extraordinary length.

Within the nucleus of NGC 3256 are many clouds of hot, glowing gas, probably heated up by young, hot stars. But exactly what event triggered the formation of so many young stars and the production of the streamers is not clear.

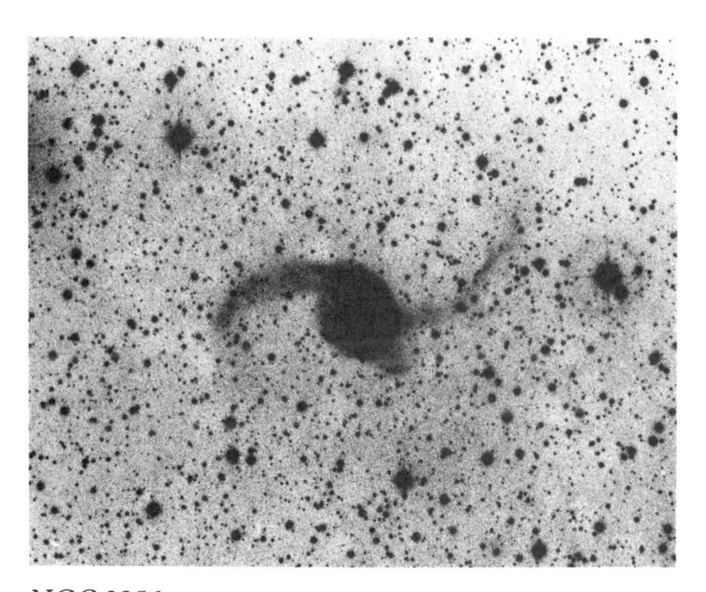

NGC 3256 AAT

NGC 4650 A

An unexplained galaxy

There are still many objects in the sky that defy explanation. Astronomers have not yet satisfactorily accounted for the shape of NGC 4650 A.

A few things are known about the galaxy. It lies in a cluster of galaxies in the constellation Centaurus at a distance of 200 million light years. The central spindle-shaped portion measures 10,000 by 25,000 light years, while the long feature running from upper right to lower left is about 140,000 light years long. This feature can be traced as a dark line where it crosses in front of the spindle to the right, but is not seen at the upper left crossing, presumably because it passes behind the spindle. Astronomers deduce that the long feature is a ring of hot stars, glowing gas, and dark dust encircling a central galaxy. The ring is seen almost edge-on. The most plausible explanation is that NGC 4650 A is a ring galaxy very like the CARTWHEEL (p. 32), but seen from within the plane of the rim. The Cartwheel is said to have formed when an intruder galaxy dropped through the center of a spiral galaxy.

This explanation fails to account for the shape of the spindle of NGC 4650 A. Theories of ring galaxies demand that the ring forms in a disk of gas which would otherwise have produced spiral arms. But spiral galaxies do not contain nuclei shaped like this one, which is elongated at right angles to the disk. Moreover, there is a very good reason why spiral galaxies do not look like this: if the spindle is rotating on its long axis, as the presence of the ring suggests, the configuration is unstable. Stars at the ends of the spindle should fall back towards the center of the galaxy, and in a short time by astronomical standards the nucleus should revert to the conventional shape. Since the hypothesized intruder is not identifiable it must be far away and we must be seeing the ring galaxy long after its disturbance. Why the nucleus of NGC 4650 A has not reverted is not understood.

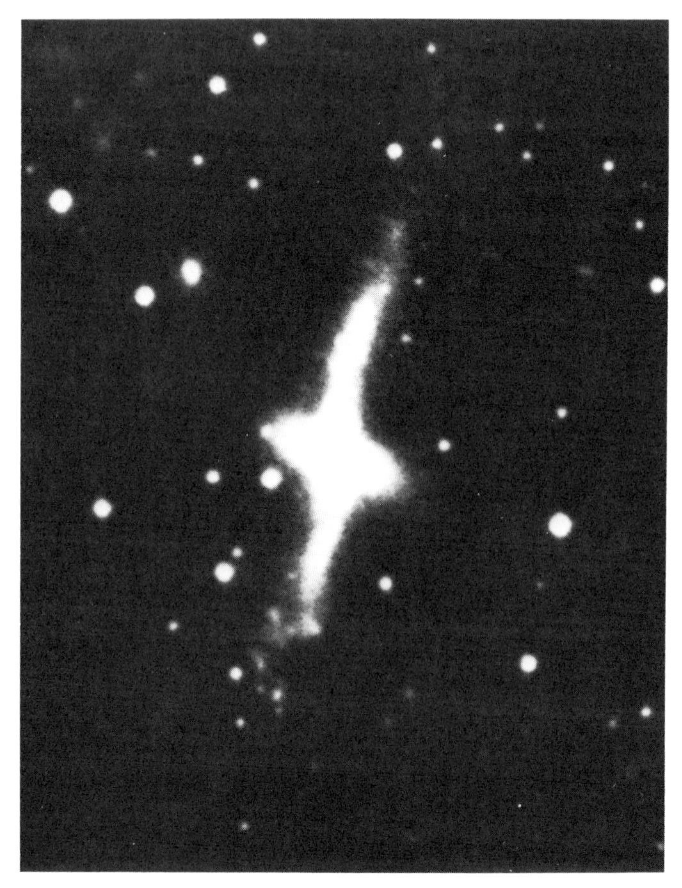

NGC 4650A UKSTU

Giant Radio Galaxies

Messier 87

A giant among galaxies

When, in 1781, Charles Messier discovered the faint, luminous patch which was to become number 87 in his catalogue, it must have appeared as just another nebula in a portion of the constellation Virgo which seemed to be overflowing with such objects.

Today we know that in that constellation Messier was discovering only the brightest members of a great cluster of galaxies lying 50 million light years distant, and now called the Virgo cluster. M 87 lies near the middle of the Virgo cluster, and it was recognized even in Messier's time to be one of the brightest nebulae in the constellation.

As the main photograph shows, M 87 is an elliptical galaxy. It appears in fact almost perfectly circular, and is probably very close to being spherical. Its constituent stars are all cool, red objects; like most elliptical galaxies it contains virtually no gas.

When radio astronomers began to explore the sky, they quickly discovered that M 87 is a very strong emitter of radio waves. Indeed, it is the third brightest object in the radio sky excepting the Sun and Moon. It was the brightest radio source in the constellation Virgo, and so became known as Virgo A. (This nomenclature had to be abandoned when more sensitive radio telescopes demonstrated that each constellation contains more than 26 other radio sources—indeed there may be hundreds or thousands.) M 87 now appears in radio catalogues under a variety of guises, such as 3C 274 and PKS 1228+12, but radio astronomers the world over still know it as Virgo A.

More recently, x-ray detectors have been sent above the Earth's atmosphere: they too recorded M 87 as a strong source. It ranks high in the top 20 celestial x-ray sources, and is known as Virgo XR-1. Near the heart of the galaxy is a very strong source of energy. This heats the very thin gas which permeates the galaxy, giving rise to the x-rays, and also produces the radiation which the radio astronomers detect. The energy comes from a peculiar jet projecting from the very center of the galaxy.

The jet is shown in the inset photograph at right which shows the region which is heavily overexposed in the larger picture. It comprises a series of at least six blobs of gas, which appear to have been thrown out of the nucleus. A very fine jet projects in the opposite direction, and a third, broad fan lies to one side. Astronomers presume that these were all ejected by a single explosion which has not, however, been reliably dated. The explosion in M 87 did not disrupt the entire center of the galaxy, as in M 82, because of the almost total lack of gas in the galaxy's central regions. Thus we see the individual jets of the explosion much more clearly. In order to produce x-rays and radio waves, the jets must contain strong magnetic fields and have a temperature of 20 million °C.

Recent optical observations of M 87 have shown that the galaxy contains a small core of unusual properties. Within the central 300 light years is a massive object which emits little or no light. Its effect can be detected because the stars in the central regions are orbiting much faster than they should, and are therefore being influenced by the gravitational pull of something heavy. The massive object weighs five billion times as much as our Sun. There are several possible explanations for the nature of this central object; some astronomers speculate that a black hole has formed here.

Black holes are predicted to occur when too much mass is crammed into too small a volume. The force of gravity near to a massive, compact star can be so large that for an object to leave its surface, the object would need to travel faster then the speed of light. In other words the star's escape velocity is larger than the speed of light. Since nothing material can travel faster than light, nothing can leave such a star. Even light itself, according to the General Theory of Relativity, would be pulled back into the star. Thus it deserves the name *black hole*.

A black hole might form from the implosion of a star in a supernova (as CYGNUS X-1 may have done). Or it might form by the amalgamation of many stars in a particularly crowded star cluster (like M 15), or in the center of a galaxy. Scenarios like the latter are in the astronomers' minds as they get to grips with the idea that there may be a black hole in M 87.

By measuring the velocities of stars at the outer edges of M 87, astronomers can determine the total mass of the galaxy. They find a value of 3000 billion times as much as the Sun, or about 6000 billion billion billion tons, which makes M 87 the most massive galaxy known.

Messier 87 with (inset) its jet KPNO

Even this enormous mass may be an underestimate by a large factor. X-ray astronomers can estimate the mass required to retain the gas which emits the observed x-radiation. Unless the galaxy weighs almost 100,000 billion times as much as the Sun, the gas should blow itself away by its own radiation. Astronomers cannot reconcile these two estimates of the galaxy's mass, and further study of the x-ray source may lead to a reduction of the higher estimate. At present, however, we can only speculate that M 87 is much larger than it appears on these photographs, and that its outer portions are purely gaseous and contain few stars.

Many tiny, faint objects cluster around M 87 to form a halo, some of which can be seen on the main photograph. Each of these is a globular cluster containing thousands of stars, crowded together into a ball like a miniature of M 87 itself. Globular clusters are known in our Galaxy, and will feature later in this book. But whereas our Galaxy contains a couple of hundred examples, about 800 have been photographed in the halo around M 87. When we make allowance for the globular clusters which are hidden by the galaxy itself, and for the smaller examples which cannot be detected, we estimate that there are several thousand globular clusters attached to M 87.

The apparent diameter of M 87 is about 7 arc minutes, and if the halo of globular clusters is included this figure is raised to 9 arc minutes, almost one-third the apparent diameter of the Sun. This corresponds to a real diameter of 120,000 light years. Some disk-shaped galaxies have larger diameters, but are quite thin; M 87, being spherical, has the largest volume of any galaxy which has been measured.

NGC 253

Spiral in the Sculptor Group

The nearest group of galaxies to us straddles the southern constellation of Sculptor, at a distance of ten million light years. As in most small groups, the brightest galaxies are all spiral. It contains many small galaxies as well, including the SCULPTOR DWARF IRREGULAR GALAXY and some LOW SURFACE BRIGHTNESS GALAXIES.

The most distinctive of the spiral galaxies in the Sculptor Group is NGC 253, seen on this photograph. It appears elongated because we view it from only 17° above the rim. If we could see it face on instead, it would appear half the diameter of the Moon. This corresponds to a diameter a little over 40,000 light years. On this photograph, the upper left end is approaching us and the other end receding, and the galaxy is inclined so that its upper right edge is nearer to us. Dark clouds along the nearer edge are clearer because they are silhouetted against the stars of the galaxy which lie beyond.

Although it appears quite regular, NGC 253's nucleus has undergone an explosion, similar to that in M 82 (p. 34), but less violent. The highest ejection speed of gas from the explosion is 120 km/sec. The remaining nucleus, about 500 light years across, contains a sizeable mass of dust and gas. The core radiates almost one-half the energy of the whole galaxy, or about 100 billion times the energy output of the Sun. The nucleus emits most of its energy at infrared wavelengths, which makes NGC 253 the third brightest infrared galaxy in the sky.

NGC 253 AAT

3C 449

An active galaxy

In 1932, an American radio engineer, Karl Jansky, discovered that radio reception on Earth was affected by a source of interference which was strongest every 24 hours or so, peaking four minutes earlier each day. Jansky realized that this behavior implied that the source lay beyond the Earth's atmosphere, and even beyond the Sun and Moon. Something out there was giving off strong radio waves. By timing the cycle of interference, Jansky was able to locate its source in the sky. He showed it to be coming from the direction of the nucleus of our Galaxy. In fact, Jansky had discovered the first radio galaxy—our own.

The GALACTIC CENTER, although 30,000 light years away, is very much nearer to us than the nuclei of other galaxies. If all galaxies emitted radio waves no more intensely than our own, we should know of very few of them, even using today's highly sophisticated radio telescopes. Some galaxies, however, are far more powerful radio emitters than ours, and were detected when the very first radio telescopes were built. In addition, some nebulae within our Galaxy emit radio waves of sufficient intensity to appear as comparably strong sources. Enough interest was generated by these early discoveries to guarantee that radio astronomy would become established. Today it contributes greatly to our understanding of the Universe.

Radio emission from galaxies arises in distinct patterns. All spiral galaxies, for example, contain nebulae which emit radio waves and which outline their spiral arms. Some galaxies, like our own, emit from their nuclei. Others release radio waves from two lobes considerably outside the visible galaxy. In the latter two cases the radio emission is caused by what is called synchrotron radiation.

Physicists can generate synchrotron radiation in the laboratory, though in only infinitesimal quantities compared to celestial radio sources. The requirements are a magnetic field and a supply of fast-moving electrons. Magnetic fields are not hard to come by: the Earth, for example, has a field which is sufficiently powerful both to swing a compass needle and to power synchrotron emission. Galaxies, too, have magnetic fields which can often be detected by their effect on starlight—the fields polarize light, just as the lenses in

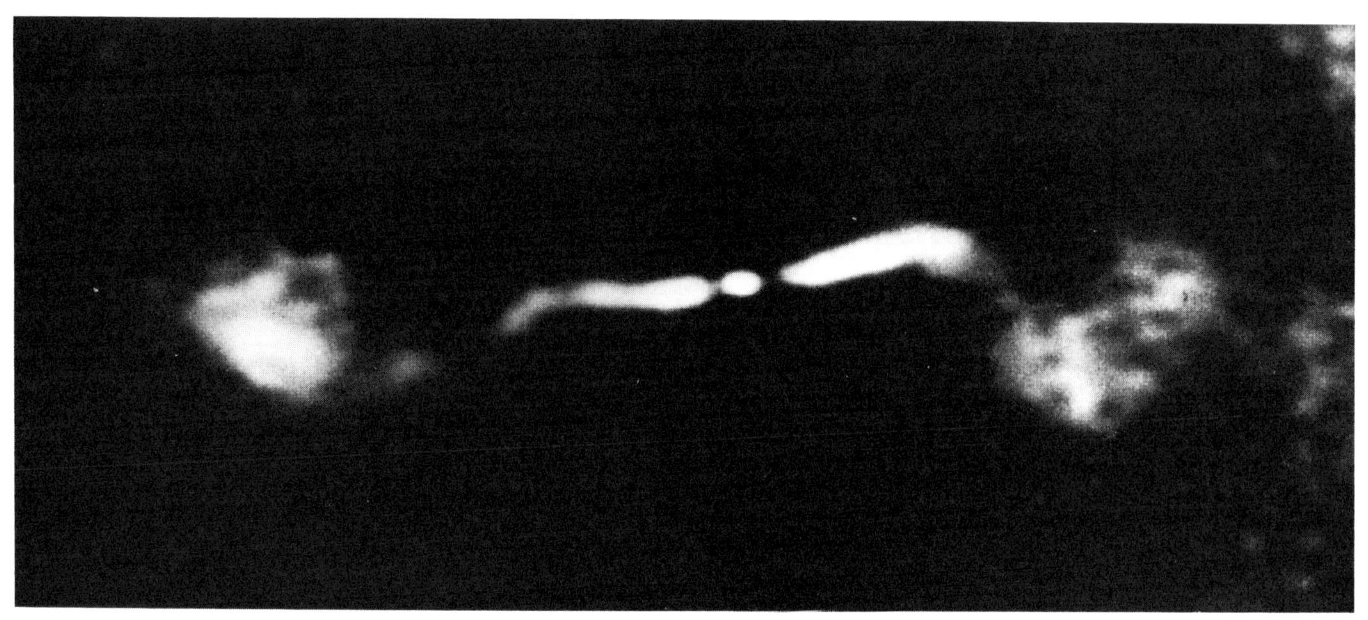

Radiograph of 3C 449 MRAO

a pair of Polaroid sunglasses do. Finding magnetic fields is therefore no problem, but finding a suitably plentiful source of electrons to generate synchrotron radiation is something which keeps astronomers in work. Uncountable numbers of electrons are continuously being injected into a radio galaxy's magnetic field. So somewhere at the heart of each radio galaxy is a violent explosion, not understood, which releases electrons at a prodigious rate.

We have already described the violent and explosive events that occur in the nuclei of some relatively ordinary galaxies. In the next pages we will illustrate even more violent specimens. In each case, the nature of the energy source at the center of it all remains a mystery.

It is a fair generalization to say that galaxies whose radio emission comes primarily from the nucleus are usually spiral, while galaxies with two lobes of radio emission are generally elliptical. The spiral galaxies are either quite conventional, including some that have already been illustrated, or else are optically distinguished and will feature in later pages. We begin by showing a typical two-lobed radio galaxy.

The ordinary-looking elliptical galaxy at the center of the *photo*graph is the origin of a source of radio emission. The other picture is a *radio*graph, made from observations with a radio telescope and shows pictorially the regions of the galaxy which emit radio waves. The galaxy lies on the central small weak radio source, on the line joining the two large, intense radio lobes. Both pictures are to the same scale. The galaxy nearer to the stronger source, which is the case in all two-lobed radio sources. Moreover, the ratio of separations of the two lobes from the galaxy is the same as the ratio of their radio strengths—that is, the stronger the lobe, the nearer it is to the galaxy.

We can draw an analogy with a balance. If the radio lobes were two balls whose weights were proportional to their radio strengths, the galaxy would lie at the point of balance of a rod joining them. Since the strength of the radio emission may well depend on the mass of material in the lobes, this is a meaningful analogy. It seems fairly certain that the two lobes were ejected from the galactic nucleus in some way. In fact, the radiograph shows that the radio emission gets more diffuse further from the center, as if being sprayed out. In a balanced ejection the heavier lobe would have traveled slower and would therefore be seen nearer to the galaxy. This is what we see.

The dimensions of the radio lobes are very great. The underlying galaxy is of conventional size. The

Photograph of 3C 449 POSS

largest radio galaxy known, 3C 236, measures 20 million light years from end to end.

In some of the two-lobed radio galaxies like 3C 449 a third source appears on the radio maps, coincident with the galaxy. New techniques have recently allowed astronomers to make very detailed maps of these galaxies. In many cases the central source is resolved into yet another pair of lobes, much more compact, and presumably ejected more recently. This is evidence that multiple explosions, separated by perhaps millions of years, occur in the nuclei of galaxies. A remarkable discovery in the case of these "double-double" radio galaxies is that the inner and outer lobes are both separated along the same line. Material from both explosions therefore left the nucleus of the galaxy along the same line. But we expect the nucleus to take part in the rotation of the galaxy, and maybe even to spin faster than the rest of the galaxy. In the many millions of years between explosions, the nucleus and the galaxy should have rotated considerably.

Why, then, does the nucleus prefer to explode in only one direction? And how does it remember its preferred direction of explosion?

Astronomers suspect that the ejection of radio lobes occurs along the rotation axis—in other words towards the north and south poles of the nucleus. This is the only direction which does not change as the nucleus rotates. It has not yet been possible to determine whether all radio galaxies rotate about an axis through the radio lobes, because the galaxies are mostly rather faint. The two pairs of lobes of the bright galaxy CENTAURUS A do point along that galaxy's poles, however.

Cygnus A

The brightest radio source in the sky

In a remote corner of the constellation Cygnus lies an insignificant little galaxy which would not have attracted the attention of astronomers had radio astronomy never developed. But this galaxy is known throughout the astronomical world. It is Cygnus A, the brightest radio source in the sky.

This photograph shows that Cygnus A is both small and faint, and the reason for this is its great distance, fully one billion light years away from us. The great distance makes its radio brightness all the more impressive. Cygnus A is radiating energy at the rate of 10^{38} watts, as much as would be released by the detonation of a nuclear bomb the size of the Earth several times a second. The origin of the energy is unknown.

Since it takes part in the expansion of the Universe, and is at great distance, Cygnus A is receding from us at a very great speed: 15,000 km/sec. If an astronaut

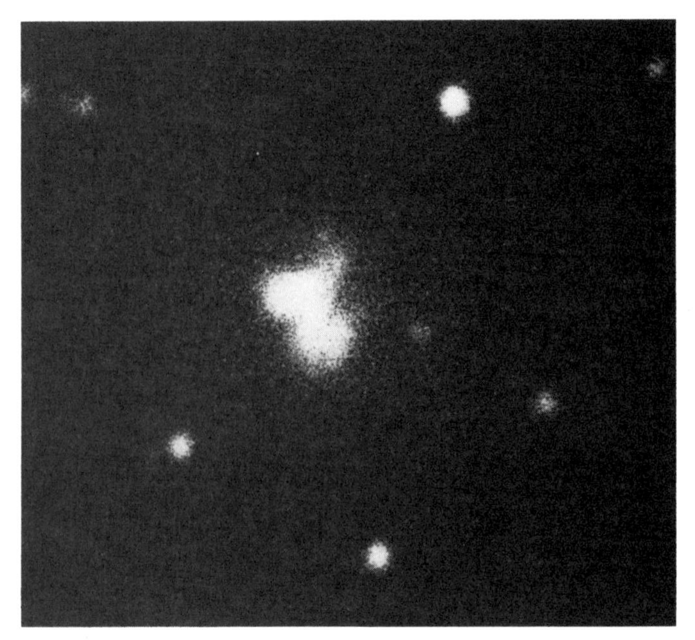

Cygnus A *Hale*

could travel at such a speed, the round trip to the Moon would take under a minute.

As a result of the release of energy in the nucleus, the gas in the galaxy is highly excited. It glows like a discharge lamp, all its chemical constituents radiating their own particular colors. The atoms in the gas are also made to dash around violently—speeds of 100 km/sec are not uncommon.

Cygnus A was the second radio source to be identified with a visible object, as long ago as 1948. Palomar Observatory astronomers Walter Baade and Rudolph Minkowski found that Cygnus A lies in a cluster of galaxies, and from the double appearance seen on this photograph, they suggested that two of the cluster members were seen in collision. This, they argued, was the origin of the energy. More recent observations have shown that Cygnus A is, like most radio galaxies, a single object. If better photographs could be taken, we might find that Cygnus A resembles CENTAURUS A (p. 41)—a galaxy cut by a surrounding ring of dust.

Radio astronomical maps of Cygnus A show it to be a classic radio galaxy. It has two bright lobes, each some 300,000 light years from the galaxy. A weaker source is exactly coincident with the galaxy. Thus Cygnus A appears to be extreme rather than unique. The radio lobes lie off the edge of our illustration.

Fornax A

A triple radio source

The brightest radio source in the small southerly constellation of Fornax has been identified with the

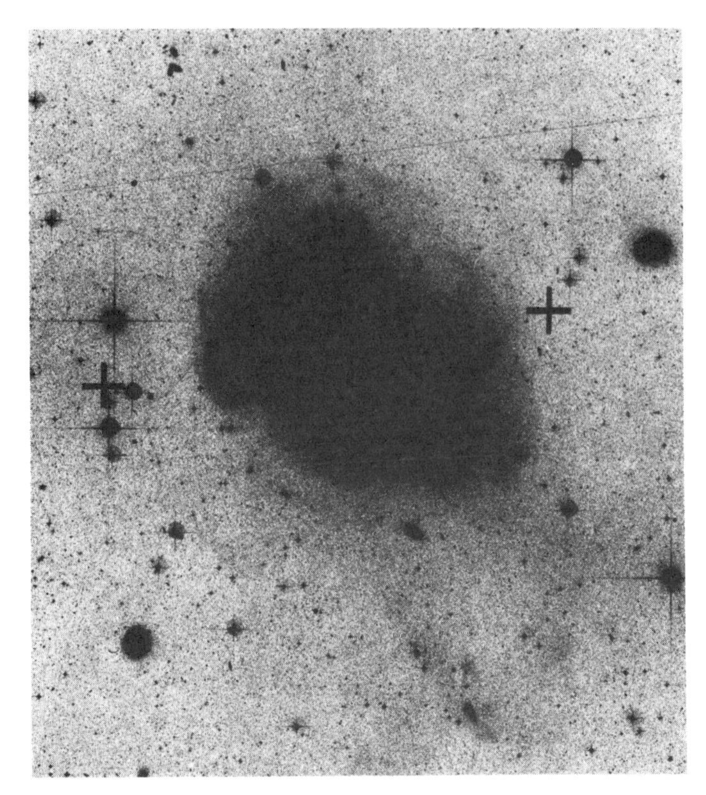

Fornax A UKSTU

larger of the two galaxies in the short-exposure positive photograph (white stars). The larger galaxy's catalogue number is NGC 1316, and the galaxy at the top of the picture is its companion NGC 1317.

The central regions of NGC 1316 pictured here are devoid of spiral structure. However, the dust lanes visible within the galaxy suggest that it is not an elliptical, and the classification of lenticular has been adopted. Unlike most lenticular galaxies, however, the faint outer reaches of NGC 1316 have a totally irregular outline, as seen on the deep-exposure negative photograph (black stars). This reveals that NGC 1317 seems to lie within the boundary of NGC 1316; however, the structure of the outer material is not disturbed near NGC 1317. This suggests that NGC 1317 is not actually involved, and instead lies either in front of or behind the radio galaxy.

The radio source is triple. Radio lobes lie one to each side of NGC 1316, and their centers are marked by crosses. A weak third source is coincident with the nucleus of the galaxy, confirming the identification between Fornax A and NGC 1316. Radio sources in lenticular galaxies are rather rare, and most sources with radio properties like Fornax A are associated with elliptical galaxies.

NGC 1316 lies rather over 100 million light years from us. The radio lobes extend 20 arc minutes on either side of the galaxy, which corresponds to a distance of at least 500,000 light years from their parent galaxy.

Fornax A AAT

NGC 5128

Centaurus A

Had Charles Messier, the 18th-century cataloguer of nebulae, lived farther south, NGC 5128 would have had an M-number like so many other prominent galaxies. But it lies in the southern constellation of Centaurus, and is difficult to explore from European observatories. The bright galaxy's title of Centaurus A results from it being the most intense radio source in that constellation. At a distance of 16 million light years, it is the nearest of the giant radio galaxies.

Following the discovery of the Centaurus A radio source, astronomers in Australia sought an optical identification. In 1952 Bernard Mills was able to pinpoint the radio source with sufficient accuracy to confirm earlier suspicions of its identity with NGC 5128. At that time NGC 5128 was thought to be a peculiar nebula within our own Galaxy. Only when Palomar astronomers Walter Baade and Rudolph Minkowski studied NGC 5128 was its extragalactic nature revealed.

The large photograph shows the principal features. Most of the light comes from an almost spherical collection of stars which is classified as a giant elliptical galaxy. The galaxy is slightly elongated along a north-south line. Across its waist, trending east-west, runs a dark lane of dust and gas, intricately woven into streaks and bands. Baade and Minkowski saw this dark band as the arms of an otherwise unseen spiral galaxy, and believed Centaurus A was a collision between an elliptical and a spiral galaxy. The energy generated by such a collision could be released as radio waves, they argued.

Today astronomers believe that the band of dark material is a girdle which completely encircles NGC 5128, and that there is but a single galaxy in the system. Moreover, measurements of the motions of stars in the galaxy show that it rotates so that the dust lane forms the equator. Thus although basically an elliptical galaxy, NGC 5128 has a disk comprising gas and dust, and maybe some stars; it therefore has some affiliations with spiral galaxies, like Messier 104, whose dust-cut image appears on p. 16.

Radio maps of Centaurus A show it to be extremely large, fully 6° by 10° in the sky, about 250 times the area of the Moon's disk. The principal areas of radio emission are two lobes extending roughly to north and south, and lying pretty accurately at right angles to the dark girdle. In Centaurus A, therefore, we have evidence that the radio lobes lie at the poles of the galaxy. Moreover, it was shown in 1961 that Centaurus A is a double-double radio source. A second pair of radio sources lies on the same line as the large lobes: these are much smaller, and lie near the northern and southern edges of the visible distribution of stars. It seems that two explosions have occurred at the center of NGC 5128, each ejecting a pair of radio sources out of the galaxy's poles.

Can we find optical evidence for the explosions which are believed to have ejected the radio lobes? Several astronomers have unearthed clues which seem to point to at least one explosion in the galaxy. Observing from Cerro Tololo Inter-American Observatory, the Dutch-Canadian astronomer Sidney van den Bergh found large numbers of young, hot stars in the dust girdle and near the center of NGC 5128. Such stars are not normally found in elliptical galaxies, and their concentration indicates that some violent event occurred perhaps 30 million years ago, precipitating their formation. John Graham, also at Cerro Tololo, recently discovered a jet 130,000 light years long stretching north into the radio lobe. The jet was a discontinuous chain of blobs and streaks which appear to be clouds of gas made luminous by the hot stars they contain. As in the case of the stars around the galaxy's central regions, the hot stars must be no more than a few tens of millions of years old.

These pieces of evidence focus our attention on the very heart of the galaxy whence the explosions might have originated. Unfortunately, this lies behind the

Centaurus A AAT

dark dust lane, and is thus difficult to study. The effect of the dust, like smoke from a chimney, is to redden the light from the nucleus behind it. A deep orange spot of light can be seen at the very center of the galaxy, and this is visible on the color photograph (Plate 1). This is a giant cluster of hot stars and gas with a total luminosity two billion times greater than that of our Sun, and thus brighter than many entire galaxies. This, however, is not the true heart of NGC 5128. At the center of this cluster of stars is a compact source of radio, infrared and x-radiation. Here broods a dense and very luminous object, yet another of the mysterious galactic nuclei whose nature we are as yet unable to fathom. The x-rays show it to be a variable object, ranging in its energy output by a factor of three or four over a period of typically five years. This variability is not well-determined, since its x-ray emission has been known only since 1970. On top of this slow variation there are short flares in the x-ray brightness, occurring about once a month. Do these flares represent individual large stars or clusters of stars being swallowed by a black hole? This is a particularly intriguing possibility, but one which will be very difficult to prove correct. The energies involved in Centaurus A are so high that gamma radiation, more energetic even than x-radiation, has been detected from the galaxy.

Because it is quite close, Centaurus A also gives us a chance to examine the extended lobes of radio galaxies. For example, they provide the first case of lobes from which x-radiation has been detected. The x-rays seem to originate from the very extremities of the radio lobes, and as the sensitivity of x-ray observations increases NGC 5128 may be shown to exceed 10° in length.

The larger photograph suggests that the galaxy generates visible radiation at least some distance into the radio lobes; this reproduction shows the faintest of Centaurus A's features recorded on a plate taken with the U.K. Schmidt Telescope in Australia. The total length of the visible nebulosity here is 1.2°. The reason for all the visible radiation is not known. Throughout the radio lobes there are patchy clouds of hydrogen gas, quite unseen except for their influence on the radio spectrum of the lobes. These probably are part of an enormous halo of gas around the whole galaxy.

NGC 5128 is receding from us at 450 km/sec. Its total brightness is 100 billion times that of the Sun, and the radio lobes are three million light years in total dimension. If they were ejected 30 million years ago, as suggested by the fact that there are still hot stars in the galaxy's central regions, material must have been ejected at 15,000 km/sec—5% of the velocity of light.

NGC 1265

A galaxy with a radio wake

In the constellation of Perseus lies a distant, loose cluster of galaxies of which several members have quite distinctive properties. The centerpiece of the cluster is the unusual galaxy NGC 1265. In our illustration we have superimposed a radio contour map onto a photograph of NGC 1265. The contours indicate the regions of intense radio emission just as contours indicate hills on a geographical map.

The radio map of this galaxy shows a different looking structure from the normal double-lobed radio galaxies illustrated on the previous pages. Blobs of radio emission stretch out in two almost parallel streams rather reminiscent of the wake of a boat. NGC 1265 is the best-studied example of the very small number of so-called head-tail radio galaxies having a head at the position of the visible galaxy and a tail trailing from it.

The Perseus cluster lies some 340 million light years from us. It is distinguished by having the largest spread of velocity between its members of any known cluster. Thus while the average member is receding from us at 5200 km/sec, some member galaxies have velocities differing from this figure by over 2000 km/sec. NGC 1265 is one of these, with a total recession velocity of 7400 km/sec. Compare this with the velocity attained by a satellite orbiting the Earth, a mere 8 km/sec. Somewhere near the center of the Perseus cluster lurks something very heavy. It is much heavier, indeed, than all the stars of the member galaxies we can see; something which exerts so strong a gravitational pull that galaxies such as NGC 1265 fall in orbit around it very quickly.

The fact that NGC 1265 is moving rapidly through the cluster gives us a clue to its radio structure. Many astronomers believe that the tail really is a wake generated as the galaxy moves through the gas of the cluster. If NGC 1265 were not moving, they argue, we would see a series of pairs of radio lobes, all ejected more or less along the same line, much as in 3C 449. Because of its motion, the multiple lobes have been dragged back behind it to form the radio tail.

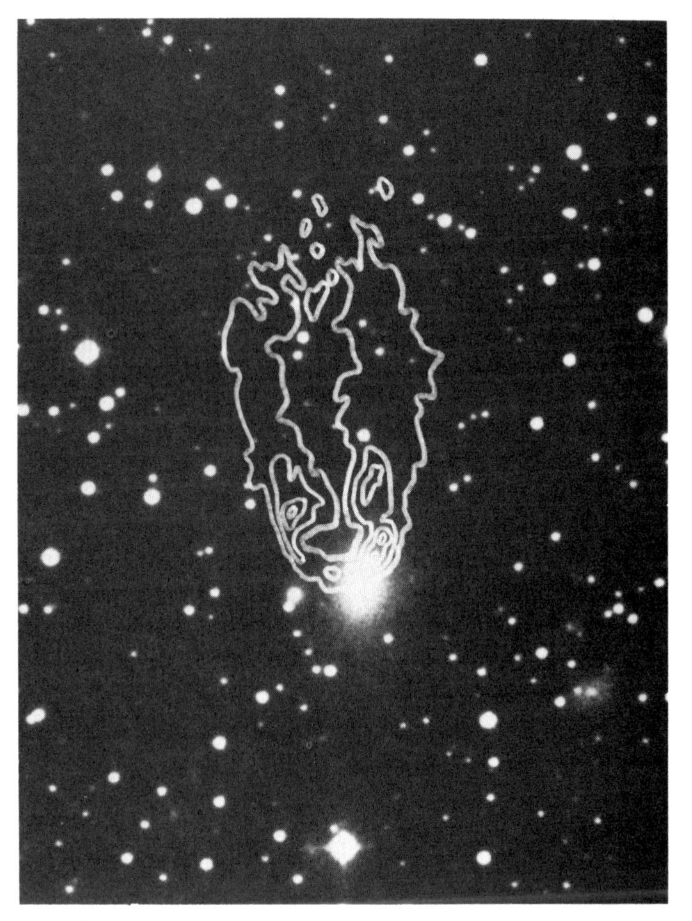

NGC 1265 POSS

Seyfert Galaxies

Energetic spiral galaxies

In 1943, the American astronomer Carl Seyfert drew attention to a small number of distinctive galaxies

NGC 3783 AAT

NGC 1566 AAT

which he had noted. All were spiral galaxies with very small, bright nuclei. Spectroscopic observations showed that within the nuclei of these galaxies was gas, excited by something very hot or energetic, and moving at very high speeds. Compare this with normal spiral galaxies, which have large nuclei of cool stars, generally with little gas. Whereas the *stars* in the middle of spiral galaxies orbit at speeds of a few hundred km/sec., the *gas* in Seyfert's galaxies moves at up to 5000 km/sec. It appears that this is not merely an orbital motion, but reflects either an infall or an outflow of material.

Ignored like the works of so many artists, the scientific paper in which Seyfert listed six galaxies with these properties languished for two and a half decades. At that time, so little was known about normal galaxies that it was not useful to study unusual specimens. In the late 1960's, however, more galaxies were recognized to have the same properties. Better telescopes and better equipment were available to study these objects, and currently they receive a great deal of attention.

Today, galaxies of this type are known as Seyfert galaxies, and about 200 are known. This number is growing rapidly. Seyfert galaxies account for a significant proportion of all large spiral galaxies, and are relatively easily found because they are intrinsically brighter than normal galaxies. Two types are recognized: Seyfert 1 galaxies are the brighter, have the faster-moving gas, and emit synchrotron radiation. As described earlier (p. 38), synchrotron radiation accounts for the radio frequency output of some galaxies. For synchrotron radiation to be dominant at optical wavelengths, however, much more energy must be released as electrons in a strong magnetic field. In Seyfert 2 galaxies, the gas moves more slowly and the light comes from hot stars. The distinction between the two types of Seyfert galaxies is not entirely clear cut; some galaxies have been known to display alternately Seyfert 1 and Seyfert 2 characteristics.

By careful research, astronomers are beginning to unravel the mysteries of Seyfert galaxies. The gas, they have learned, exists in individual blobs or filaments which are moving very rapidly within a region only a few light years across. These gaseous filaments do not completely fill all of this space. On the contrary, they occupy about one-millionth; they probably move outwards through a void which previous gas has swept clear. Amid these filaments is a minute nucleus within which vast amounts of energy are generated and converted into synchrotron radiation. A typical bright Seyfert galaxy radiates several million million times as much energy as the Sun. Although unable directly to measure the size of this nucleus, astronomers infer

NGC 1068 POSS

that it has a diameter no more than about 10,000 times that of the Sun, and maybe much less.

What is the source of all this energy? Most astronomers prefer to sketch diagrams of Seyfert galaxies with a question mark at their centers. Some are prepared to suggest that, like M 87, a black hole has formed at the center of each Seyfert 1 galaxy, and possibly within Seyfert 2 galaxies too.

The overwhelming force of gravity, which prevents even light from escaping from a black hole draws into the black hole any further material which comes within its influence. In his wildest dreams, Tolkien never imagined so terrifying a creature as a black hole at the nucleus of a galaxy, invisible, yet all-engulfing,

NGC 4151 POSS

swallowing into oblivion stars far larger than our Sun. Although the black hole itself would not give out energy, the stars and gas falling into it would be compressed and heated and would release large amounts of energy before they became engulfed. It is quite feasible that the energy so released accounts for that produced in the nuclei of the Seyfert galaxies illustrated here.

The radio galaxies described earlier were all elliptical; Seyfert galaxies are all spirals. Radio galaxies and Seyferts both have unusual energy sources in their nuclei. This fact suggests that they are the same type of object. Astronomers currently believe that the two phenomena are closely related, and that Seyfert galaxies would be radio galaxies if they had no gas in their nuclei. Spiral galaxies do contain gas, which absorbs the energy output by the nucleus, producing the Seyfert galaxy spectrum. If the gas were not present, the energy could escape to excite giant radio lobes.

X-Ray Galaxies

Like Seyfert galaxies?

The galaxies on p. 46-47 are members of a type only recently discovered. Until satellites orbited above the Earth's atmosphere, the existence of objects in the sky which outpoured radiation of very high energy—x-rays—was unsuspected. Only in the mid 1970's were some of the sources of x-radiation identified as galaxies. When one considers the distances of these galaxies, typically several hundred million light years, the energy emitted from their nuclei must be enormous. Thus astronomers expected the x-ray galaxies to resemble Seyfert galaxies.

There certainly are some similarities beween x-ray and Seyfert galaxies. Both seem to have small, bright nuclei with a large gas content. Most or all of the x-ray emission comes from this nucleus. Both resemble spiral rather than elliptical galaxies. Moreover, many Seyfert galaxies have also been found to emit x-rays.

But if there are similarities, there are also features which distinguish x-ray galaxies from Seyferts. One difference is apparent here: they are rather irregular and contain more dark dust clouds. It is thought that the presence of large amounts of dust enhances the galaxies' x-ray emission by complicated physical processes which cannot easily be reproduced on Earth. The irregularity may indicate that more violent and energetic events occur in their nuclei, and this would be consistent with the generation of more intense x-radiation.

All this suggests that x-ray galaxies are more energetic versions of the Seyferts. But one piece of

NGC 1365 AAT

NGC 5506 AAT

NGC 7582 AAT

NGC 2110 POSS

evidence contradicts this belief. Whereas in the Seyfert galaxies, gas is forced to move around at very great speeds, in these x-ray galaxies no fast-moving gas is observed.

Because the x-ray galaxies were so recently discovered, they are not yet well understood by astronomers. Every year many more are located; each helps to elucidate what processes are going on in their nuclei.

Quasars

Galaxies masquerading as stars

Radio galaxies, Seyfert galaxies, x-ray galaxies: all of these appear to be conventional galaxies with overly luminous nuclei. Suppose a galaxy has a nucleus so bright that it outshines the stars in the galaxy. Light from the nucleus would hide the light from the stars. The galaxy would be so bright that it would be visible at much further distances than conventional galaxies. Indeed, because such a power output is liable to be rare, we would have to look out to large distances in order to survey enough volume of the Universe to see many examples, just as you would probably have to

4U 0241 & 61 POSS

Sagittarius and center of our Galaxy

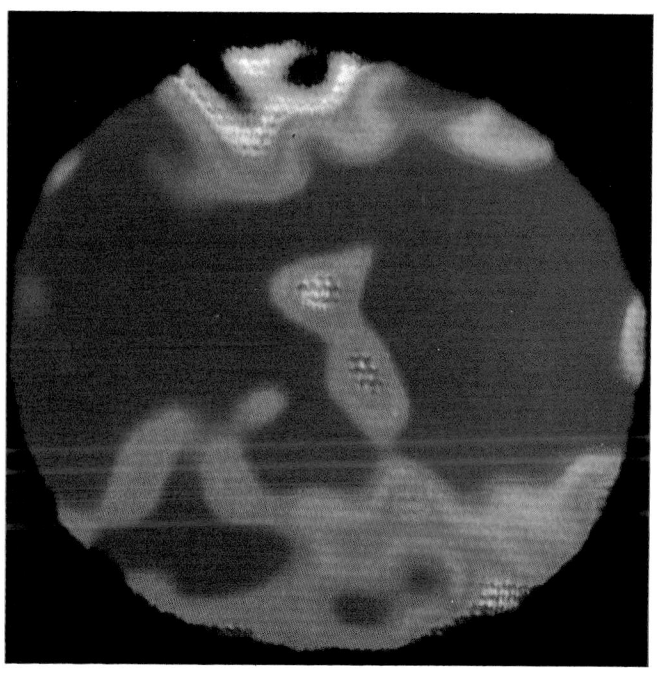

South Celestial Pole

Face of Betelgeuse

PLATE 5

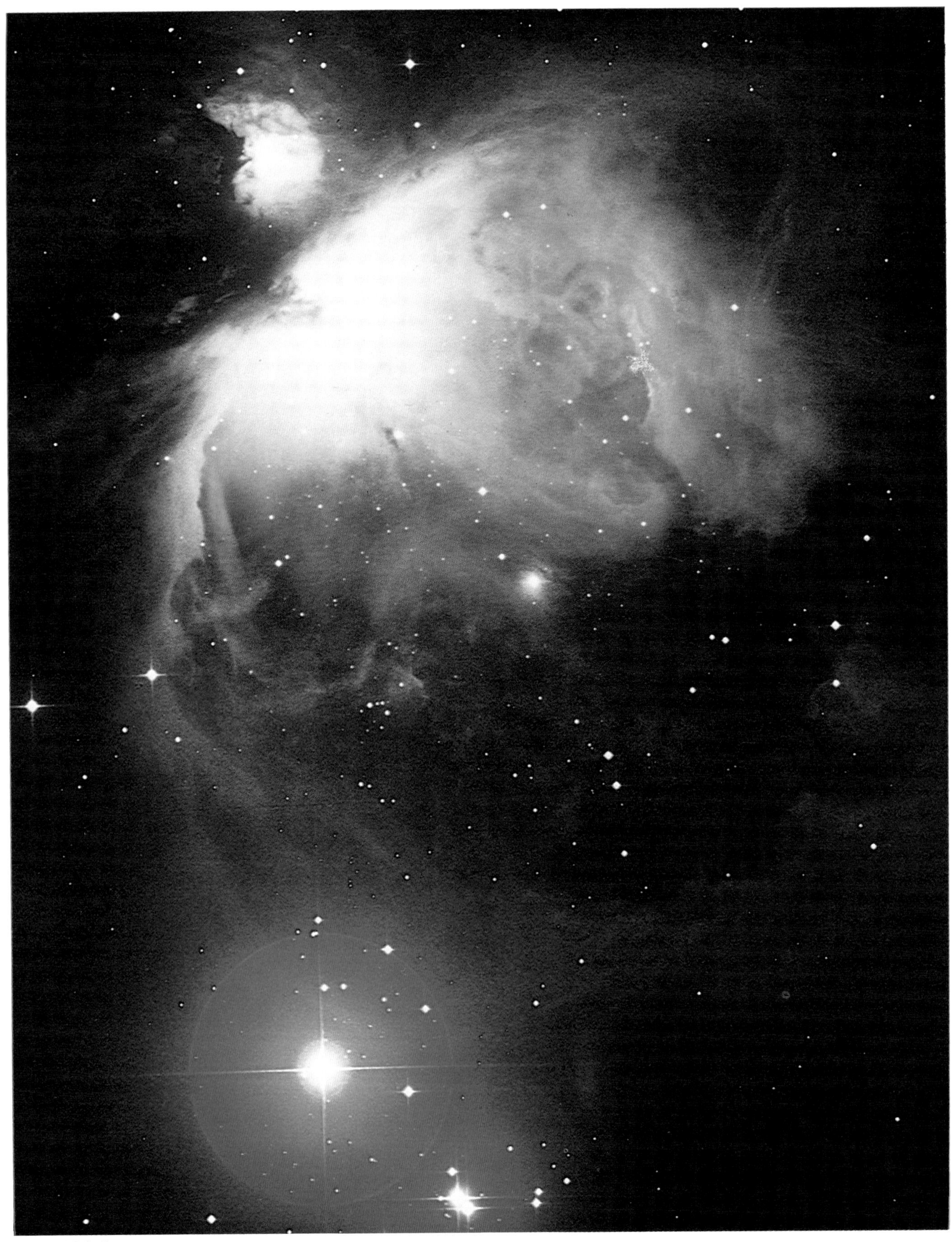

Orion Nebula

PLATE 6

Rosette Nebula

PLATE 7

Trifid Nebula

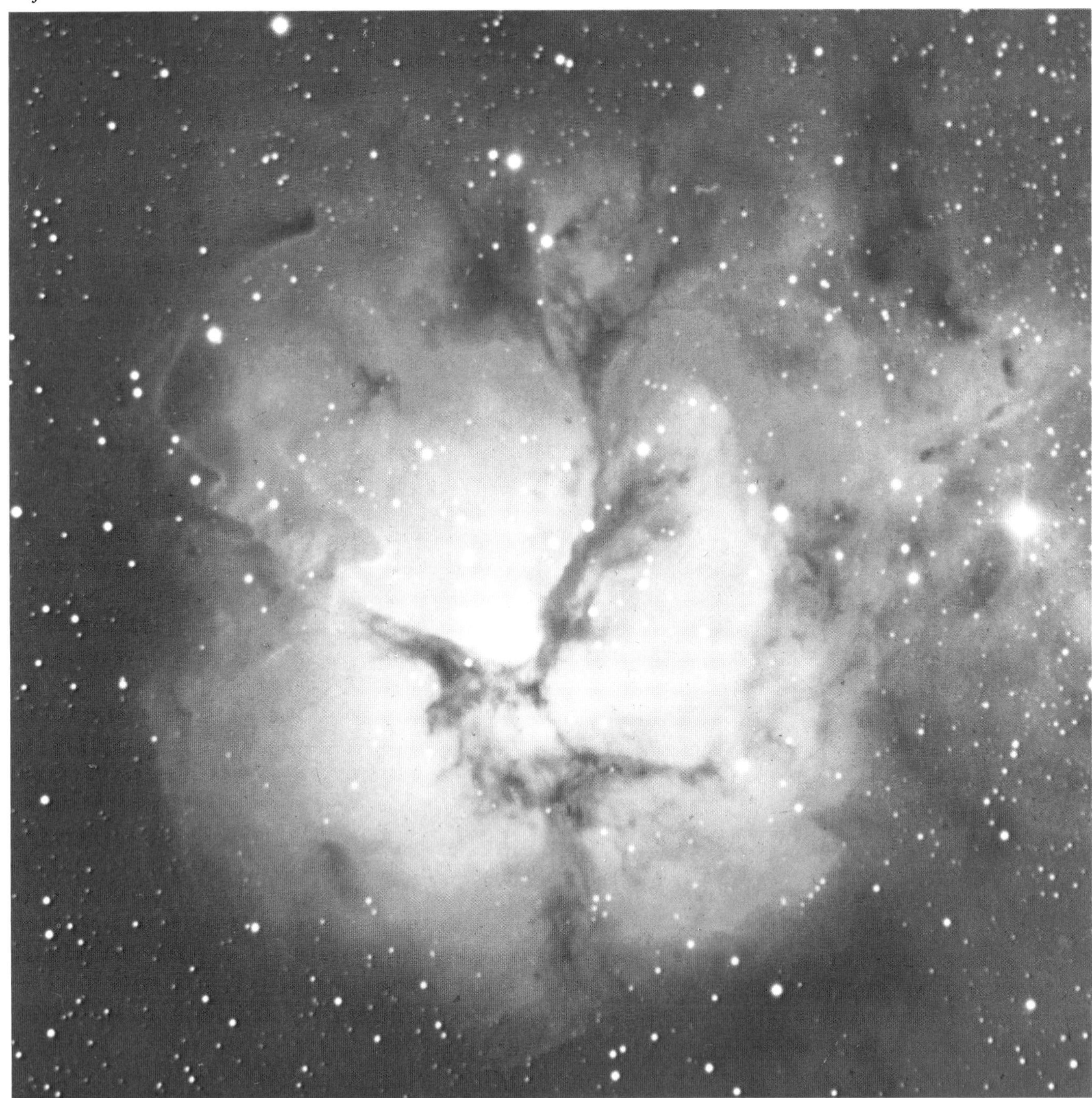

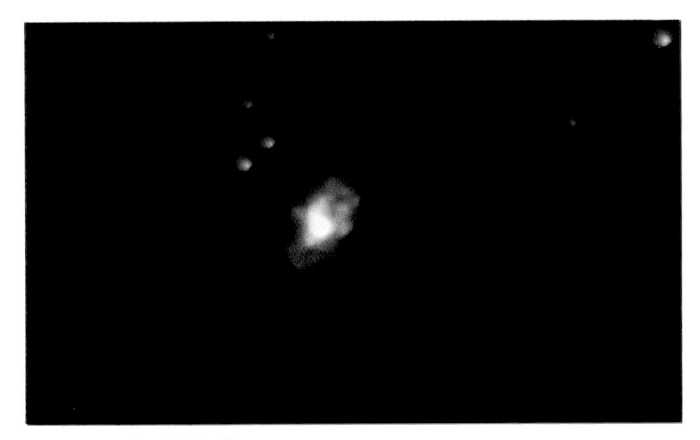

Homunculus Nebula

PLATE 8

search a wider area of the U.S.A. to find a high-powered Lamborghini automobile than a low-powered Volkswagen. Thus galaxies with excessively bright nuclei might well lie much further away than most galaxies.

Perhaps the majority of astronomers believes that galaxies with such bright nuclei have been found. They are quasars.

What are the properties of quasars which lead to this view? First, quasars do not look like galaxies. Their very name was derived from *quasi-stellar*, signifying that quasars appear starlike on photographs such as these. Astronomers find that, despite apparently coming from stars, quasars light is shifted towards the red end of the spectrum. The redshift closely resembles that found in galaxies. In the case of galaxies and clusters of galaxies, redshift is proportional to distance according to the well-known Hubble relationship (p. 11). If we adopt the same proportionality, we find that quasars lie at much greater distances. The *nearest* known quasar (2S 0241+622) is almost a billion light years away, while the most remote (OQ 172) lies 20 times as far away. By comparison, few galaxies have been recognized more than two billion light years from us. Although lying at much greater distances than galaxies, quasars appear equally bright and must therefore emit much more energy.

The first quasars to be discovered, in 1963, were singled out from true stars by their radio emission. Again if redshift can be linked to distance, their radio output far exceeds that of normal galaxies. These days, astronomers can detect quasars which have no radio emission, and indeed the majority seem to be radio-quiet. By this very fact, such quasars are more difficult to discover: this explains why most quasars known are radio-emitting. Recent results also indicate that many quasars strongly emit x-rays. Optical astronomers find that the physical conditions in quasars very closely resemble those in Seyfert galaxies. If quasars are Seyfert galaxies with extremely luminous nuclei, the

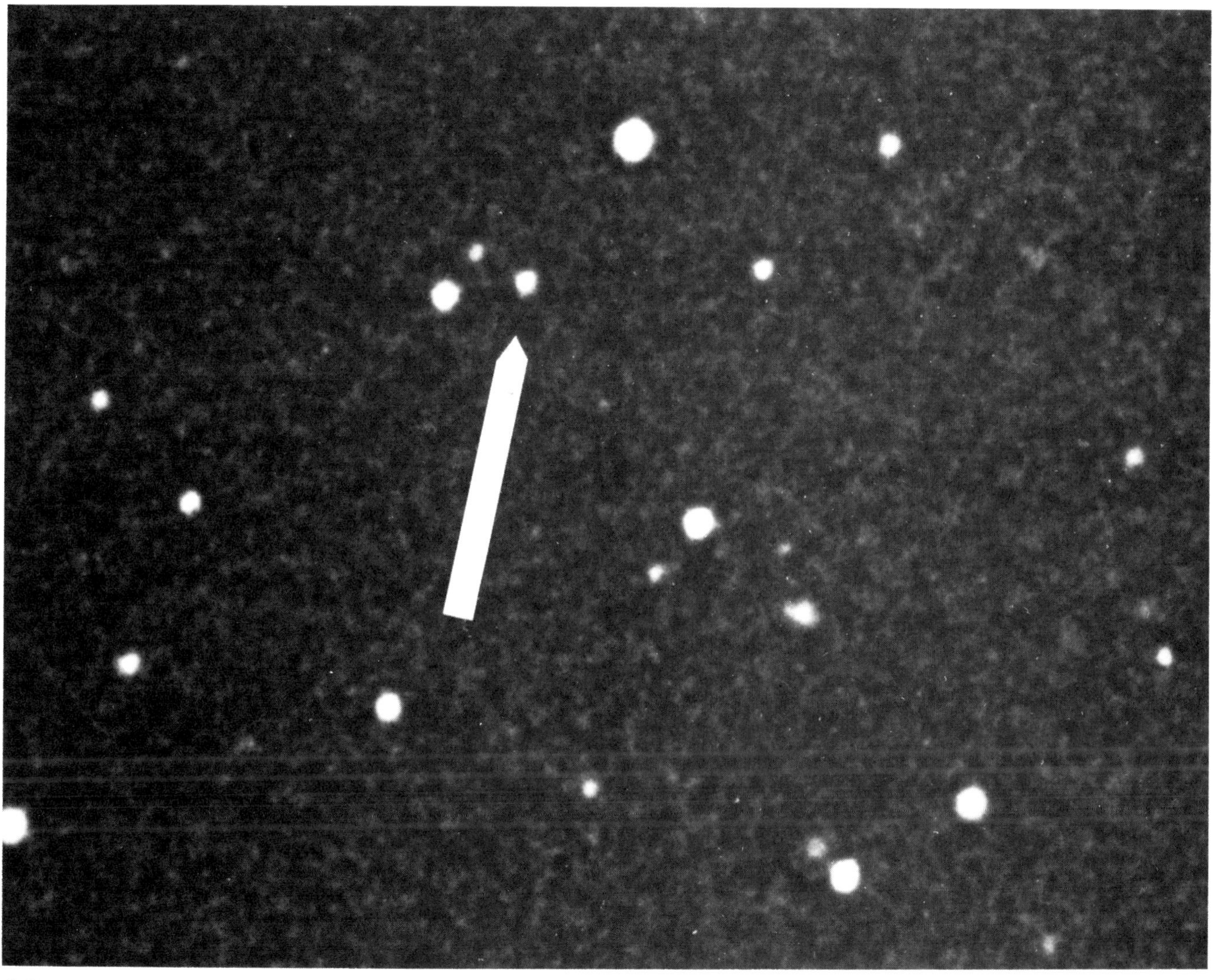

OQ 172 POSS

glare from each nucleus would be enough to obscure all the other stars in the galaxy, thus rendering the quasar stellar in appearance.

If we restrict our attention to objects at high redshifts, our sample is composed entirely of quasars. It is therefore easy to be misled into believing that quasars are very common in the Universe. This is not so: normal galaxies far outnumber quasars, but are too faint to be seen at distances corresponding to large redshifts. We get a more realistic estimate of the proportion of galaxies that are quasars by examining that portion of the Universe which lies near the Sun. Suppose we were to list the million nearest galaxies to us. That list would include less than half a dozen quasars. Moreover, as quasars go these would be extremely faint specimens. Only one galaxy in a hundred million is a high-luminosity quasar.

Not all astronomers accept this interpretation of quasars. If quasars are truly remote, as their redshifts suggest, the luminosity of quasars are so great that there is some doubt whether any mechanism can generate so much energy. But redshifts can arise in ways other than by velocity of recession, according to the theory of relativity. And even if the redshifts are a result of recession, quasars may not obey Hubble's law. Some astronomers therefore argue that quasars are no more distant than conventional galaxies. It has even been suggested that they are stars within our own Galaxy.

The nearest known quasar is 2S0241 +61, which was discovered as recently as 1978. Since it does not emit much energy at radio wavelengths, this faint stellar object in its crowded star field had attracted no attention. In 1978, the object was found to coincide with an x-ray source; only then was it shown to have the optical properties of a quasar. The distance of 2S0241 +61 is 800 million light years.

At the outermost edge of the observed universe lies OQ 172, a quasar discovered in 1973 by Lick Observatory astronomer Joe Wampler and his collaborators. The object had first been detected as a radio source during a survey made by a team in Ohio. Optical observations by Wampler's team confirmed it to be a quasar. The distance of OQ 172 is not known with any great accuracy, since at such remoteness our interpretation of distance depends on how we envisage the Universe to be constructed. We do not know, for example, whether Hubble's law which relates distance and recession velocity is still valid. If it is, the distance to OQ 172 is nearly 18 billion light years. Since 1973 astronomers have been seeking quasars at greater distances without success. Many examples have come to light that lie almost as far away as OQ 172, and a more distant specimen may be found at any time (even before this book goes to press). The failure to better the record set by OQ 172 is prompting astronomers to ask whether we have, at last, reached the real edge of the observable Universe.

3C 273

Faster than light?

The brightest quasar in our skies is numbered 273 in the third Cambridge catalogue of radio sources. For several years it had been known only as a strong radio source in the constellation of Virgo, with nothing to distinguish it from other similar sources. In the early 1960's radio telescopes were incapable of pinpointing sources with adequate precision to identify their visible counterparts.

But by lucky chance, the Moon periodically passes in front of 3C 273, cutting off the radio radiation. By timing exactly when this occurs, it is possible to determine quite precisely where the source is. This experiment was first performed in 1963, in Australia, by Cyril Hazard using the Parkes 64-meter radio telescope. Hazard and his collaborators showed that 3C 273 comprised two sources. The fainter, 3C 273B, coincided with the moderately bright starlike object at the center of this photograph and at the time was thought to be the first true radio star. 3C 273A lay almost 20 seconds of arc away. Photographs such as this one revealed a faint jet stretching from the stellar object to the brighter radio source.

Spectra of the stellar object were quickly taken, by Palomar Observatory astronomers Maarten Schmidt and Bev Oke. They found 3C 273 to have a spectrum like that of a Seyfert galaxy, but to be receding from us at almost one-sixth the speed of light. It was clear that this was no ordinary star emitting radio waves, and it soon became known as a quasar—a quasi-stellar radio source.

If, as most astronomers believe, quasars obey Hubble's law, 3C 273 lies at a distance of about three billion light years. This in turn implies that it is more than a factor of ten more luminous than the brightest known Seyfert galaxies.

Interest in 3C 273 was maintained by a sudden surge of its radio output, amounting to a factor of three, over the period 1965 to 1967. Some still-unexplained explosion occurred in the quasar. The explosion arose within a space less than one light year across, yet its energy output was greater than that of any known galaxy. In 1969 an x-ray detector flown on an Aerobee rocket detected 3C 273 as a source of x-radiation; it was the first quasar to be found as an x-ray source, and caused a further stir of excitement because few astronomers at that time expected quasars to emit so much radiation at so short a wavelength.

But it was in 1971 that the most striking discovery was made. Radio astronomers had by then developed

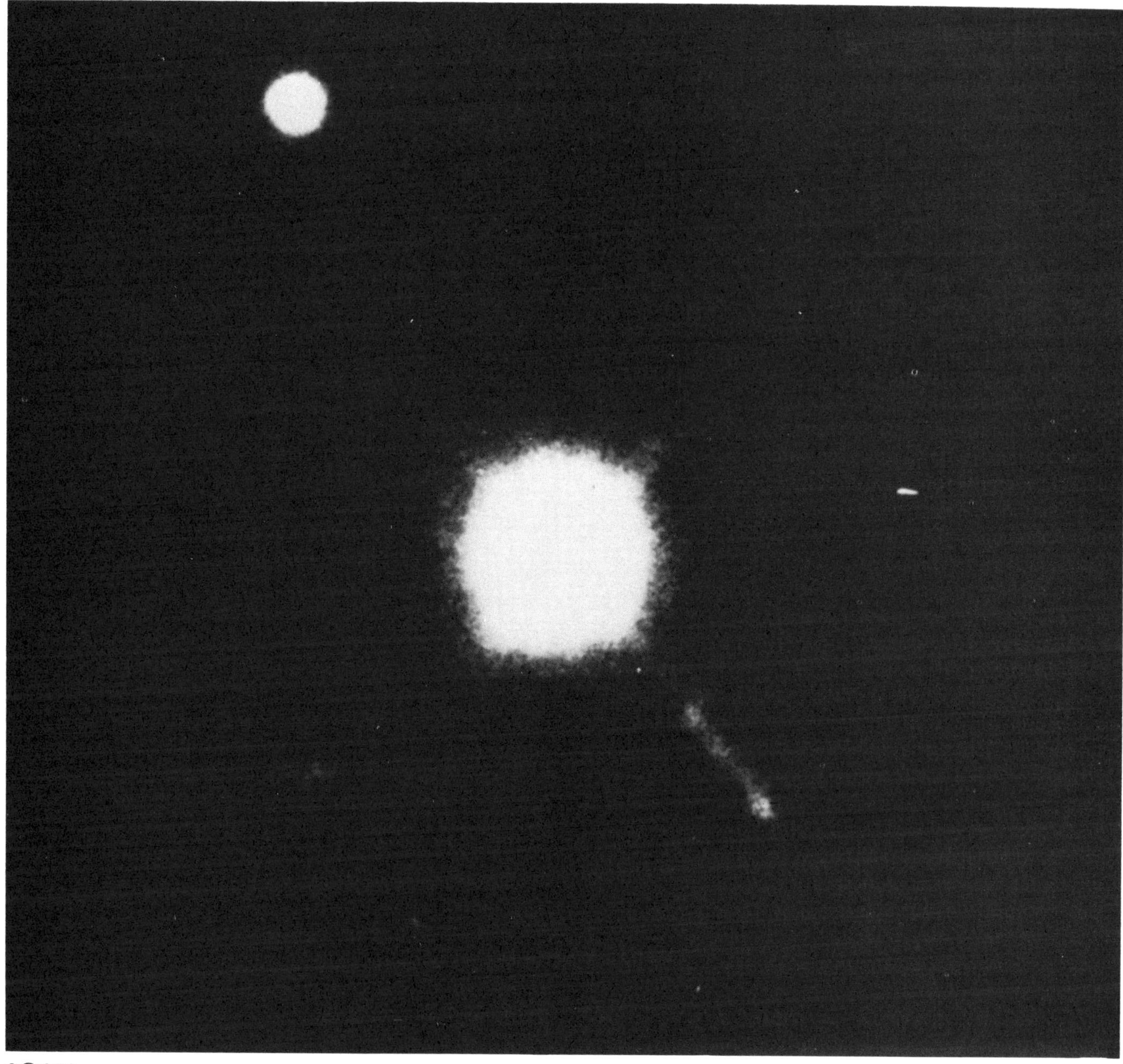

3C 273 KPNO

techniques for studying sources on a scale of a fraction of a second of arc. They accomplished this by combining the signals from radio telescopes almost on opposite sides of the Earth, and thereby simulated, in certain respects, a telescope 12,000 km in diameter.

It was already known in 1971 that 3C 273B was smaller than two-thousandths of an arc second in diameter, or about 30 light years. As the techniques improved, astronomers found that 3C 273B was itself double on a scale smaller even than 30 light years. Then in 1971 came a measurement which showed the two components to have moved apart. It appeared that 3C 273 was developing a new jet before the eyes of modern astronomers. But the astounding observation was that the speed of ejection of the new jet (assuming 3C 273 is indeed at the distance which Hubble's law indicates) is three times the speed of light!

For about seventy years scientists working in all fields of physics have accepted the reality of Einstein's precept that nothing can move faster than light. How could theory and observation be reconciled?

Today we know of several quasars whose component sources are separating at an apparent speed in excess of that of light. We can attempt to explain the effect in several ways. We can say that Einstein was wrong; but so great a body of data supporting the theory of relativity and its related laws is not so lightly set aside. We can say that quasars are not so distant as their recession velocities suggest; and some rebel astronomers believe this is the case. We can even argue that the laws of physics are different in the remote parts of the Universe where quasars lie, but we say that with little conviction.

Astronomers are still debating the best explanation of these velocities apparently in excess of the speed of light. Currently the majority view is that we happen to be viewing the relevant quasars from a favored direction. It transpires that if the components are moving very nearly, but not exactly, along our line of sight, they can *appear* to separate at velocities several times that of light. This curious effect comes out of the mathematics of the same theory of relativity which prohibits *real* motions faster than the speed of light!

So is 3C 373 developing a new jet in addition to its present one? If we assume that 3C 273A and the visible jet were ejected in a similar way to the new jet, we can date the old explosion to about 20,000 years ago. It seems unlikely that explosions arise only once every 20,000 years, and yet that one should have occurred within a few years of the discovery of 3C 273. More likely, the event which produced 3C 273A was unusually energetic.

BL Lacertae Objects

Quasars masquerading as variable stars

Quasars are the most luminous single objects known; for that very reason they are uncommon. One step down in luminosity from quasars is a class of objects which is even rarer— the BL Lacertae objects.

The first BL Lac objects to be located had previously been classified as stars in our own galaxy (the designation BL Lac is that of a variable star in the constellation Lacerta), and their true extragalactic nature was demonstrated only in the late 1960's. The photographs show the two best-known examples, BL Lacertae and AP Librae.

The characteristics of BL Lac objects are diverse. All vary irregularly in their light output, hence the variable star designation. All emit strongly in the radio domain. BL Lac itself was identified because it is listed in the Vermilion River Observatory catalogue of radio sources. Also, all are bright at infrared wavelengths. Their light is polarized, though not so strongly that the polarization could be detected by wearing polarizing sunglasses and tilting one's head from side to side. Finally, BL Lac objects are distinguished by having featureless spectra at visible wavelengths. When their light is dispersed through a prism, no bright or dark features can be made out. Thus we are prevented from learning much about them, even by this technique, the most powerful tool available to astronomers. The lack of features is

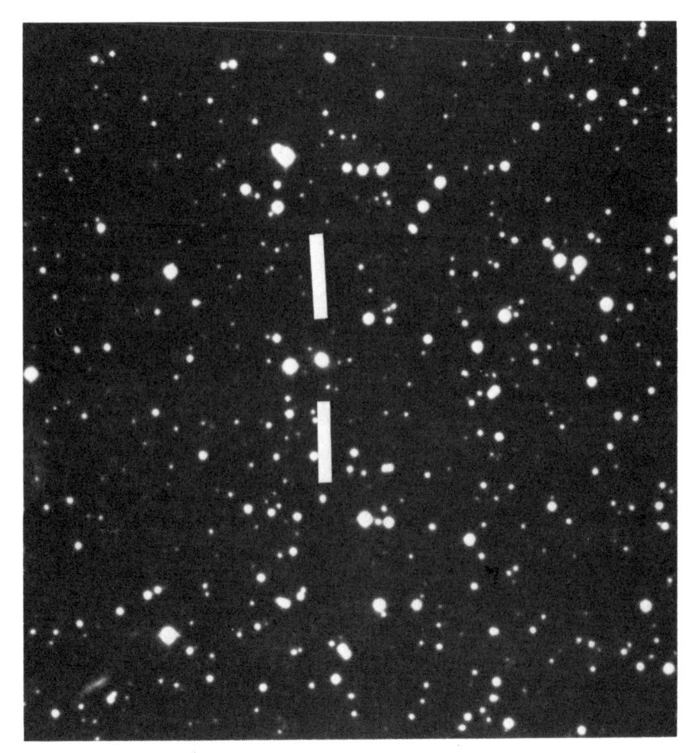

BL Lacertae POSS

probably due to an absence of gas in BL Lacertae objects and this must be a clue to their nature.

BL Lacertae and AP Librae can both be seen to be fuzzy, like galaxies. In each case, the fuzz is made up of the stars of a large galaxy lying at a great distance from us. Typical distances of the known BL Lac objects are a billion light years. Like quasars, they appear to be galaxies in which something very energetic is happening to the nucleus. In fact, since they are obviously the bright nuclei of galaxies many astronomers view them as a kind of missing link between QUASARS (so bright they outshine any stars in their parent galaxy) and most galaxies (in which stars outshine the nuclei).

Naively we expect the second most luminous class of objects to be commoner than the most luminous. Since BL Lac objects are much rarer than quasars, some astronomers believe that they are short-lived phenomena. If their life expectancy is shorter than that of quasars, we are less likely to see them since there will be few at any one time. If this argument is correct, they may represent very young or very old forms of quasars or Seyfert galaxies.

Local Group of Galaxies

The Andromeda Galaxy

Most distant object visible to the unaided eye

If we lived amid a rich cluster of galaxies, our sky

AP Librae UKSTU

would be filled with its members. Galaxies would be easily seen through binoculars, or even to the unaided eye, in many parts of the sky. Plainly this is not the case. However, it seems that we do inhabit a sparse collection of galaxies, which is referred to as the Local Group.

Our Galaxy is the second or third largest member of this group, and several other galaxies which do look fairly bright in our skies also belong. Three of these were catalogued by Messier, one of the first astronomers to list the fuzzy objects he had seen in his explorations of the sky. These are now referred to by their Messier numbers, M 31, M 32 and M 33.

The Local Group is about 500,000 light years in diameter, and many of its members are concentrated into two associations, centered on M 31 and our Galaxy. Unlike some of the groups illustrated earlier, there is little evidence that major tidal interactions have thrown streamers of stars out of any of its members. However, our Galaxy is gravitationally influencing some of its neighbors.

Currently about 30 galaxies are recognized as members of the Local Group, but there may be more, particularly among the LOW SURFACE BRIGHTNESS GALAXIES currently under investigation in Australia. Four Local Group members are spiral galaxies, M 31, M 33, Maffei 1 and our Galaxy; the remainder are dwarf elliptical and irregular galaxies. Some of these will be introduced in the next few pages. The first is the Andromeda Galaxy, M 31 (Plate 2).

The Andromeda Galaxy is a large spiral galaxy greatly tilted to our line of sight so that it appears elliptical. We view it from an angle of 13° above its edge. The central part appears quite bright, and was noticed as a small, luminous patch in the sky by the unaided eyes of Arab astronomers of the 10th century. M 31 was discovered from the murky skies of Europe only with the aid of a telescope, by Simon Marius in 1612. He saw the central quarter degree of the galaxy; on modern deep photographs it can be traced out over an area 1.25° by 4.1°:20 times the apparent area of the full Moon. The radius of the circular disk of the Andromeda Galaxy is about 80,000 light years, and its distance is 2,300,000 light years.

The bright, central nuclear bulge of M 31 is surrounded by patchy spiral arms, particularly prominent to the west (left of photo on p. 54) where dark lanes of dust are silhouetted against the bright background of the nuclear bulge. The west side is therefore the nearer; the east side is more distant although it looks closer because it is brighter. Measurements of the speed of both extremities of the Andromeda Galaxy have shown that one end is approaching us and the other receding: the sense of rotation is such that the Andromeda Galaxy's seven spiral arms are trailing back.

The total mass of the Andromeda Galaxy is equivalent to 300 billion Suns and it emits as much light as 13 billion Suns. Most of the light is emitted from the central bulge, which contains old red stars, but most of the mass resides outside the central bulge, in the spiral arms. These contain younger, blue stars and the dust and gas from which they have recently formed. The difference between the inner bulge and the outer spiral arms of M 31 led Walter Baade in 1944 to recognize the existence of two populations of stars. If Baade had had a color picture like Plate 2 he would not have needed to work so hard to establish this idea.

M 31 has two close companion galaxies, M 32 (NGC 221) and NGC 205, as well as two more distant companions, NGC 147 and 185. M 32 was discovered by Le Gentil in 1749. NGC 205 was known to Messier, although not included in his catalogue for some reason, and its discovery is often credited to Caroline Herschel in 1783. Much smaller than M 31, the companions orbit the main galaxy like planets orbiting the Sun; M 32 is currently moving in front of its parent galaxy. Their presence seems to have warped M 31 out of true, for its spiral arms do not all lie in the same plane but are noticeably kinked in places. Both M 32 and NGC 205 are small elliptical galaxies.

Messier 33

Spiral galaxy in the Local Group

Another spiral galaxy in our Local Group is Messier

Messier 31 *Hale*

33. Not only does this accompany M 31 and M 32 in Messier's catalogue, but it lies near the M 31 system in the sky. Moreover, M 33 is only slightly more distant from us: 2.7 million light years. Thus M 33 must be quite close to the giant spiral galaxy M 31, and some astronomers consider it a satellite to M 31. However, no bridge of gas has been found to link the two.

With a mass 30 billion times that of the Sun, M 33 is a large spiral galaxy. This photograph shows it to have a small nuclear bulge and well-developed spiral arms. The arms contain a large quantity of gas, and some of this lies in compact clouds illuminated by very bright stars. The largest of these individual clouds of gas contain enough material to build a million stars like our Sun.

NGC 6822

A dwarf galaxy

NGC 6822 is a small galaxy. Unlike the gargantuan assemblages of stars which were illustrated in the preceding photographs, here is a galaxy in which individual stars are easily photographed, and bright

Messier 33 KPNO

NGC 6822 AAT

clouds of glowing gas can be recognized. One might even suspect that the stars of this galaxy could be counted, but a photograph such as this reveals only the very brightest specimens, and the stars probably number hundreds of millions.

Individual stars can be seen not merely because they are isolated from their neighbors, but also because they appear bright. This is not a peculiarity of the stars themselves, but an effect of the proximity of NGC 6822. Lying at a distance of about two million light years, this is one of the closest galaxies to us, and is a member of our Local Group. The brightest stars output 500,000 times as much light as our Sun, and are comparable to the brightest stars in most larger galaxies.

No symmetrical pattern exists to NGC 6822: the galaxy is irregular and about 10,000 light years across. There is a prominent bar of bright nebulae and stars across its top in this photograph. The nebulae are clouds of gas which are made luminous by hot stars in their midst, and so are miniatures of the giant irregular galaxies. Such nebulae are called H II regions; later photographs will show in much more detail some of the H II regions in our Galaxy. This nomenclature refers to their principal constituent, hydrogen. The Roman numeral II signifies that the hydrogen is in a state of ionization, in other words that the hot stars continually cause the single electrons to be stripped off the hydrogen atoms. The gas glows when the electrons drop back into place. Where no hot stars ionize the gas, it is in its neutral state, referred to as H I. Neutral hydrogen cannot be detected on photographs, since it does not glow, but it radiates at a wavelength of 21 cm, and is thus amenable to study by radio astronomy. In NGC 6822, radio astronomers have found that a curving band of H I extends from each end of the bar of H II regions. The H I resembles loose spiral arms. Possibly a photograph taken 100 million years hence, when the gas in these arms has condensed to form stars, would show NGC 6822 to be a small spiral galaxy.

H II regions in our Galaxy comprise 93% hydrogen, 7% helium and only traces of all other elements. The 15 H II regions in the bar of NGC 6822 contain even smaller proportions of the other elements. This indicates that the galaxy is young, for hydrogen and helium are only very gradually converted by stars into heavier elements. Young is, of course, a relative term. The first stars of NGC 6822 formed hundreds of millions of years ago, but this is a small fraction of the age of many galaxies, including our own.

Because NGC 6822 is a member of our Local Group, its velocity is not determined solely by the expansion of the Universe, but also by its own motion within that group. As a result, NGC 6822 is one of the very few galaxies which are moving towards us, at a speed of about 60 km/sec. Although this sounds extraordinarily rapid, the scale of even the Local Group of galaxies is such that at the time of the hypothetical photograph 100 million years hence, NGC 6822 would still lie at nine-tenths of its present distance.

Carina Dwarf Elliptical Galaxy

Faintest galaxy

Dwarf elliptical (dE) galaxies are faint collections of stars having an elliptical distribution, and the Carina Galaxy is the latest to be found. None was known until in 1938 the American astronomer Harlow Shapley discovered one in the constellation Sculptor, and a second in Fornax a year later. A total of seven nearby dE galaxies is known, of which this one in the constellation Carina seems to be the least populous. It was discovered by Edinburgh astronomers Russell Cannon, Tim Hawarden and Sue Tritton in the course of the photographic survey of the southern sky by the U.K. Schmidt Telescope. It lies at a distance of 500,000 light years, and is about 3000 light years across.

The Carina dE Galaxy is visible as a very sparse peppering of faint stars barely seen at the center of this photograph between the numerous brighter stars of our own Galaxy. The presence of foreground stars is the major obstacle to the discovery of further dE galaxies. Indeed, this difficulty in finding and studying dE galaxies is illustrated by the fact that one, Leo I, lies behind the bright star Regulus, and is rarely photographed successfully since light from this star vastly outshines it. Nevertheless, it is believed that dwarf galaxies are the most common kind of galaxies in the Universe.

None of the seven nearby dE galaxies contains discernible traces of dust or gas, though they must once have contained these for stars to have formed within them. The evidence suggests that star formation in the dwarf elliptical galaxies occurred long ago and promptly ceased. It is difficult to identify individual objects in distant galaxies. Despite this, globular clusters have been noticed in some dE galaxies, and variable stars resembling the RR Lyrae stars in our own Galaxy have been discovered. These are both aged objects, typical of a population in which star formation has long since ceased. What happened to the gas which must have been left over from the formation of these oldsters? Perhaps it was swept from the dE galaxies as they passed by, or even through, our own much larger Galaxy. This must have first occurred soon after the formation of the dE galaxies.

The tidal force of a large galaxy like our own on a dwarf elliptical is huge. One dE galaxy, that in Ursa

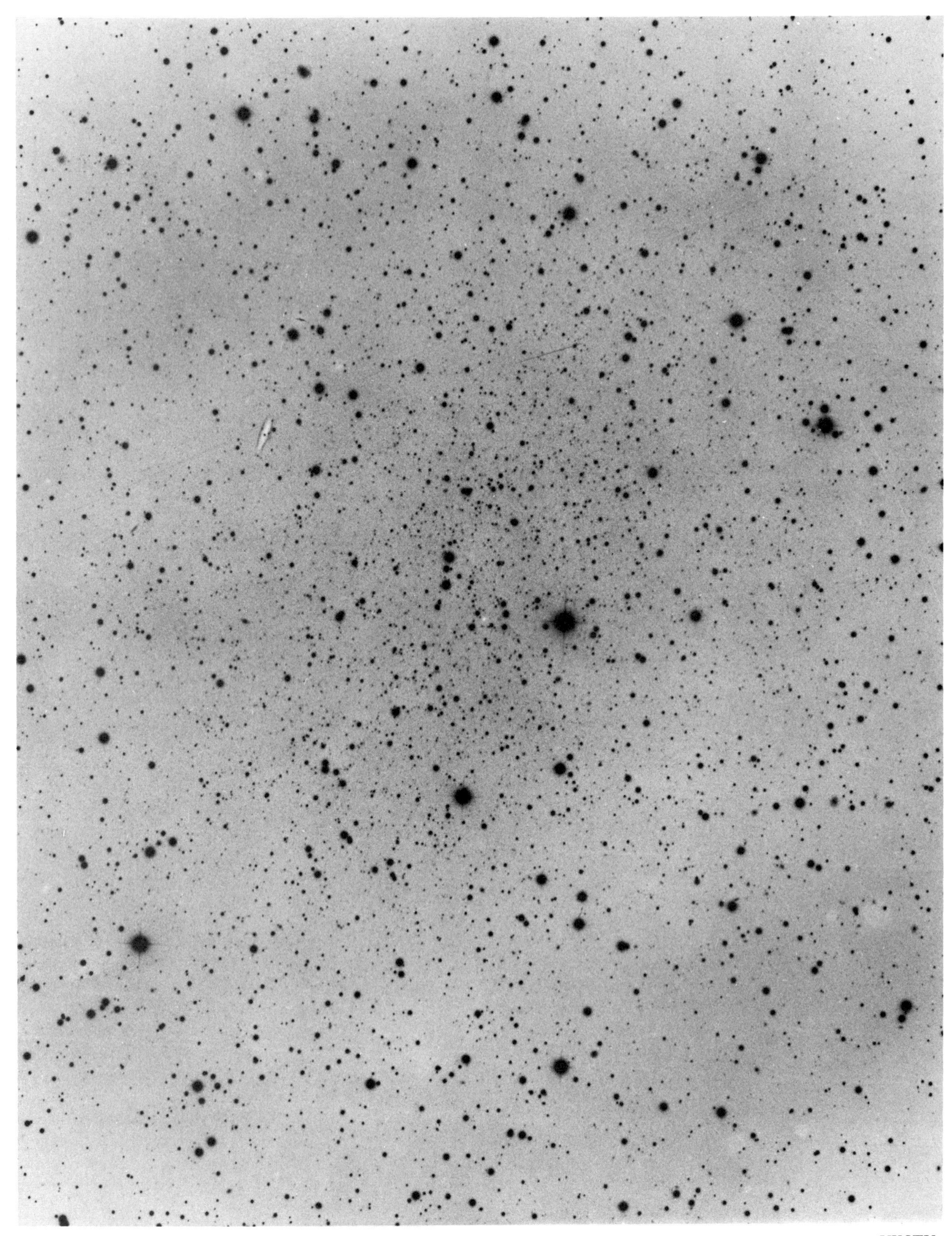

Carina dwarf galaxy UKSTU

Minor, is on the verge of total disruption. If it is now approaching us, it must be for the last time: its stars will mostly be captured by our own Galaxy on the next pass.

Maffei 1

The hidden giant

This faint smudge is a member of the Local Group, and is potentially one of the largest and brightest galaxies in the sky. It is named in honor of its discoverer, Paolo Maffei, who in 1968 noticed two such objects on deep photographs of the northern sky. Maffei 1 is the brighter of the two.

On photographs taken through a blue filter, Maffei 1 is not seen at all. This is a red-filtered photograph, which shows it weakly. Photographs in infrared light show it more clearly, but still fail to reveal any structure to the object. In order to study Maffei 1, astronomers had to use infrared detectors which are sensitive to radiation beyond the range of photographic emulsions.

A group led by Hyron Spinrad showed in 1971 that Maffei 1 is a galaxy, either an elliptical like those illustrated on p. 15, or perhaps a lenticular like NGC 1533 (p. 22). If Maffei 1 is lenticular, however, it differs from NGC 1533 in having no detectable neutral hydrogen. Spinrad's team estimated that Maffei 1 lies about four million light years from us, and is therefore close enough to be a member of our Local Group of galaxies. It weighs about 200 billion times as much as our Sun, which makes it one of the largest members of the Group. It is probably escaping from the Group, as it is receding from our Galaxy at over 150 km/sec.

Maffei 1 POSS

Why is so large and nearby a galaxy so faint? The clue lies in its position in the sky, which is deep within the Milky Way. We view it through our own Galaxy which, like some of the spiral galaxies illustrated earlier, is streaked with dust. The dust dims the light from Maffei 1 more than one hundredfold. If Maffei 1 did not by chance lie behind the galactic dust, it would be a bright and spectacular galaxy in our sky, easily visible in a pair of binoculars.

Maffei's second object is also a galaxy, lying beyond the Local Group, and thus made fainter by distance.

The Magellanic Clouds

Large and Small satellites to our Galaxy

Those who live south of the Tropic of Capricorn never see a cloudless night sky. Even on the clearest of nights two small luminous clouds can be seen in the far south. They are always there, circling with the fixed stars, sometimes high in the sky, sometimes low on the southern horizon. Not clouds of water vapor, they are star clouds. In fact they are the brightest galaxies in the sky after our own.

They are perpetually below the horizon as seen from northern latitudes and, although known and described in the oral mythology of Australian Aborigines, Kalahari Bushmen and Pacific Islanders, they were new to the 15th-century European explorers who first sailed to the south. The Spanish explorer Magellan's name became associated with the two galaxies after the record of his round-the-world trip of 1518-20 was published by the historian Pigafetta. The two galaxies are known as the Magellanic Clouds.

The two galaxies have rather different apparent sizes, so are distinguished by the words Large and Small. Since *Large Magellanic Cloud* is a cumbersome title, the appreviations LMC and SMC are invariably used among the astronomical fraternity, and will be adopted in this book.

The Magellanic Clouds lie quite close to us on the cosmic distance scale. The LMC is 160,000 light years away; the SMC rather farther. This makes the LMC the closest galaxy to us, at a distance less than three times the diameter of our own Galaxy. In fact both

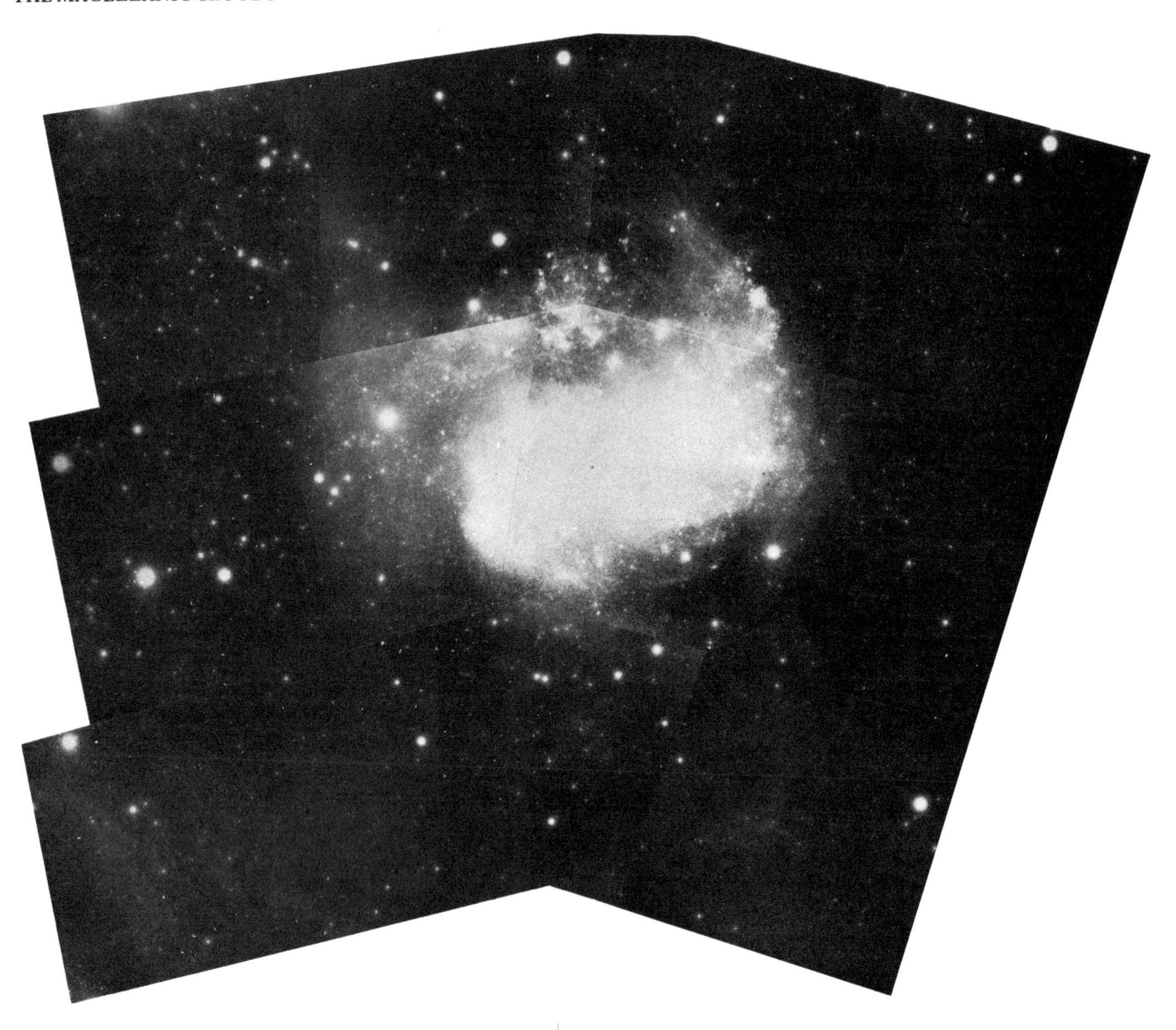

Large Magellanic Cloud (mosaic of six photographs) UKSTU

Magellanic Clouds are orbiting our Galaxy, almost hand in hand, just as the Moon orbits the Earth; they are satellite galaxies. In the same way M 32 and NGC 205 are satellites to the ANDROMEDA GALAXY though they follow unrelated orbits around their primary.

The Small Magellanic Cloud, the SMC, is pictured twice in this book. One photograph, in color, appears in Plate 3. It shows a multitude of mostly blue stars and a few nebulous red patches. The stars are only the very brightest members of the galaxy; myriads of much fainter, white, yellow, orange and red stars are unrecorded. The red nebulae are clouds of gas, glowing with the characteristic color of hydrogen gas ionized by hot stars, as in the case of the ORION NEBULA, Messier 42, described on p. 103 and Plate 6.

The black and white photograph shown here records many of the fainter stars in the SMC. These crowd together so as to almost fill the central region of the galaxy, and in this respect the photograph rather resembles the naked-eye view of the SMC. However, the photograph also records the outer regions which are quite unseen by eye or even through moderate-sized telescopes.

Compare these two photographs with the illustrations of various types of galaxy given earlier. Clearly the SMC is neither spiral nor elliptical: it is classified as an irregular. Now compare the color photograph with that of the ANDROMEDA GALAXY (Plate 2). The stars of the SMC resemble the blue Population I stars in the arms of spiral galaxies. The bright red Population II stars found in elliptical galaxies and in the nuclear bulges of spiral galaxies are hardly represented in the SMC. Ionized hydrogen nebulae too are found only in the arms of spiral galaxies, and of course in other irregular galaxies.

Population I stars are young; so are the nebulae. Thus the color photograph immediately reveals that the SMC is a young galaxy. Its age is not well known, but is probably about two billion years, only one-quarter the age of our Galaxy.

Not all the stars of a galaxy form at the same time. Star formation is a continuing process. This is particularly evident in spiral galaxies, whose nuclear bulges contain stars of far greater antiquity than their arms. At any moment we see in a galaxy some stars which were among the first formed, and whose chemical compositions reflect the initial chemistry of the galaxy, together with some stars of more recent vintage.

The chemistry of a galaxy changes with time. This is because stars derive their energy by nuclear processes, converting hydrogen into heavier elements. As stars die they eject some of the products of their combustion into their parent galaxy, and hence enrich it with the elements they made. Carbon, nitrogen and oxygen figure prominently in the exhaled materials. The stars which formed later thus contain more of these "heavy" elements. The later in its life cycle we view a galaxy, the greater will be the proportion of elements heavier than hydrogen and helium which it contains.

This theory of galactic enrichment, which describes how the chemistry of galaxies changes, predicts that a galaxy as young as the SMC will contain few of the heavy elements. Measurements of the chemical compositions of the stars and nebulae of the SMC confirm this prediction. The proportions of several elements have been measured in various of the stars and nebulae of the SMC. All measurements confirm that there is only about one-tenth as much of these elements in the

Small Magellanic Cloud UKSTU

SMC as in our Galaxy.

The Large Magellanic Cloud, the LMC, so dominates its companion that it receives most of the attention lavished by inquisitive astronomers on our Galaxy's satellites. In part the dominance is caused by its proximity, for the LMC is about 30% nearer. But the LMC is also the larger galaxy, containing a greater number and variety of stars and nebulae, and including intrinsically more luminous specimens of most types of object. Indeed, the LMC is a fair-sized galaxy, with a largest dimension in excess of 30,000 light years; this is double the size of the SMC, and nearly half the diameter of our own Galaxy.

As for the Small Cloud, we illustrate the LMC by two photographs, one in color (Plate 3) and one in black and white. The color picture reveals the greater number of nebulae within the LMC, and indeed more lie outside the bounds of this photograph. More significant, perhaps, is the prominence of yellow stars in the LMC when compared to the SMC. In part this reflects the fact that we see fainter stars because they are closer to us. But the yellowness is also due to a greater proportion of Population II stars in the LMC's central regions. Thus we can tell that the LMC is an older galaxy than its companion, at least in its inner regions.

The color photograph suggests that the LMC, like the SMC, is an irregular galaxy, and in many catalogues that is the galaxy type quoted. However, there are the beginnings of spirality in the outer parts of the LMC, and these can be seen in the mosaic of black and white photographs. The LMC appears to be maturing from irregular youth into a barred spiral. The elongated distribution of Population II stars maps the bar.

Among the variety of studies of the LMC being undertaken by astronomers, determination of the chemical composition is probably the most important. The proportion of heavy elements is higher in the LMC than in the SMC, as astronomers expect for a more aged galaxy. Indeed, the elements oxygen and nitrogen are nearly as abundant in the LMC as they are within our own Galaxy, and much the same appears to be true of other elements. Moreover, there is some evidence that the proportion of heavy elements is greater near the central bar than it is in the outskirts. This too is consistent with theory, since more stars have died in the older and more crowded central regions than farther out, and thus more of the elements they release have accumulated there.

Of particular interest in these studies has been the ability to map the changing proportion of heavy elements with time. This has been possible in the LMC (and to a limited extent in the SMC) by means of some clever astronomical archaeology. Astronomers of the Mount Stromlo Observatory in Australia have been particularly active in pioneering this work.

The secret lies in the study of clusters of stars within the LMC. From the distribution of different types of stars within a cluster it is possible to determine its age. In essence, Population I stars evolve into Population II stars towards the ends of their lifetimes. Population II stars are thus like the teeth of a sheep: a count of them unambiguously determines when the entire cluster condensed from its primordial gas cloud. The present chemical composition of certain types of the stars in that cluster is a good measure of the composition of the gas whence it came. Each cluster therefore yields a value for the proportion of heavy elements at the particular time of its genesis. If many clusters of diverse ages are examined, the progression of galactic enrichment can be mapped out.

This work is still underway, and there is some disagreement among different groups of astronomers working along these lines. The preliminary results suggest that the enrichment of the LMC with heavy elements did not proceed gradually. In at least one part of the galaxy there seems to have been a rather sudden increase in the abundance of heavy elements. The reasons for this are not obvious, and astronomers are still trying to understand it.

The Magellanic Stream

Wake of the passing Clouds

In the section depicting interacting galaxies, several distorted and misshapen specimens were illustrated where two galaxies were making a close mutual pass. Since the Magellanic Clouds lie within three diameters of our Galaxy, and may once have been closer, it is reasonable to ask whether there is any evidence for interaction in the triple system.

Because of the small masses of the Magellanic Clouds, the disturbance to our Galaxy has been slight. But our Galaxy certainly has influenced the Clouds. It has even been suggested that the LMC and SMC were once a single galaxy which became rent in its encounter with ours. However, this suggestion is compromised by a variety of observations.

Notwithstanding their separate identities, the LMC and SMC are linked by a bridge of invisible neutral hydrogen gas, detectable only by radio telescopes. Both galaxies contain considerable amounts of gas which has yet to form stars, and the bodies of gas which surround the two are clearly joined, so that the Clouds appear as two condensations within a single envelope. Moreover, it has been known for some time that a spur of gas projects beyond the LMC in a direction away from the SMC.

Over the last few years, Mt Stromlo astronomer Don Mathewson has mapped the distribution of hydrogen

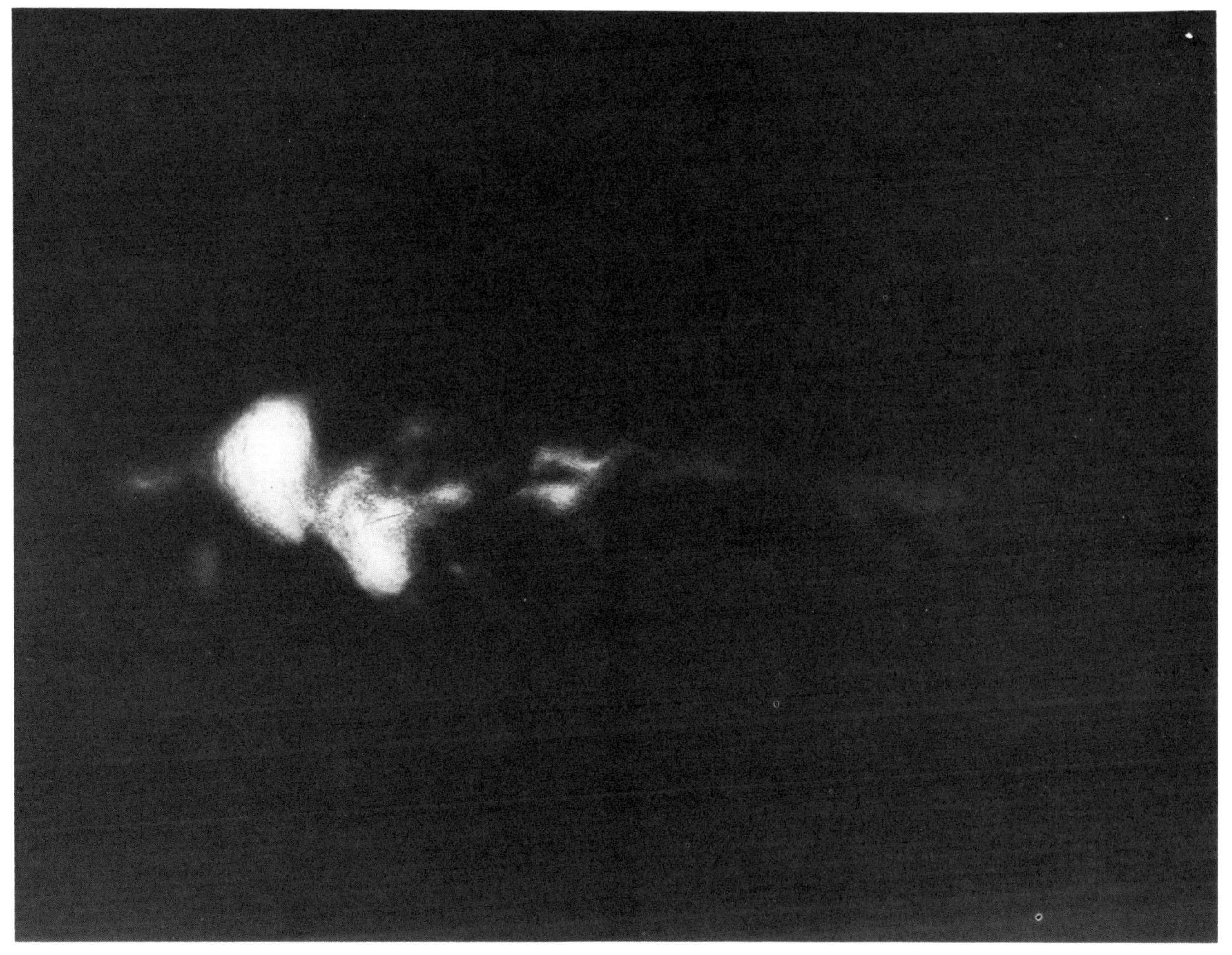

Magellanic Stream DAA

around the Magellanic Clouds. He finds that the spur projecting from the LMC is the densest portion of a long tail of hydrogen which stretches much of the way across the sky until it disappears behind the hydrogen of our Galaxy. Mathewson calls this tail the Magellanic Stream. He and others have searched for stars associated with the Magellanic Stream, but without success. It seems that the gas is too tenuous for many stars to have formed in it.

The Stream was probably extracted from the outer parts of the Magellanic Clouds at a time when there were few stars involved in it. Without stars the Magellanic Stream cannot be self-luminous, and it would not even appear on optical photographs taken from other galaxies. We can illustrate it here only by a sketch map showing the irregular distribution and long jet formed of gaseous hydrogen, which constitutes the Magellanic Stream.

There is some debate among astronomers as to the origin of the Magellanic Stream, and even as to its reality. An interpretation based on gravitational interaction seems natural, but the measured motions of the gas do not quite fit those predicted by computer simulations of the interaction. This may simply indicate the inadequacy of the computer models. Or it may point to an entirely different origin of the Magellanic Stream.

The Tarantula Nebula

The heart of the LMC

Because the Magellanic Clouds are irregular galaxies, they have no obvious nuclei. However, the incipient spiral structure of the Large Cloud induces many astronomers to seek a nucleus to that galaxy. At one end of the LMC's bar lies a great nebula known as the Tarantula Nebula (Plate 4). If the LMC has a nucleus, this is it.

The Tarantula is an H II region. Within it, hydrogen has been ionized by the energy of starlight. The characteristic red of recombining hydrogen is

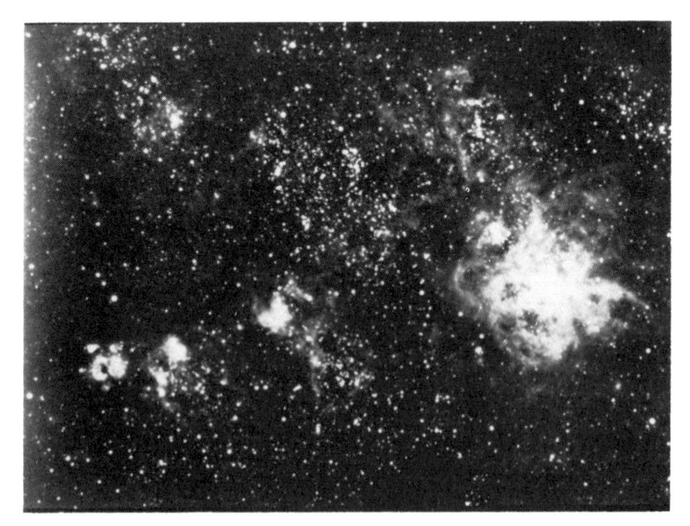

Tarantula Nebula UKSTU

moderated slightly by the other colors emitted by hydrogen and by other elements, most notably a green due to oxygen, to which, however, color film is not equally sensitive. The nebula appears brighter than the stars because it has harvested its luminosity from their abundant but invisible ultraviolet radiation, converting it to visible light.

Before the hot stars formed, perhaps a few million years ago, the hydrogen was not ionized, and was therefore not luminous. Slowly, over the countless millions of years that the LMC had been evolving, gravity tightened its grip on the giant hydrogen cloud, shrinking blobs of it until stars condensed from them. One by one the stars began to shine, as do street lights in a city when dusk falls. Each star released ultraviolet radiation to ionize the hydrogen.

Today sufficient stars are shining within the Tarantula to illuminate a region of gas 4 billion billion kilometers across, almost 900 light years. At the distance of the LMC this amounts to half the apparent diameter of the Moon. The total mass of ionized hydrogen in the nebula is about 500,000 times the mass of the Sun.

As luck would have it, little of the nebula lies on our side of the newly-formed stars; all that does has been ionized, and we see through it to the stars and beyond. But on the far side the situation is different. Behind the Tarantula Nebula lies an enormous cloud of neutral hydrogen waiting to be ionized. The neutral gas cannot glow: it is dark and unseen. It also contains small grains of dust, which accounts for less than 1% of the nebula's mass. So widely spaced are the dust grains that a volume of nebula equal to the volume of the Taj Mahal would contain only two or three grains. Yet so vast is the scale that this minuscule proportion of dust is still sufficient to render the nebula opaque. We cannot see the neutral nebula, nor can we see through it. Evidence of its existence comes from radio observations which detect neutral hydrogen by its radiation at a wavelength of 21 cm. Optical evidence of material is also provided by very faint wisps of ionized hydrogen stretching more than 1° from the Tarantula Nebula.

The picture of the Tarantula Nebula that emerges from this description may seem complicated, but it is by no means a full description of the discoveries astronomers have made in observing this fascinating object. In particular, a large number of old, red stars have recently been found within it. These are Population II stars, and must have been formed many hundreds of millions of years ago. They may number among the first stars to have formed within the LMC. Now they are nearing the end of their lifespan and their place is being taken by a new generation of young, hot stars. The new generation was probably born because the explosive death of one of their predecessors sent a ripple through the nebula, triggering the collapse of the gas into star-sized blobs. Radio astronomers find evidence of stellar explosions (supernova remnants) in the Tarantula Nebula.

If for the lifetime of the LMC stars have been forming within the Tarantula Nebula, it has a strong claim to be called the galaxy's nucleus. Even today the Tarantula Nebula is the largest accumulation of glowing hydrogen in the galaxy. It seems we must accept that the LMC has an off-center heart.

The LMC in detail

What galaxies are made of

Because of its great proximity, on a galactic scale, we may study the LMC in far greater detail than any other galaxy except our own. We see more clearly what galaxies are made of. We can, for example, study individual stars. When we do so we find a great range of different types: some hot and blue, others cool and red. Some stars vary in brightness; others shine with constancy. Some travel alone through the LMC, others in pairs, and yet others as part of a congregation, the sort of extended family which astronomers call a star cluster.

In the LMC we also see gas clouds, like the Tarantula Nebula just described. H II regions are irregular clouds of gas glowing because hot stars within them feed energy into the gas. Supernova remnants are expanding bubbles of gas, the relics of stars whose lives ended explosively. They shine because they are energized by bumping into stationary gas as they expand. Other gas clouds are not illuminated; we see them because the dust grains they contain dim more distant stars. These examples of galactic smog are appropriately known as dark clouds.

The photographs in Plate 4 illustrate some of the objects found in the LMC. Several clusters of stars are seen in the upper right photograph, the images of their member stars blending together towards the centers of the clusters. A group of nebulae is seen in the lower photo. Most of the nebulae are H II regions like the TARANTULA NEBULA or the ORION NEBULA in our own Galaxy. Dark clouds obscure some parts of these nebulae. A filigree bubble, perhaps a supernova remnant, surrounds a loose peppering of bright stars in a stellar association at lower right.

There are many examples of objects within our Galaxy whose distances we do not know. If we can identify similar objects in the Magellanic Clouds, whose distance we do know, from their relative brightnesses we can get the distances of the local examples, and hence better understand the structure of our Galaxy. The best illustration of this is that of the Cepheid variables, like DELTA CEPHEI (p. 95). The Cepheid variables in the LMC all lie at more or less the same distance from Earth. This made possible the discovery that the longer the variation period of the star, the brighter it actually is. If we know how bright it actually is and how bright it appears in the sky, the star's distance can then be worked out. In this way astronomers can determine the distances of all other Cepheid variables in our own and other galaxies.

All these types of objects and more are also found within our own Galaxy (which is usually written with a capital G to avoid confusion with other galaxies). They can be studied in yet greater detail in our Galaxy, of course. Thus this is the point in the book where we forsake the more distant realms of the Universe and focus our attention on relatively local, and therefore better understood phenomena.

Our Galaxy

The Milky Way

The Milky Way is the band of light from the distant multitude of stars of our Galaxy. Though each star is too faint to be picked out by eye, together they glow visibly. Swamped by bright lights from the city, diluted by moonlight, the Milky Way can best be seen from country areas around the time of New Moon, particularly in summer and winter when it arcs from near the celestial poles to the northern and southern horizons.

The Milky Way exists because stars in our Galaxy are not equally distributed in all directions. The Galaxy is flattened like a pie and there are fewer stars visible when one looks through the thin dimension of the Galaxy than through the thick dimension. A being living on a planet circling a star in the middle of a spherical galaxy (p. 15), which would have equal dimensions in all directions, would see no Milky Way at all. His sky would be equally bright everywhere. If his sun were offcenter, the sky would be brighter on one side (towards the center of his galaxy) than on the other, but there would be no band of light. The existence of the Milky Way shows us that we live in a

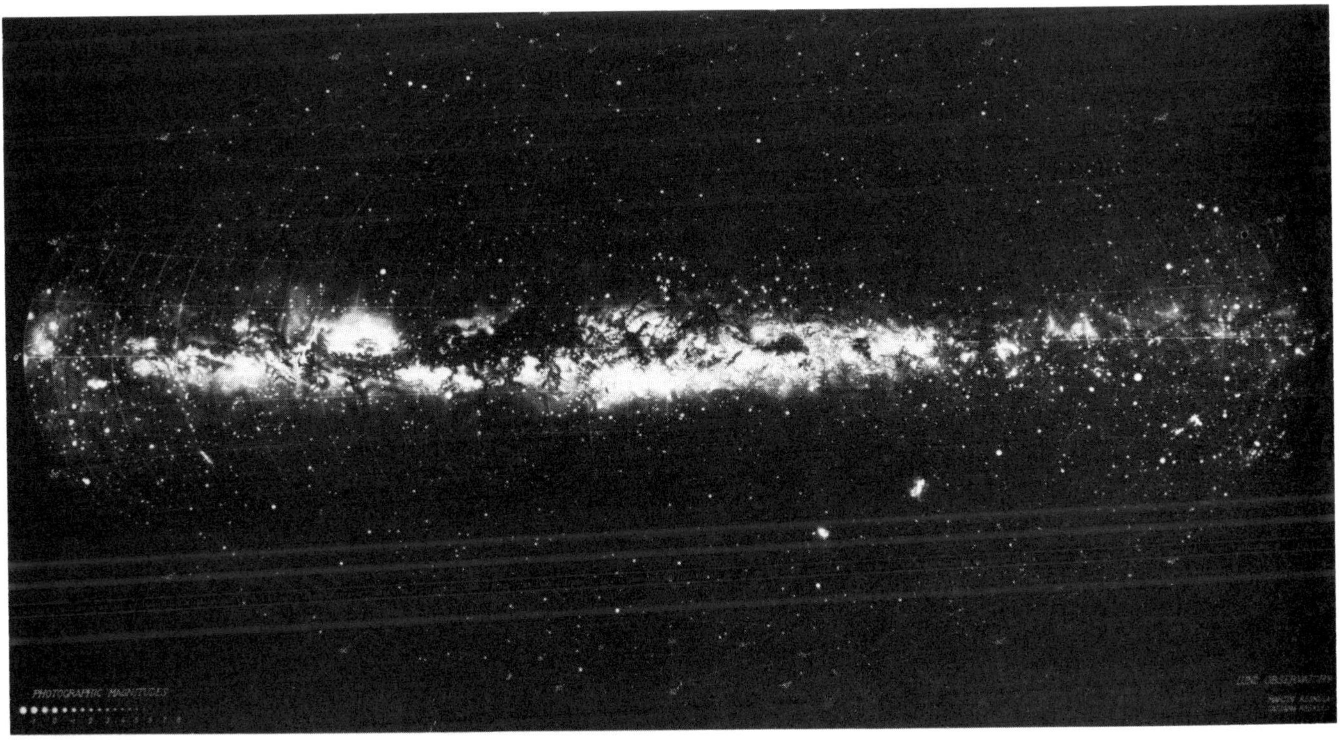

Milky Way *Lund*

very flattened galaxy like the spiral galaxies on p. 16.

Not only does the Milky Way represent the presence of stars, it corresponds to an absence of galaxies. While galaxies are found elsewhere in the sky, they are not found near the Milky Way, within the so-called Zone of Avoidance. The reason for the Zone of Avoidance is that dust concentrates in the plane of the Milky Way. We cannot see through the dust towards the galaxies which are certainly beyond it.

The Galaxy's Spiral Arms

Mapping young stars

Because we live inside the Galaxy, it is difficult to see its plan clearly. In the same way, it is difficult to draw a plan of our own city from inside it, but from a highflying aircraft the structure can be clearly seen. We can therefore see the structure of other galaxies better than that of our own. After mapping other galaxies, we can look back to our own and identify similar features in it.

From various pieces of evidence we deduce that we live in a spiral galaxy like NGC 2997 (p. 14). The evidence is first that we obviously live in a flattened kind of galaxy or we would not be able to see a Milky Way. Second, radio astronomers detect large amounts of hydrogen in our Galaxy, mostly lying along the center line of the Milky Way. No hydrogen is seen in elliptical or spherical galaxies like NGC 4564 (p. 16) or NGC 4374 (p. 15). Therefore we know that we do not live in either of those types of galaxies. Third, although we can see only the "local swimming hole" of our Galaxy, out to a distance of about 15,000 light years, astronomers are able to trace the existence of three spiral arms.

What is the best method of looking at spiral arms? Optical astronomers need to identify the bright objects which lie along the arms in order to see out as far as possible. These objects are the hot, blue stars, the Cepheid variable stars like DELTA CEPHEI (p. 95), and clusters of stars like the PLEIADES (p. 121) and the HYADES (p. 119). Bright, young stars congregate towards the central line of the Milky Way. The spiral structure of our Galaxy has a flat appearance, as if we are viewing a thin disk from a position on it. It can accurately be drawn on a sheet of paper provided that the distances to all the various objects can be determined. This can usually be done by studying the spectra or the light variations of the stars concerned, which yield clues to their true brightness. Then by using a measure of the apparent brightness as well, an estimate of their distance can be obtained.

The resulting plan of the local spiral structure of our Galaxy shows portions of three spiral arms. The central one is called the Orion Arm because it contains the stars of the constellation Orion. We live on the inner edge of the Orion Arm. It stretches halfway round the sky, from the constellation Cygnus, through Cassiopeia, Taurus and Orion to Monoceros. The reason why so many clusters of stars and nebulae are found in Cygnus and Monoceros is that in those directions we look along the Orion Spiral Arm, and see a line of star clusters and nebulae stretching away in the distance.

Another spiral arm running parallel to the Orion Arm is called the Perseus Arm because it contains the clusters h AND CHI PERSEI (p. 122), and others in that constellation. The Perseus Arm lies in the direction away from the Galactic Center.

The third spiral arm visible to optical astronomers is the Sagittarius Arm. It lies between us and the center of the Galaxy, which is in the direction of the constellation Sagittarius. The Sagittarius Spiral Arm is markedly curved even over the short length of it which optical astronomers can see. Its nearest approach to the Sun is in the direction of Scorpius and it curves away in the direction of Carina on one side and Scutum on the other. In Carina we look tangentially along the arm as it curves away from us, so the spiral arm material—nebulae, clusters, bright stars—is richer in Carina than in any other constellation.

The American astronomer Bart Bok and his associates have studied the Carina end of the Sagittarius Spiral Arm. They have confirmed in our Galaxy what radio astronomers have observed in others. The dustiest regions with the most dense gas clouds lie not along the spiral arms seen by optical astronomers and traced out by the bright stars, but on the outer edge of the optical spiral arms.

Why is the gas separated from the stars? The answer is that spiral arms, despite appearances, are not permanent features of our Galaxy.

The Galaxy is rotating with the inner regions turning faster than the outer regions for exactly the same reason that the nearest planet to the Sun, Mercury, revolves around the Sun faster than the most distant, Pluto. If the spiral arms were permanent, they would therefore long ago have wound up into a general blur. This is not the case as they are being constantly renewed. The renewing process is called a density wave. It flows around the Galaxy in spirals, compressing and releasing the gas as it passes. The theory of the density wave came from the Chinese-American astronomers C.C. Lin, Frank Shu and Chi Yuan.

The theory explains both why the spiral arms do not wind up, and why the gas and stars in a spiral arm are separated, in the following sequence of events. The arrival of a density wave compresses the gas in a region, triggers the collapse of the dust clouds and causes the formation of stars, including massive and

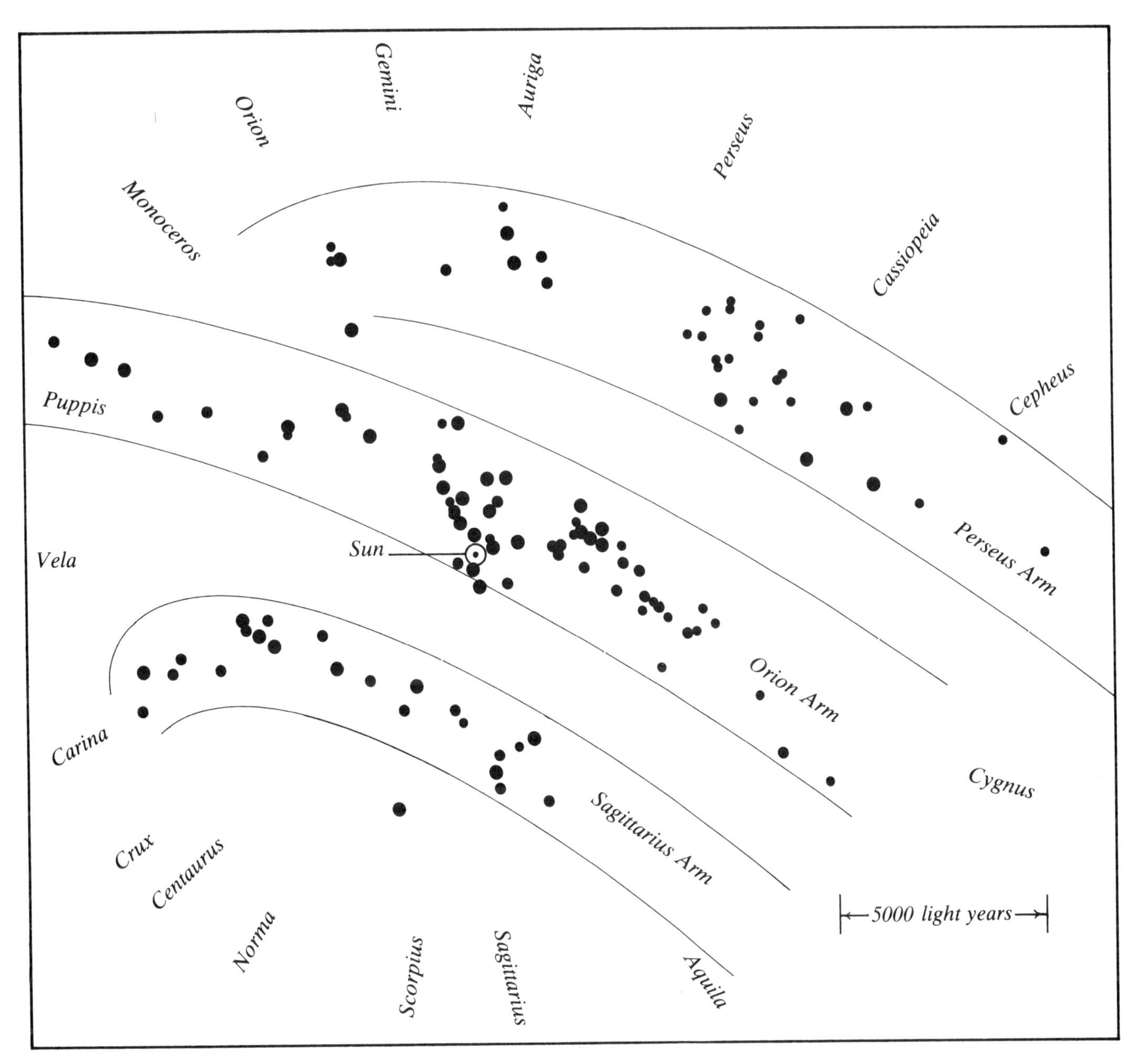

Spiral arms of the Galaxy

bright ones such as are found in clusters. The wave passes on, triggering the formation of more stars, while those which have already formed age. The massive stars age quickly: being brighter they use their available fuel faster than others do. As the wave moves on, one sees in its wake compressed, high density gas, bright young stars and faint aged stars, in that order.

NGC 55

Mimicking our Milky Way

NGC 55 is a nearby spiral galaxy, very similar in form to our own. We see NGC 55 edge-on because we happen by chance to lie in its galactic plane. This gave John Graham of Cerro Tololo Inter-American Observatory the idea to prepare a cropped picture of part of NGC 55 in order to emphasize the fact that our own Milky Way is the view of an edge-on spiral galaxy. Even astronomers who are specialists in the structure of the Milky Way have been fooled by the fake, and have tried with singular lack of success to identify the star patterns in the NGC 55 photograph (p. 68) with constellations in our own sky!

The cropped picture shows the gap between two of NGC 55's spiral arms, while the galaxy's center lies to the right. The brighter stars concentrate in clusters in the spiral arms and many can be seen where we

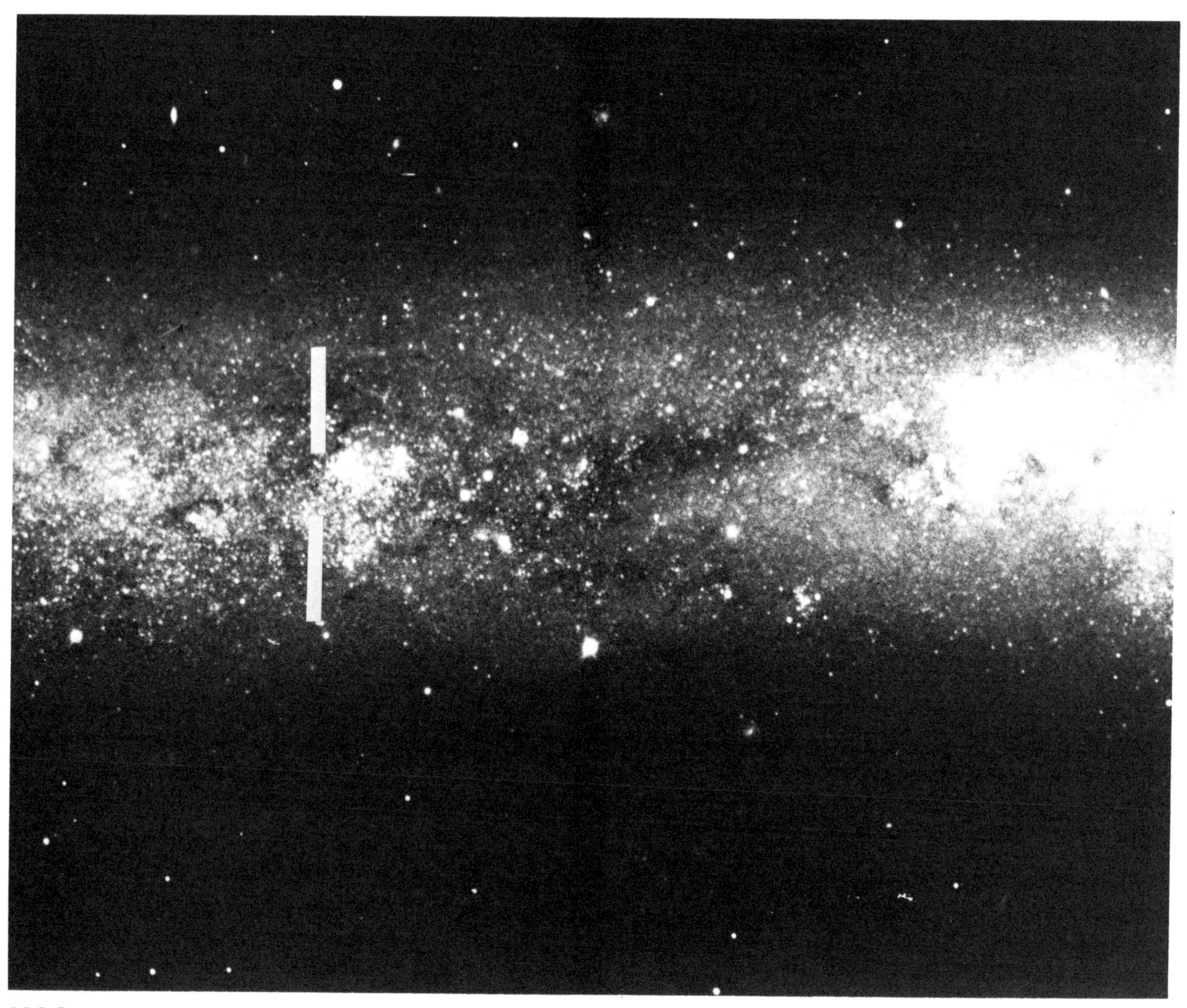

NGC 55

CTIO

happen to look along the arms. The fainter stars, not individually visible, form the gray background band in the center of which the more massive, brighter stars are embedded. If this were our own Galaxy, the Sun would be at the point indicated, Sagittarius would be to the right, Cygnus beyond the Sun, in the back of the picture and Cassiopeia to the left. The GALACTIC POLE in the constellation Coma Berenices, near to the COMA CLUSTER (p. 12) of galaxies, would be at the top.

Galactic Center

A massive black hole?

Our Galaxy is spinning about its center, which is 28,000 light years away towards the constellation Sagittarius. At the actual position of the Galactic Center is the extremely compact radío source known as Sagittarius A West, also a strong source of infrared radiation. As evidence of its compactness, measurements with radio interferometers show that it is smaller than the solar system.

We show here a photo of the direction of the Galactic Center, on which is superimposed a contour map made in infrared, revealing the bright patch which is the very center of our Galaxy. We have to show an infrared map, since our view of the Galactic Center is hidden by a heavy cloak of dust through which light cannot penetrate. Other galaxies such as the ANDROMEDA GALAXY also possess a starlike nucleus, actually made up of many millions of stars, mostly yellow and red giant stars. Our Galaxy may possess such a central region. However, M 31 lacks our Galaxy's compact radio core within its starlike nucleus. On the other hand, many spiral galaxies contain stronger radio cores. For example, the central source in M 81 is 10,000 times stronger in its radio power than ours. It changes intensity from day to day, signifying that it is less than one light day across. It is therefore very compact, like our own Galactic Center which is less than 80 light minutes—about the size of the orbit of Jupiter—across.

The central region of our Galaxy has always been something of a mystery to astronomers. In the 1950's radio astronomers began to look at the motions of interstellar hydrogen in the Galaxy. They soon found that although the outer parts of the Galaxy, from the Sun's position outwards, moved in a relatively orderly circular flow, the inner region (within the central 10,000 light years) was expanding as well as rotating about the Galactic Center. The expansion speed of the spiral arm at 10,000 light years from the center is approximately 50 km/sec. Other clouds of hydrogen even nearer the Galactic Center are fleeing at a rate up to four times faster than this. Our Galaxy's center is somewhat reminiscent of the SEYFERT GALAXIES and QUASARS observed in far space, or of milder exploding galaxies such as M 82 (p. 34) and NGC 253 (p. 38). The center of the Galaxy emits enough ultraviolet radiation, presumably from bright blue stars, to ionize the hydrogen in its vicinity. The ultraviolet radiation in fact drives the electrons out of the hydrogen atoms. When the electrons jump back into their atoms they emit visible light but it cannot be detected because of the intervening dust. Its presence can be inferred because ionized neon atoms embedded in the hydrogen emit an infrared spectral line at a wavelength of 12.8 microns. This infrared radiation can penetrate the dust and also, as luck has it, is not halted by the water vapor, ozone and carbon dioxide in Earth's atmosphere. It turns out that the neon embedded in the ionized hydrogen within the central two light years of the Galaxy has speeds of up to 500 km/sec, suggesting that it orbits an object of a mass of at least four million solar masses.

Combining the radio and infrared evidence, Dutch astronomer Jan Oort speculates that all the four million solar masses are contained in the small core of Sagittarius A West. To maintain the observed turmoil within the Galactic Center there must be major explosions occurring there more often than every 15,000 years. Perhaps these explosions could be supernovae, powerful enough to disrupt massive stars. Indeed the nearby radio source, Sagittarius A East, is probably a supernova remnant, which is evidence that

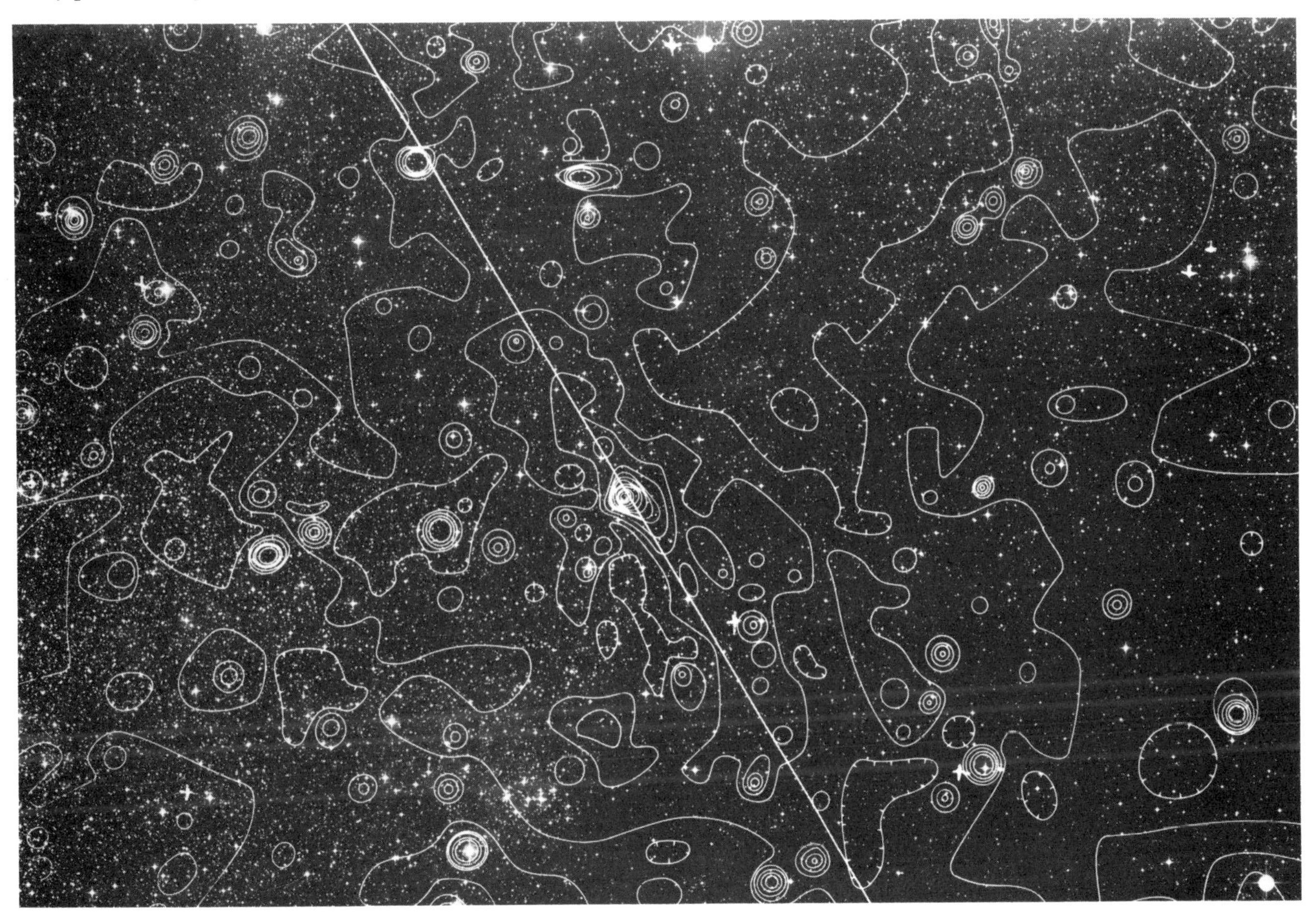

Galactic Center and infrared map UKSTU, *Becklin*

supernovae have occurred near the Galactic Center. Oort, however, believes that supernovae are unlikely to be as frequent as the theory requires. Probably the same unknown mechanism that is needed to cause the expansion of gas from the center of the Galaxy is the engine responsible for the motions in the core of the Galactic Center. "Something out of the ordinary appears to be required," writes Oort. "A massive black hole?"

Galactic Pole

The view out of the Galaxy

In contrast to the GALACTIC CENTER the Galactic Pole is not marked by any actual object. It is merely the direction perpendicular to the plane of the Galaxy. It is the axis about which the Galaxy rotates approximately every 100 million years. The North Galactic Pole is in the direction of the constellation Coma Berenices, the South Galactic Pole towards Sculptor. Because we look along the smallest dimension of our flattened Galaxy, the stars in this direction are few and far between, as can be seen in our photograph.

The direction to the Galactic Pole is of interest because as astronomers look farther and farther into space in that direction, they see progressively into the older regions of the Galaxy. The first objects encountered are hydrogen gas, bright blue stars, DELTA CEPHEI (p. 95) variable stars and galactic clusters like the HYADES (p. 119) and associations like ORION (p. 81). These objects are all of an age less than 100 million years. They concentrate in a layer 800 light years thick.

Looking beyond the layer of young objects, astronomers see the population of planetary nebulae like the RING NEBULA (p. 143), variables like the novae and short period RR LYRAE (p. 98) stars. These are objects one to five billion years old in a layer 4200 light years thick, surrounding and permeating the central slab.

Higher still above the galactic plane can be found MIRA (p. 99) variables. They are five to six billion years old and are contained in another slab 6000 light years thick. Higher still are the globular clusters (like OMEGA CENTAURI, p. 135) and longer period RR Lyrae stars, which form what is called the galactic halo. In external galaxies like M 87 (p. 36) the globular clusters can clearly be seen to swarm in a spherical halo about the galaxy. They have lost all sign of concentration to a slab.

Generally speaking, the objects which are found highest above the galactic plane in the direction of the Galactic Pole have the highest speeds: it is their speed which enables them to travel so far from the plane. Studying their speeds enables us to determine the density of the Galaxy in the region near the Sun, since we can estimate the amount of mass affecting them. A value of 4.5 solar masses in each cube ten light years on a side is found. All types of visible stars and gas account for just over half this. More than one-third of the matter is invisible, probably made up of neutron stars, white dwarfs, black holes, and miniature stars too small to be luminous.

Olin Eggen, Donald Lynden-Bell and Allan Sandage have proposed that the structure of the Galaxy, as just described, could be explained if it had collapsed from a relatively spherical form to a disk-like spiral galaxy. While the Galaxy was a sphere the globular clusters formed, and they still fall in and out of the Galaxy on cigar-shaped orbits. Now that the Galaxy is flat, recently formed objects orbit in relatively circular paths within its plane. The shapes of the parts of the Galaxy thus correlate with their age. The shapes also correlate with the composition of the objects within them. This is because throughout the history of the Galaxy, stars have been making elements heavier than hydrogen and helium in greater and greater abundance. Younger parts of the Galaxy are thus more contaminated with heavy elements.

Probing the properties of stars at greater and greater distances above the galactic plane, towards the Galactic Pole, gives information about this theory and is an active area of astronomical investigation.

Sagittarius

Center of a spiral galaxy

The center of our Galaxy lies in the direction of the constellation Sagittarius (Plate 5). The Milky Way is bright here because more stars lie in that direction. It is also the widest part of the Milky Way because the spiral arm which lies between us and the GALACTIC CENTER comes closest to us in the direction of Sagittarius. Another reason is that above and below the spiral arm we can see beyond it to the central bulge of our Galaxy, 30,000 light years away.

Laced along the Milky Way in Sagittarius is a network of relatively nearby dusty, dark nebulae. These dark nebulae generally lie within 600 light years of the central plane of the Galaxy and they form a thinner band than the stars contributing to the light of the Milky Way. The stars seen above and below the densest dust clouds are red because they are old and cool. In addition, their light is made redder by being viewed through the dust.

A sixth of the Messier objects (listed in the Appendix) lie in Sagittarius (the LAGOON NEBULA shows as a prominent red patch on Plate 5). The list includes globular and galactic clusters, and nebulae. The

Galactic Pole UKSTU

Sagittarius DFM

galactic clusters and emission and reflection nebulae are associated with the dark nebulae concentrated in the plane of the Galaxy, but the globular clusters make up a more widespread swarm about the center of the Galaxy, like bees around a hive. Even so, more than half the globular clusters known lie in the 6% of the sky within 30° of the direction of the galactic center. It was this offcenter distribution of globular clusters which, in 1917, gave Harlow Shapley the clue as to the position of our Sun in our Galaxy, now estimated at 26,000 light years from the Galactic Center.

Our own view of our Galactic Center resembles that seen by an astronomer looking at it from another galaxy lying in the plane of the Milky Way—for example an astronomer on MAFFEI 1 (p. 59). That astronomer will see our Galaxy looking like an external spiral galaxy in whose plane we lie, such as MESSIER 104 (p. 17). By looking to the Milky Way in Sagittarius, we have the clearest proof that we live in a spiral galaxy.

2. Stars and Nebulae

In the last section we showed how stars and gas exist in enormous collections called galaxies. It is time now to see how stars exist within each galaxy. It is only recently that astronomers have had large enough telescopes to study individual stars or small groups of stars in galaxies beyond our own. Thus most of our knowledge of the kinds of smaller collections of stars comes from studying the ones in our own Galaxy.

There is an immediate problem. We can say that all the stars in a distant galaxy are at the same distance and belong to that galaxy because we can see that the stars are separated from other nearby galaxies by the enormous distances of intergalactic space. But the view we have of the stars of our own Galaxy as seen from inside the Galaxy is far more confused. It has always been difficult for astronomers to disentangle chance, line-of-sight coincidences of stars from real groups. In the following section we see how it is possible to separate the real groups from the apparent.

Our first problem in distinguishing the meaningful groups from the meaningless is confusion by the sheer number of stars. But although poets talks of the "countless" visible stars, in fact less than 4000 stars can be seen at a time with the unaided eye. The Greek astronomer Hipparchus (129 BC) classified them into six magnitudes, with stars of the first magnitude being the brightest, and stars of the sixth magnitude the faintest which he could see. This scale, with modifications, is still in use today. The *Catalogue of Bright Stars*, prepared at Yale University Observatory, lists stars brighter than magnitude 6.5. Stars as faint as this might optimistically be seen when directly overhead by the keensighted in a clear desert or mountaintop sky away from city lights. The Bright Star Catalogue contains 9110 entries and the stars in it are plotted in *Norton's Star Atlas*, a popular sky map drawn up by an amateur astronomer in 1910 and still in everyday use.

With a pair of binoculars many previously unseen stars can be distinguished, perhaps to magnitude eight, a factor of four fainter. The Milky Way can be seen to be the unresolved light from numerous stars. "Stars" like Omega Centauri or h and Chi Persei can be seen to be clusters of many stars. The Czechoslovak astronomer Antonin Becvar's *Atlas Coeli* reaches magnitude 7.75 and plots about 30,000 stars. Of course, the larger the telescope the more stars can be seen. F.W. Argelander, observing from Bonn in the first half of the 19th century, systematically scanned the sky north of the equator with a small telescope and listed all the stars he found. His "exhaustive list," the *Bonner Durchmusterung*, contains 324,189 entries of stars brighter than magnitude 9.5. The southern part of the sky was observed in a similar way from Argentina to magnitude 10, and the *Cordoba Durchmusterung* lists 613,953 stars.

Stars fainter than this are not all individually listed by astronomers. Selected areas of the sky have been chosen as typical for the purposes of a stellar census; more thorough listings have been made of faint stars of particularly interesting kinds. Several atlases of all the stars in the sky, have, however, been prepared, usually photographed by the Schmidt type of telescope, invented by Estonian optician Bernhard Schmidt in 1930 to photograph a wide area of the sky at once. The German amateur astronomer Hans Vehrenburg used his Schmidt telescope in both the southern and northern hemispheres to chart the entire sky to magnitude 14; probably 20 million star images can be counted on his *Photographischer Sternatlas*.

The whole sky visible from California has been photographed by the 48-inch Schmidt telescope of the Palomar Observatory. 935 areas, each six degrees square, were photographed twice, once with red filters, once with blue; copies of the 14-inch square photos were distributed in the mid 1950's. Stars as faint as magnitude 20 or 21—that is, 400,000 times fainter than those visible to the naked eye—can be seen on the prints. There are approximately a billion star images on them. Two recently built Schmidt telescopes in the southern hemisphere, one at the European Southern Observatory (ESO) at La Silla, Chile, and the other the U.K. Schmidt Telescope Unit (UKSTU) at Siding Spring Observatory in Australia are producing atlases of the southern half of the sky. The ESO "Quick Blue" Atlas is already complete and reaches to magnitude 20.5. Photographs by UKSTU are on a new kind of photographic emulsion, IIIa-J, and reach magnitude 23. Three-quarters of this survey was complete by 1978. Altogether it will probably contain the images of up to ten billion stars. Stars of magnitude 23 appear four million times fainter than the 6.5 magnitude stars visible to the unaided eye.

The largest telescopes in the world can record the images of stars as faint as magnitude 24. Perhaps there are 20 billion such stars able to be photographed in our Galaxy, maybe half the total number it contains. The stars in the sky, although uncountable, are not countless. The sequence of pictures on p. 74-5 shows a zoom in from naked-eye stars as plotted on *Norton's Star Atlas* to the faintest stars photographable.

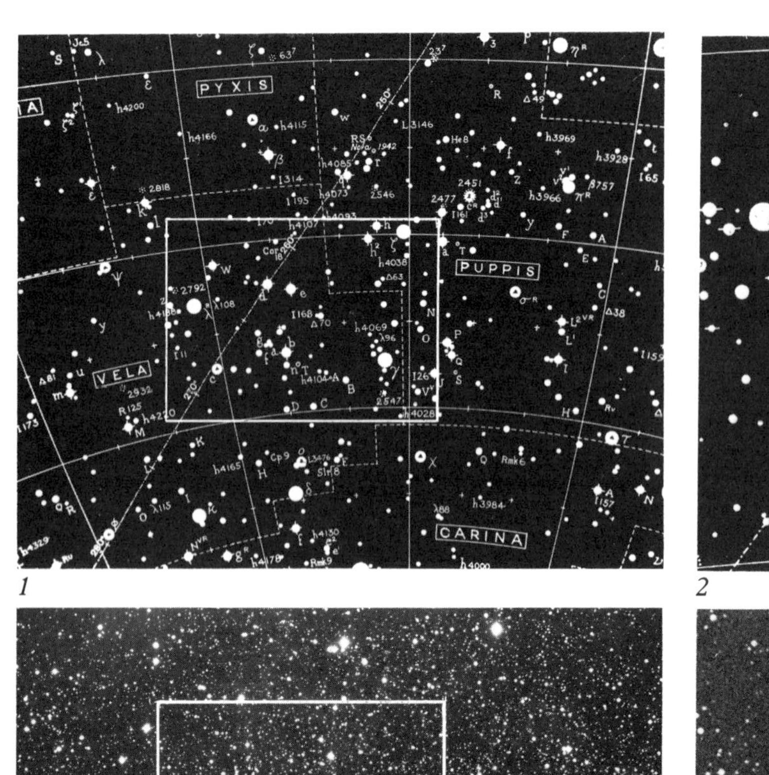

1

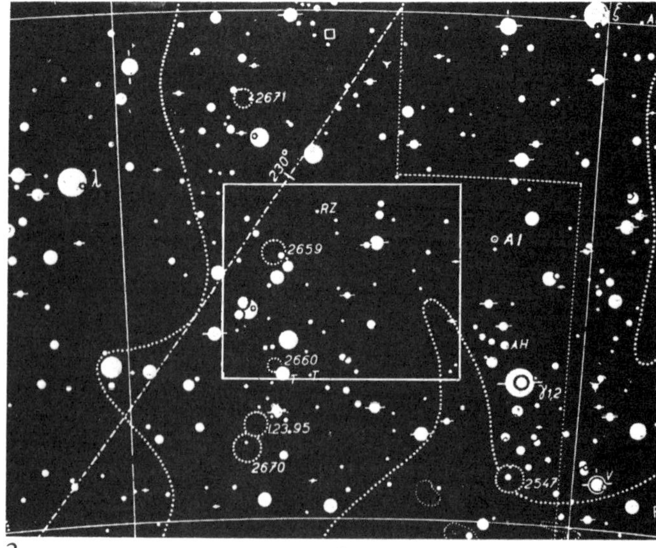

2

5

6

Zooming in to the faintest stars. Each frame, reading from left to right along the two rows, shows the area boxed on the preceding frame. The brightness covered ranges from naked-eye stars (upper left) to the Vela pulsar (lower right), a factor of one billion.

Constellations

Cassiopeia

Line-of-sight constellation

Cassiopeia is the well-known W-shaped constellation of five second-magnitude stars near the North Celestial Pole. Its stars appear almost equally bright and they are grouped together. But although they may appear to be a related group, the stars of the constellation are widely separated along the line of sight. The nearest, Beta, is 36 light years away, while the farthest, Gamma, is 980 light years distant. The distances of the remainder are: Alpha, 130 light years; Delta, 45 light years; Epsilon, 620 light years. Moreover, the group of five Cassiopeia stars are of widely different ages. The youngest is Gamma, nine million years old. The two oldest are Alpha and Beta, each about 10 billion years old.

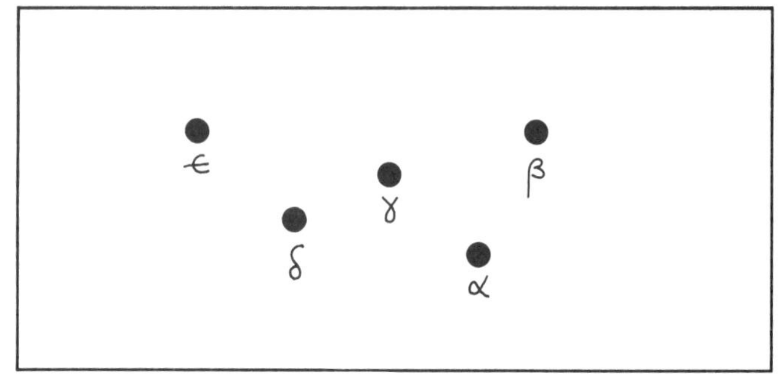

Cassiopeia

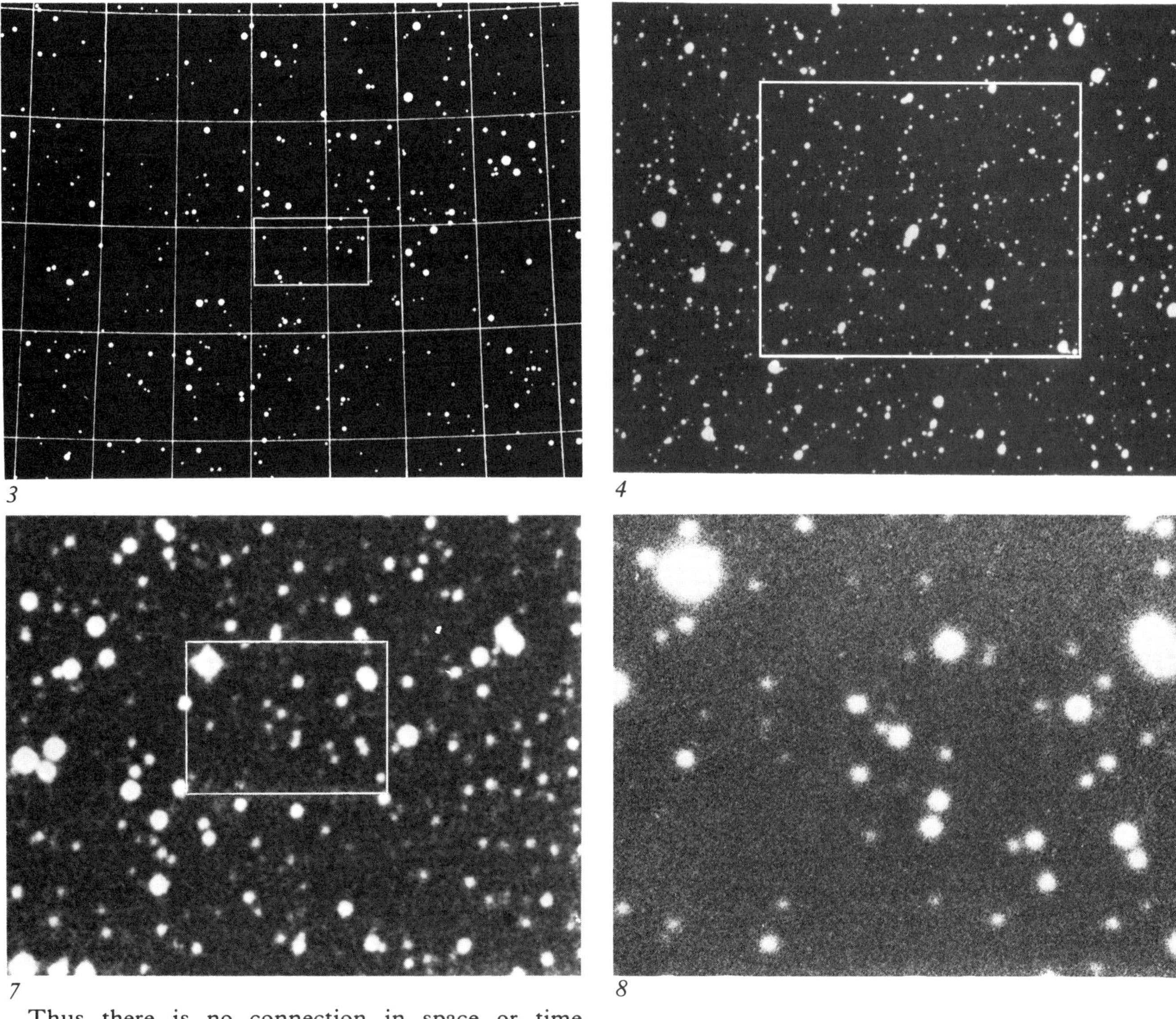

3 4 7 8

Thus there is no connection in space or time between the Cassiopeia stars. This is the case with most constellations. The arrangements of the stars which appear so natural to us here on Earth are in fact chance line-of-sight groupings and signify nothing.

Messier 73

Asterism

Messier discovered this unspectacular little "cluster" of four stars in 1780. The cluster is in the constellation Aquarius, close to the globular cluster M 72, which Messier discovered on the same night. Perhaps M 72 is the real reason why Messier recorded M 73. Although M 73 found its way into Dreyer's compendium of star clusters and nebulae, the *New General Catalogue*, as NGC 6994, it has been regarded as an asterism, a chance coincidence of four stars along the line of sight. The authors of this book do not know if this is true; one day they plan to determine the types of

Messier 73 POSS

stars concerned to see if they are related. They suspect in fact that M 73 might be a real little cluster, for the following reason. On average there are 60 stars per square degree which are brighter than magnitude 12, as are the four stars in M 73. The probability of finding four such stars by chance in a given area of sky one arc minute across (like M 73) is about two chances in a billion. However, there are 150 million such little areas on the sky, so the chances are only one in four that one such random asterism exists on the sky. M 73 could be it, but we would gamble that it is a genuine multiple star of some kind.

Messier 24

Star cloud

Charles Messier discovered this object in the Milky Way in the constellation of Sagittarius in 1764. He described it as a degree and a half in diameter and reported that it contained several stars of different magnitudes. The stars it contained were clustered in several different parts. Even Messier's own description hints the truth: M 24 has no reality in itself. The stars it contains are too spread in brightness and have no readily distinguished center. What Messier in fact saw was a star cloud of the Milky Way, a pseudo-cluster of stars spread thousands of light years along the line of sight, perceived through a chance tunnel in the interstellar dust.

The interstellar dust generally dims the light of stars behind it. But the dust is patchy. For some unknown reason it clumps in clouds typically 25 light years across: many such clouds can be clearly distinguished, projected against the star cloud. There are typically two such clouds in a line of sight 1000 light years long in the Milky Way. But even over the 30,000 light years to the central regions of the Galaxy there could be, and

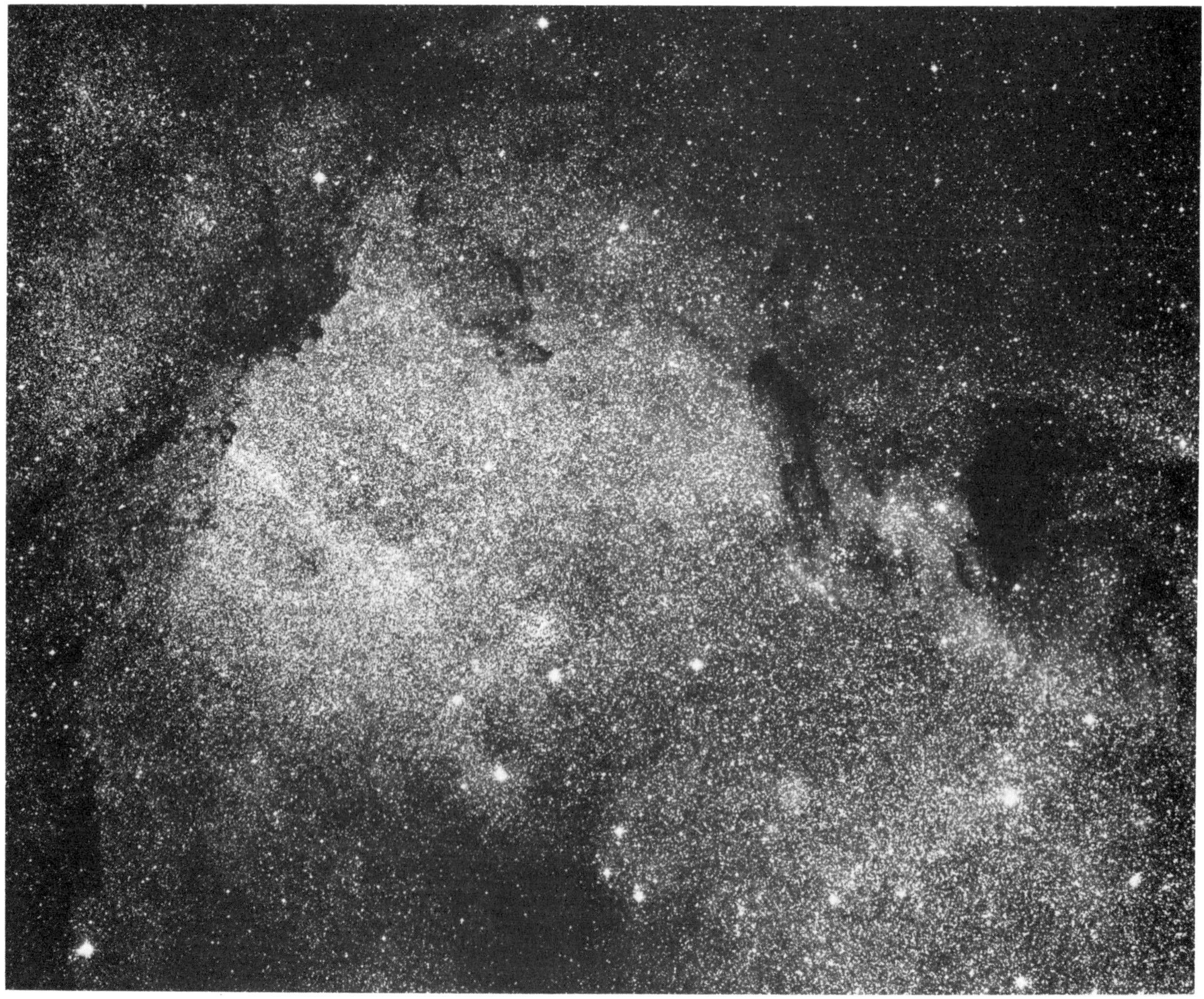

Messier 24 POSS

by chance are, clearer windows than normal in the interstellar medium. M 24 is in effect one of these windows.

These clear windows through the Galaxy have great significance in the study of galactic structure, since they make it possible to study otherwise hidden, distant regions.

South Celestial Pole

Real point, insignificant star

The South Celestial Pole, around which the southern stars circle each night, as seen on a time exposure photograph (Plate 5), is marked by the star Sigma Octantis. It is a faint and insignificant star compared with Polaris, the star within a degree of the North Celestial Pole. Of course, there is no reason why there should be any significant marker at either celestial pole, since neither pole is the axis of rotation of the sky, but the axis of the rotation of the Earth, as seen traced out by the apparent daily motion of the stars. Thus although as the photograph shows, the celestial pole is a very real point in the sky, it is of no significance when we are considering the stars themselves.

The celestial poles move slowly on the sky, as the Earth wobbles like a spinning top. The poles trace out a circle of radius 23½ degrees around a point in the constellation of Draco (North Pole) and near the Large Magellanic Cloud (South Pole). They describe this cycle every 25,700 years, the so-called period of precession.

South Pole AAT

Ursa Major

Moving group

In 1869 the British astronomer and popularizer R.A. Proctor called attention to the fact that the five central stars of the Big Dipper (the Plough) shared a common motion. Our figure shows their displacement after 100,000 years. Three years later William Huggins, the great English pioneer of spectroscopy, pointed out that these five stars, which have approximately the same brightness, have similar spectra. Thus they are moving together, are the same distance (so their light is diminished to the same degree) and are of similar age. Unlike the majority of the constellations, therefore, these Dipper stars are really related, rather than being in happenstance line-of-sight coincidence. They constitute five members of a kind of loose cluster of stars named, after the Latin name of the Dipper constellation, the Ursa Major moving group. They and five fainter stars in the Ursa Major constellation are all grouped about 75 light years from the Sun.

In 1909, Danish astrophysicist Ejnar Hertzsprung noticed that 37 Ursae Majoris and Alpha Coronae Borealis, two stars in widely separated parts of the sky, shared the same motion as the Ursa Major moving group. This led him to make a systematic search over the whole sky among the stars whose velocities had then been measured, in order to find other members of the group. He found eight new members, including to his and everyone's surprise, Sirius, which lies in almost the opposite direction to Ursa Major.

Over many years the efforts of astronomers have made more data available on the velocities of stars. More members have been added to the Ursa Major moving group. It now numbers 100 possible members, of which Sirius is the nearest to us and also the brightest. These 100 stars surround the Sun and are heading towards a point in space on the border of Sagittarius and Microscopium at a speed relative to the

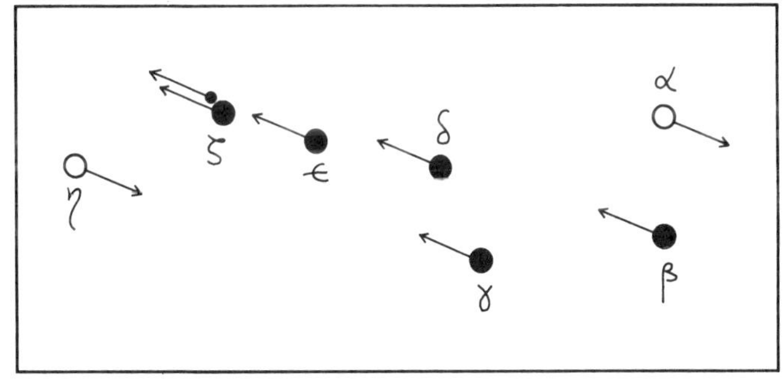

Ursa Major after 100,000 years

Sun of 18 km/sec. They seem to be about the same age as the PLEIADES star cluster, in the range, say, 50 to 100 million years old. The Pleiades itself is the nucleus of another moving cluster, as are the HYADES and PRAESEPE star clusters.

Why do moving groups exist? Evidently all the stars in a moving group were born at much the same place and time. Their parent gas cloud, having been jostled about the Galaxy by random encounters with other star clouds, had some motion of its own, some small deviation from the general rotational motion about the Galaxy. When the gas cloud fragmented, the stars which it formed "remembered" their parent's motion, continuing to move in a group around the Galaxy, just as the fragments of a bursting skyrocket continue the motion of the rocket through the air. During their motion around the Galaxy, some members of the moving group encounter other random stars and are swung off course. Thus members of a moving group "evaporate" in time, and the group dissipates. Old

36 Ursae Majoris BD

groups are difficult to identify with certainty, although Olin Eggen, now of the Cerro Tololo Inter-American Observatory, believes that a moving group containing the star 61 Cygni could be comparable in age to the oldest galactic cluster M 67, say five billion years old. If so the 61 Cygni moving group survives as the oldest living fossil of galactic rotation.

36 Ursae Majoris

Common proper motion

Some double stars are recognized because over the course of time the motion on the sky of one around the other is noticed. But the lifetimes of astronomers and even records about astronomy are short, and there could be pairs of stars in space so widely separated that their orbits around each other take many millions of years to complete. Such stars are recognized to be fellow travelers by the fact that they journey together in space, coursing along on parallel trajectories. They are called common proper motion stars. (A star's proper motion is called that, not because it has any moral virtue, but because the motion is intrinsic to the star itself. The concept excludes an apparent motion such as the way a star rises and sets in the sky because of the rotation of the Earth.) 36 Ursae Majoris is a fourth-magnitude naked-eye star near the bowl of the Big Dipper (or the front of the Plough as it is called in Britain). It moves across the sky by 18 arc seconds every century, in a due westerly direction—that, in other words, is the value of its proper motion. Two arc minutes away is a ninth-magnitude red dwarf star (BD +56° 1450 is its number). Both are visible at the center of our picture. BD 56° 1450 moves in a parallel course to 36 Ursae Majoris, although separated from it in space by at least light days. More than one degree away is a third common proper motion star to this multiple group, an eighth-magnitude yellow dwarf star at least three-quarters of a light year from the other two. It takes about 15 million years to orbit the other two. It is therefore little wonder that the curvature of this star's motion about the other pair cannot be noticed and the proper motions of all three stars appear parallel.

Gould's Belt

Tilt of a nearby galactic fin

The brighter stars are not scattered uniformly over the sky: they tend to lie along a band known as Gould's Belt. Discovered by Sir John Herschel (1847) and studied by B.A. Gould (1879), Gould's Belt is close to but not identical with the Milky Way. Winter observers in either hemisphere can see Gould's Belt with their own eyes. In the northern sky the bright stars concentrate in Orion and Taurus and lie off to one side of the Milky Way. In the southern hemisphere the Lupus and Centaurus constellations lie off to the other side. The tilt between Gould's Belt and the central line of the Milky Way is 16°. The figure shows a diagrammatic representation of the Milky Way and its centerline as determined from radio astronomy measurements. Each bright star in the sky is individually plotted on the picture. The brighter stars tend to lie below the Milky Way line in the left and right quarters of the figure and above it in the central half.

Closer study by Harlow Shapley and Annie Cannon in 1924 showed that the tilt was a property of the easily seen hot blue stars. About a third of the stars which are visible to the naked eye are hot, blue stars, including virtually all the stars of the constellation Orion. Hot, blue stars are massive stars. They are profligate with their nuclear fuel, which runs out quickly. They never, therefore, live to a ripe old age, but die as relative youths. Gould's Belt, then, is a feature of stars that are both young (less than say 100 million years old) and nearby (within, say, 600 light years). The more distant hot, blue stars tend to lie closer to the Milky Way.

According to the late Australian astronomer Colin Gum, Gould's Belt represents a fin projecting from the

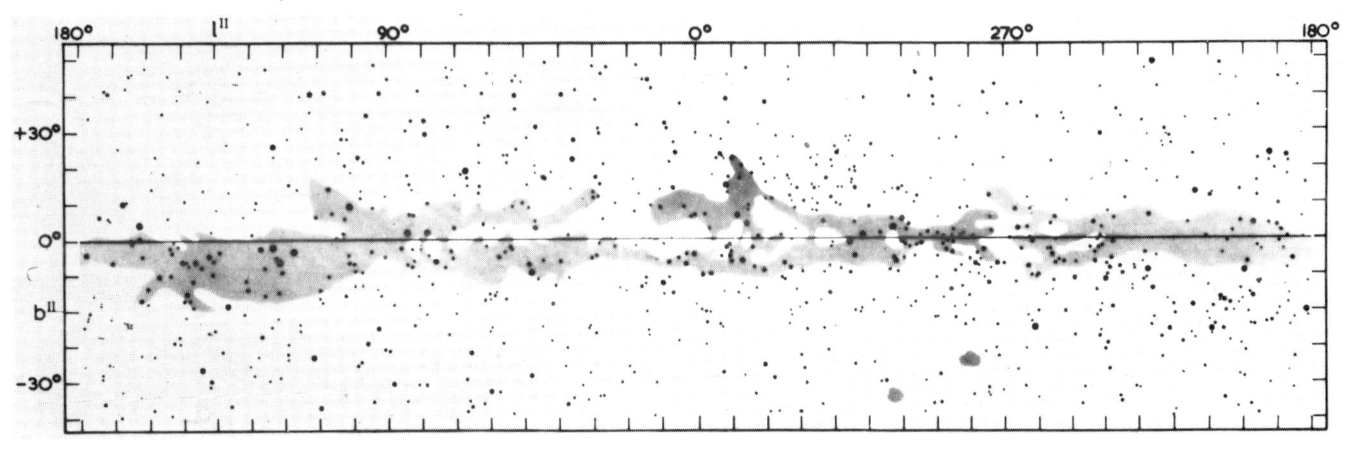

Gould's Belt

Orion's Belt and Sword *Roberts*

nearest spiral arm of our Galaxy, stuck out from the plane of the Milky Way at an angle. In time, as the Milky Way rotates, this locally tilted fin will smear out and disappear, like a transient eddy in a whirlpool. It would be natural, then, to expect Gould's Belt to be apparent only in stars which are younger than the time for the Galaxy to make one rotation, namely 100 million years. Gum's suggestion fits well.

Orion

Association in time and space

Unlike most constellations such as CASSIOPEIA (p. 74), which are meaningless line-of-sight coincidences, the constellation Orion is a group of stars associated in time and space. The mythological figure which the constellation represents is The Hunter, with three stars figuring as his Belt, and three his Sword hung from it. These six stars and five at Orion's head, shoulders and feet form the main stars of the constellation. Our picture is by Isaac Roberts, British pioneer celestial photographer, and shows the Belt and Sword. Of the prominent stars in the constellation, only Betelgeuse (Alpha Orionis) is an odd man out, being a red supergiant at much greater distance. The stars appear bright to us, first because they are relatively nearby (1330 light years), and second because they are intrinsically luminous.

Rigel (Beta Orionis) is a blue supergiant star. Many of the other stars in the group including all three of the stars of the Belt of Orion (Delta, Epsilon and Zeta Orionis) are giant stars.

As stars exhaust their nuclear fuel, they paradoxically brighten for a short time. Since the stars of the Belt are brighter than the stars of the Sword, one might expect that the Belt stars have used most of their nuclear fuel while the Sword stars are still, as it were, in the prime of life with plenty of fuel in reserve—and this is indeed the case. The Belt stars are about five million years old whereas the Sword stars are about one or two million years old. The youngest stars of all are found near the TRAPEZIUM (p. 89) and the ORION NEBULA (p. 102).

Reddened Star Cluster

Penetrating interstellar fog

The right-hand photograph on page 82 shows a previously unknown globular cluster whose photograph is, in this book, published for the first time. It is in fact faintly visible, but unrecognizable as a cluster of stars, in the photograph on the left; the two photos being of the same area of the sky, which is near the Galactic Center in the constellation Sagittarius.

Why is it that so many more stars, including those of the new globular cluster, are visible in the right-hand photo than in the left? The photo on the left was taken through a filter passing blue-green light. The photo on the right was filtered to show only infrared radiation. Infrared is no different from light, except that its wavelength is longer and human beings cannot see it.

The extra long wavelength of infrared, two to three times the wavelength of visible light, enables infrared to pass easier through dust in interstellar space. The stars visible in the left-hand photo are nearby. The extra stars visible in the right-hand picture are much farther away; they are screened from view in blue-green light by the copious amounts of dust in this part of the Galaxy.

How can longer wavelength radiation penetrate dust easier? Think of waves of water in your bath. Small wavelength water waves, such as are made by wagging a finger rapidly on the water's surface, are reflected by an object which is larger than their wavelength (such as your knee) and do not pass it easily. Longer wavelength waves, made by sliding your body back and forth in the bath, pass easily around your knee, reforming almost undisturbed on the other side. Now, a grain of interstellar dust is of size typically a half of a micron (a micron is a millionth of a meter, so half a micron is just under one-twenty-five-thousandth of an inch). Blue light waves are smaller than this and are affected more by an interstellar dust grain. Infrared waves are twice its size, at least, and pass it by easier. By photographing the Galaxy in infrared radiation many more stars are revealed, as in our example. Infrared reveals what is beyond the interstellar smog.

Double and Variable Stars

Sirius

Apparently brightest star

If we could judge by the most prominent example of a star which we can see, our own Sun, we would guess that most stars were single and we would in fact be wrong. The majority of stars exist in twos and threes. Only 15% of all stars are single.

Why is it that our impression of what is typical is wrong on this occasion? The reason is that the orbit of a planet around a star in a multiple star system is likely to be erratic, due to the ever-changing gravitational tugs of the other stars in the system. The planet may not even be able to orbit its star more than a few times before being snatched away from its parent sun. Even if this does not occur, the planet is likely to move close to and far from its sun and as a con-

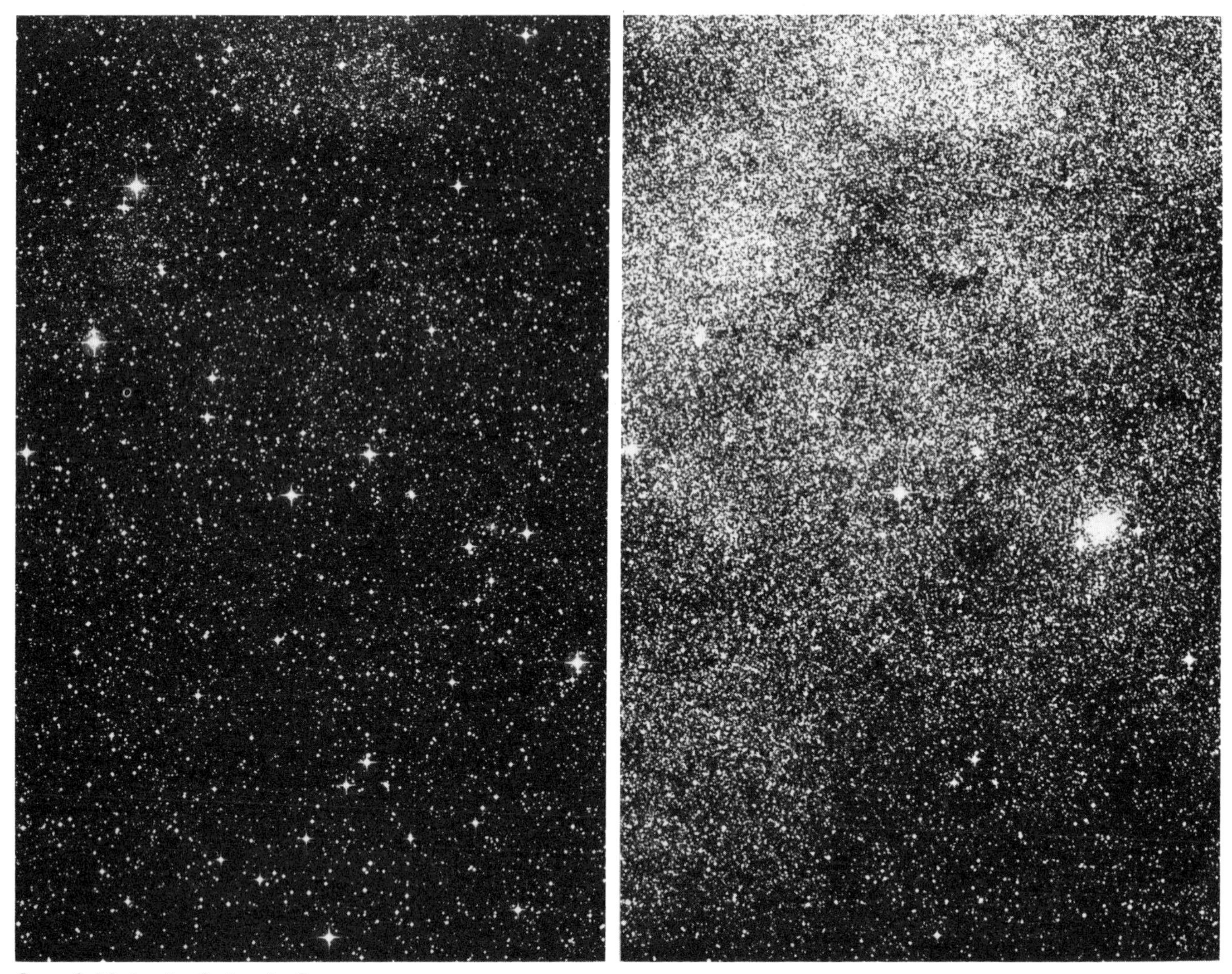

Star fields in the Galactic Center UKSTU

sequence its temperature will change drastically from searing heat to frozen cold. The prospects for the development of life on such a planet are not good.

It may be then that the fact that we are here to perceive our Sun implies that it is single. What is there to say about the majority, multiple stars?

Our first example of a multiple star is the Dog Star, Sirius. Sirius is the brightest star in the sky: Alpha Canis Majoris. It is a brilliant white star, although, curiously, classed with the red stars by Ptolemy (AD 144), who may have been thinking of its scintillating appearance when rising. It is the sixth nearest star system, only 8.7 light years away. Its closeness is betrayed by its angular motion across the sky. In the past 1350 years its motion, generally towards the south west, has been equal to the diameter of the Moon.

In 1834, Prussian astronomer F.W. Bessel noticed that Sirius does not move uniformly but swings from side to side. It appeared to be deviating back and forth by about two arc seconds with a period of 50.09 years. In 1844, Bessel wrote that he was convinced that Sirius is a binary system, consisting of a visible star, Sirius A and another, invisible star, Sirius B.

The orbit of Sirius B was worked out before the star was actually seen. It was detected in 1862, during the test of a new 18-inch telescope by the son of American telescope-maker, Alvan G. Clark. The separation of the two ranges up to a maximum of 11 arc seconds. This was last attained in 1973. The difference in brightness of the two stars is almost 10 magnitudes, and B is therefore 10,000 times fainter than A. From time to time, various observers have thought that they could glimpse a third star in the system, but no evidence of a third star can be found in Sirius' motion.

The mass of Sirius A is 2.2 times that of the Sun. The mass of B is nearly the same as that of the Sun (0.94 times). How can a star of the mass of the Sun be as faint as Sirius B? For a long time this was a mystery. According to A.S. Eddington, noted English astrophysicist, the message of the Companion of Sirius, when it was decoded, ran "'I am composed of material 3000 times denser than anything you have

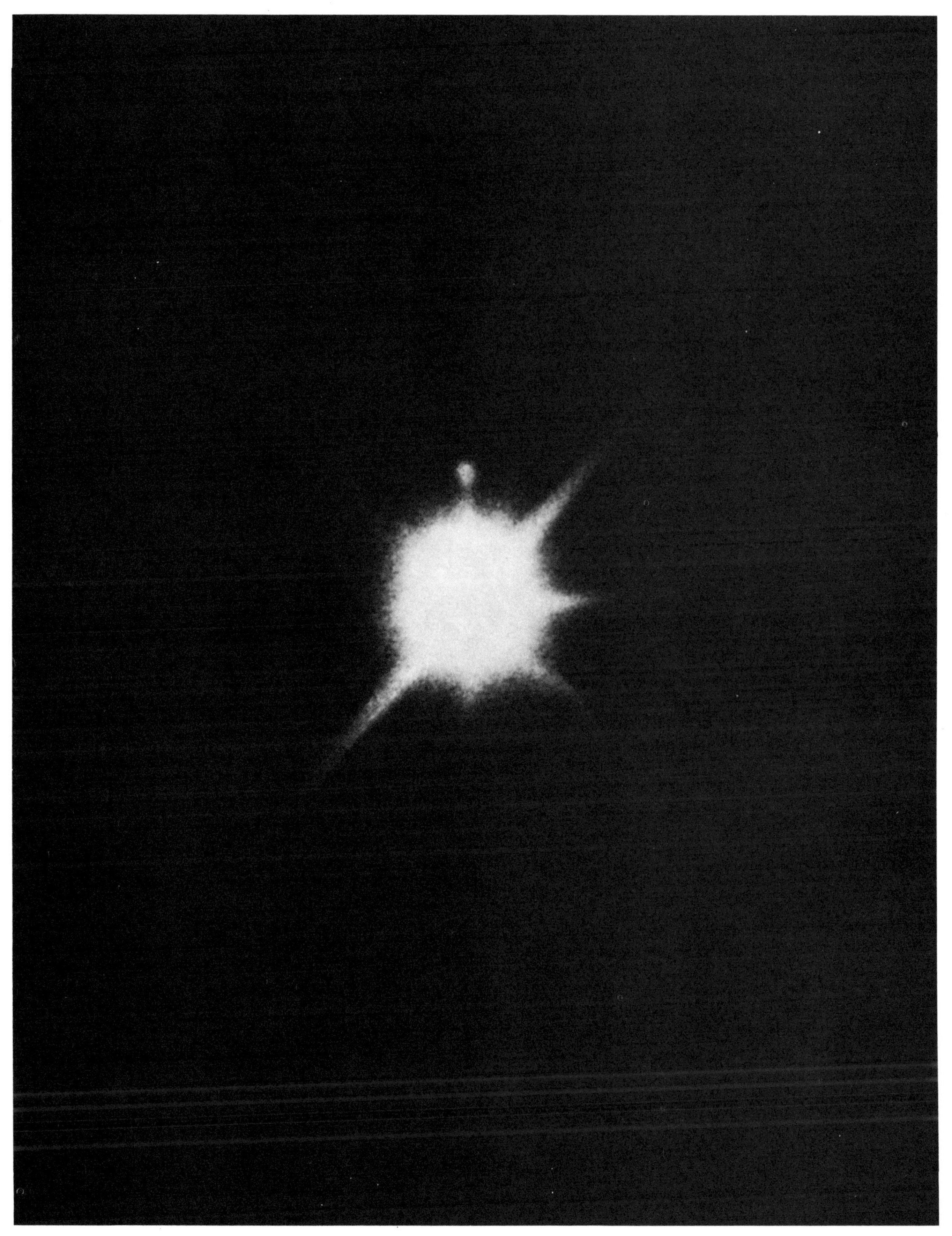

Sirius A and its faint companion, B *Lick*

come across; a ton of my material would be a little nugget that you could put in a matchbox.' What reply can one make to such a message? The reply which most of us made in 1914 was — 'Shut up. Don't talk nonsense.'"

The answer which should have been made, and which Eddington proposed in 1924, came from checking the astonishingly high density of Sirius B with the recently discovered gravitational redshift. Albert Einstein had shown that when radiation left the surface of a very compact star, its wavelength would shift because of the star's large force of gravity. The light from Sirius B did reveal a redshift of the expected size. This was recently confirmed in 1973 when Sirius B could be clearly seen again. Thus the density of Sirius B had to be accepted. Faint hot stars like it were named "white dwarfs."

In 1940 the leading Indian astrophysicist S. Chandrasekhar explained the structure of white dwarfs and why they are so dense. Later it became clear that they were formed at the end of a lightweight star's lifetime, after it had thrown off a planetary nebula like the RING NEBULA. Thus the full explanation of Sirius B came more than 100 years after the first hint of its presence was discovered by Bessel.

Proxima Centauri

Nearby star

The second nearest star to the Earth (the Sun is the nearest) is Proxima Centauri. It is the faintest of three stars in the Alpha Centauri triple system, and was discovered by R. Innes in 1915. Alpha Centauri—not to be confused with Centaurus A, the radio source—is one of the brightest stars in the sky. Its popular name

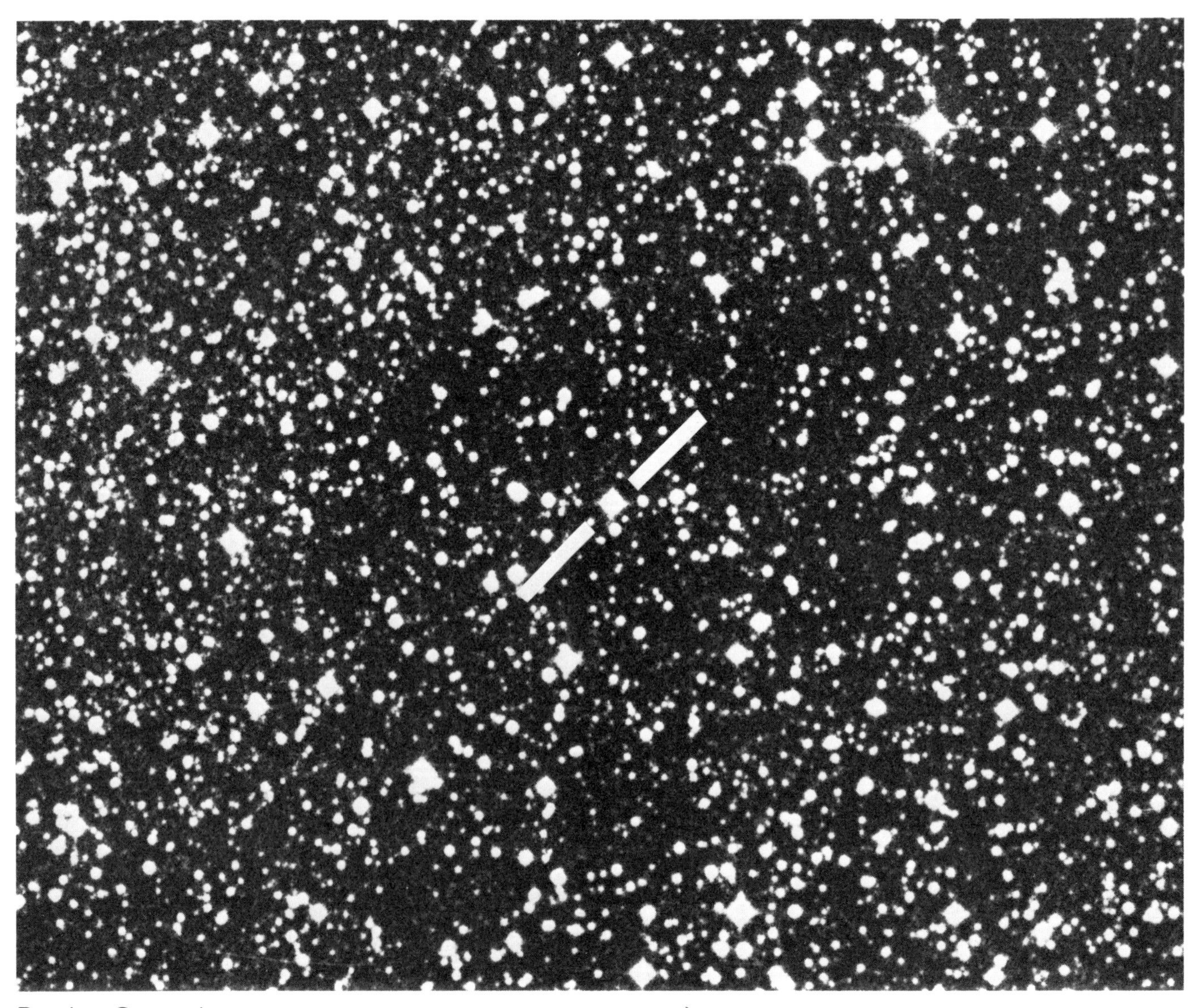

Proxima Centauri UKSTU

is Rigil Kent.

The period of orbit of the two brightest stars, which make up Alpha Centauri itself, is 79.92 years and they are separated by approximately 20 arc seconds in the sky. More than 2° from the Alpha Centauri pair is Proxima, slowly orbiting them with a period measured in millions of years. Proxima is 4.2 light years from Earth, and 0.19 light years from Alpha Centauri. It is a variable star, also called V645 Centauri. It shows sporadic flares occurring from hour to hour, almost doubling its brightness for about 6% of the time.

Alpha Centauri was one of the first six double stars to be discovered. Together with 61 Cygni, it was the first to have its distance determined, by Bessel in 1838. The method he used was that of trigonometrical parallax, in which the position of the star was seen to change as viewed from each extremity of the Earth's orbit, much as the position of a finger against a background changes when viewed by each eye in turn.

The modern value for the parallax, which is half the angular shift as seen from the two extremities, is 0.76 arc seconds. This is equivalent to the shift in position of a needle at a distance of seven miles when viewed by each eye alternately, and corresponds to a distance of 4.3 light years, just farther than Proxima.

Centaurus X-3

Binary x-ray star

Centaurus X-3 was the first clear proof to astronomers that x-ray sources occurred in double-star systems. X-rays from the stars do not penetrate the Earth's atmosphere, but x-ray sensitive telescopes have been sent above most or all the atmosphere in balloons, rockets or satellites. At the end of the 1960's theoretical astronomers were beginning to think that the x-ray stars which had been glimpsed during the brief flights of rocket and balloon-borne telescopes were binary stars, but no observational proof was then available. In 1972 the Uhuru satellite group of x-ray astronomers, using their satellite-borne x-ray telescopes, were able to observe x-ray stars for more than a few minutes or hours. They found that x-rays from the source Centaurus X-3 disappeared every 2.087 days. The disappearance was extremely regular and abrupt (see lower panel in figure). It was clearly due to a small x-ray source passing in orbit behind a large star with a well-defined edge.

The Uhuru x-ray astronomers also found on close inspection that the x-rays from Centaurus X-3 were rapidly pulsing with a period of 4.8 seconds. This was reminiscent of the hitherto only other pulsing x-ray star then found, the CRAB PULSAR (p. 152). The pulsation period of Centaurus X-3 was not perfectly regular, but got shorter and longer in synchronism

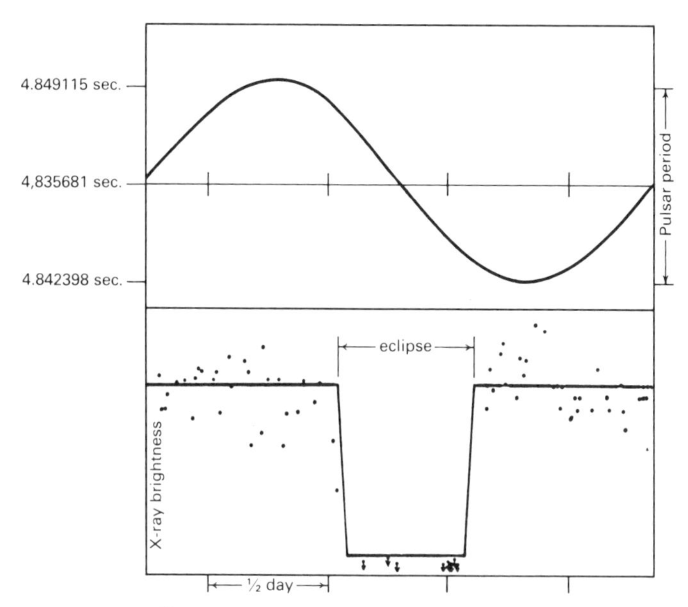

Centaurus X–3

with the orbit, every 2.087 days (upper panel of figure). This behavior results from the doppler shift of the x-ray source, with the pulse period lengthening as it receded from Earth. The pulses disappeared briefly while the star was eclipsed, and their period then shortened as the star approached Earth. In every way the x-ray facts fitted the notion that the x-rays originated from one component of a double star. It remained to find the visible object which the x-ray emitting star orbited if indeed it was bright enough.

After x-ray astronomers pinpointed the x-ray source on the celestial sphere, Polish astronomer W. Krzeminski found a star there whose light varied with the same period as the x-ray source. Astronomers believe that the x-ray emitting star distorts the optical star into an ellipsoid—that is, a stubby cigar-like shape. The ellipsoid spins about its short axis, presenting first the broad side and then the short side to Earth, making its light vary by about 10% as seen from Earth.

Studying the spectrum of the optically visible star, and using the measured periods, astronomers can calculate several details about the system, which is some 25,000 light years away. The orbit of the two stars is tilted to our line of sight by no more than 10°. The mass of the visible star is 20 times that of the Sun and that of the x-ray emitter is two solar masses. The 4.8 second pulsations are connected with the rapid rotation of the two-solar-mass star: it is a pulsar somewhat like the CRAB PULSAR.

Cygnus X-1

Black hole candidate

Cygnus X-1 is a bright x-ray star in the constellation

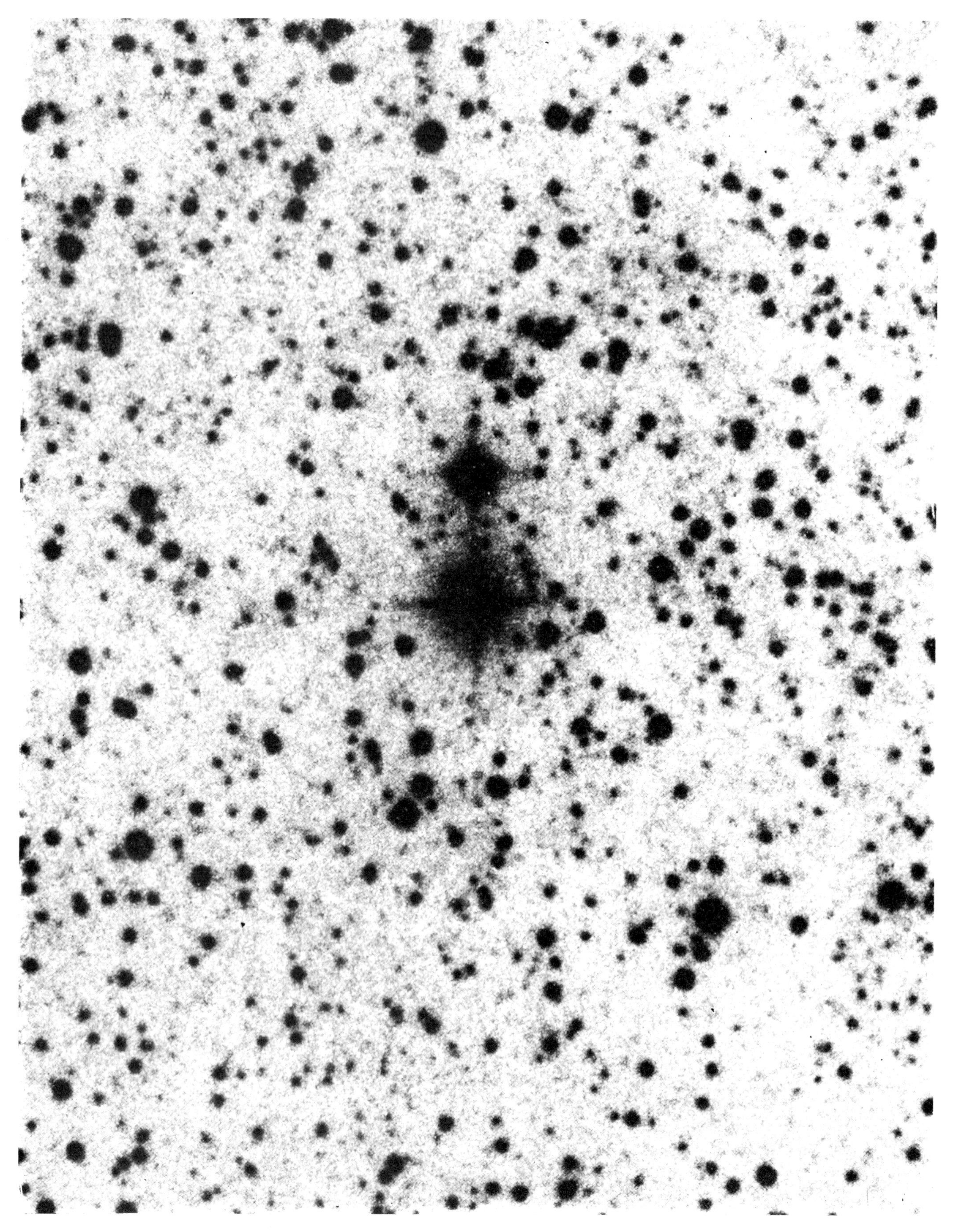

Cygnus X–1 POSS

Cygnus, and it is a strong candidate for a black hole.

A black hole is a star so dense and therefore with so strong a gravitational pull that nothing may leave its surface, because to do so would require it to travel faster than the speed of light. "Of all the objects which one can conceive to be traveling through empty space," wrote Remo Ruffini and John Wheeler in 1971, "few offer poorer prospects of detection than a solitary black hole."

Although solitary black holes may be undetectable, Soviet astronomer Iosif Shklovsky showed in 1967 how to find a black hole which was orbiting an ordinary star. When the star began to expand in the course of its evolution, it would dump gas from its atmosphere onto the black hole. As the gas fell into the black hole it would be heated, compressed by the black hole's force of gravity. The temperature of the gas would reach millions or billions of degrees, and would emit x-rays.

According to Shklovsky, to find a black hole, look among the x-ray stars. Find one associated with an otherwise ordinary but evolved star, which has expanded enough to overflow onto any companion. Show that the star is indeed a double star. As a clincher, show that the invisible companion star cannot be a neutron star or white dwarf. This last part is necessary because it turns out that gas falling onto a neutron star—a star even denser than a white dwarf, so that all its material is compressed into neutrons—will also be heated to very high temperatures. In fact a marshmallow falling onto a neutron star will release the energy of a World War II atomic bomb, and be heated accordingly. But, according to theoretical astrophysicists, a neutron star or white dwarf cannot be more massive than about two solar masses. Thus if the invisible x-ray emitting star is larger than three solar masses, the only object of which astronomers know which could be responsible is a black hole.

Cygnus X-1 was first seen briefly by rocket-borne x-ray sensitive telescopes in the 1960's. It was one of the x-ray stars studied by the Uhuru satellite in 1971. This satellite was the first to carry an x-ray telescope and for about a year it surveyed the celestial sphere for x-ray sources. Uhuru x-ray astronomers located 400 x-ray sources more precisely than ever before, including Cygnus X-1. Radio astronomers could now survey the area, knowing where to look, and one day found that a radio source had appeared in the area which was not there before. At the same time that the radio source appeared, the Cygnus X-1 x-ray source dramatically changed its character. Thus the two objects, radio source and x-ray source, must relate to the same object. A modern determination of the position of the x-ray source Cygnus X-1 is that it lies within an ellipse containing the brightest star on the photograph.

The radio astronomers were able to pinpoint the object much more accurately than the x-ray astronomers. Having narrowed it down, they consulted star charts and found an optically visible star known as HD 226868 at the location indicated by the radio telescopes. HD 226868 is the brightest star within the x-ray ellipse. Optical astronomers then determined that the star is a supergiant, and that it circles an invisible companion, as betrayed by the doppler shift of its spectral lines. Astronomers could not be sure if they were seeing the full motion of the star (orbit edge on) or a small part (orbit inclined to the line of sight), and so could not tell the mass of the invisible star. They could, however, deduce that the companion was at least six times more massive than the Sun. This is twice the mass of the most massive neutron star possible. Thus Cygnus X-1 fitted the description of a black hole which had been outlined as an explanation four years before the existence of a problem.

In 1978 another x-ray binary system, V861 Scorpii, was found, very like Cygnus X-1. In the case of V861 Scorpii astronomers have the extra information that the x-ray object disappears every orbit, eclipsed by the supergiant star. Thus the orbit has to be almost edge-on. This enables a better determination of the mass of the invisible object than in Cygnus X-1. It is about eight solar masses, far in excess of the mass of the heaviest possible neutron star. V861 Scorpii could be a better black hole candidate than Cygnus X-1.

Barnard's Star

Largest proper motion

The great American observer E.E. Barnard noticed in 1916 that the star BD +4° 3561 had a large angular speed across the sky. It is, in fact, the star of largest proper motion: 10.3 arc seconds every year. At present centrally located in the constellation Ophiuchus, Barnard's Star will move 10°, into the constellation Hercules, in 3600 years. Its motion in 10 years is easily seen on the pair of Lick Observatory photographs (Barnard's Star is arrowed).

Barnard's Star is the third nearest star system, just 5.88 light years away: its distance is known very accurately, with an uncertainty of just one light week! It is approaching Earth, coming within 3.75 light years in AD 11800, when, because of its closeness, it will appear about one magnitude brighter than its present magnitude of 9.5.

In its course through space, Barnard's Star wobbles from side to side over a range of 0.04 arc seconds. Like the wobbles of SIRIUS (p. 81), they are attributed to perturbations caused by unseen companions. Unlike the case of Sirius, however, no companion has been seen. The reason became clear when analysis of the mass of the unseen companion was possible. Sophis-

Motion of Barnard's Star *Lick*

ticated calculations suggest that Barnard's Star has at least two invisible companions, orbiting with period of 12 and 26 years, and with masses 0.7 and 1.15 times the mass of Jupiter respectively. Thus the companions are planets rather than stars. The planets are at Jupiterlike distances of three and five astronomical units from their sun (one astronomical unit is the average distance between the Sun and the Earth).

Mizar and Alcor

Double, double, double

Mizar is the star also called Zeta Ursae Majoris, the central star in the handle of the Big Dipper (the Plough), 88 light years from Earth. Mizar was the first double star to be discovered by telescopic observation, by G.B. Riccioli in 1650. It was also the first to be photographed, by the pioneer American astronomer W.C. Bond in 1857.

The separation of the two stars (Zeta^1 and Zeta^2 Ursae Majoris) is 14.4 arc seconds with the fainter towards the south. They can easily be seen as a pair through small telescopes. Their period, the time they take mutually to orbit each other, is at least 20,000 years. The brighter, Zeta^2 is itself double, and was the first binary or double star found by means of a spectroscope. It was discovered by E.C. Pickering in 1889, who found that the spectral lines of the apparently single star periodically doubled, as the result of the doppler shifts of two stars, one approaching and one receding. Both stars have nearly equal mass and brightness, and their orbital period is 20.53860 days.

The fainter star, Zeta^1, is also a spectroscopic binary with a period of 175.55 days, but only one spectrum from the pair of stars can be distinguished in a spectroscope because the brighter so much outshines the fainter. In this case, all that can be seen is a periodic shifting of the spectral lines.

Mizar BD

Both the pairs of stars which comprise Mizar are orbited at an apparent distance of 12 arc minutes by Alcor, a star which can easily be seen near Mizar with the naked eye, and which is visible near Mizar, brightest image on this picture. The period of Alcor around Mizar is some ten million years. Finally, since the radial velocity of Alcor is variable, there is some suspicion that it too may be a spectroscopic binary, although no period has been determined.

This remarkable multiple star system thus consists of possibly six stars in a hierarchy of double star orbiting double star, orbiting double star.

Trapezium

Disintegrating multiple star

The Trapezium is the star Theta^1 Orionis, which was discovered by Nicolas Peiresc in 1611. As he examined the central star of Orion's Sword, he found that there were "tightly grouped small stars embedded in a bright white cloud." The nebula he referred to is the ORION NEBULA (p. 102). The four brighter stars of the group form the Trapezium. The name is derived from their pattern and they are separated from each other by 10 to 20 arc seconds. With a moderate sized telescope, two fainter stars can be seen nearby. Our short-exposure photograph of the Orion Nebula is designed to show the region which is normally heavily overexposed, as in Plate 6. The Trapezium stars are the four brightest.

In 1931 R.J. Trumpler photographed the Trapezium and found that it was embedded in a cluster of fainter stars known as the Orion Cluster. These stars are not easily seen visually for two reasons: first, because they are confused by the brightness of the Orion Nebula; second, light from them is very much diminished by absorption by interstellar dust. Russian astronomer P.P. Parenago found that many of these faint stars are young variable stars, still forming out of the dust and gas of the region. The age of the Orion Cluster is estimated at four million years.

The Trapezium could be a very young system of stars indeed, a suggestion put forward by Parenago and colleague V.A. Ambartzumian in 1953. They had in mind the fact that the multiple stars which can survive the longest without disruption are those in which a pair of stars is orbited by a third star much farther away, and in which that triple system is orbited by a fourth at a still greater distance and so on. This is a so-called hierarchical multiple star (like the one to

The Trapezium *Lick*

which PROXIMA CENTAURI (p. 84) belongs), and a system such as this is stable and can last indefinitely. In the Trapezium, however, all the stars are approximately equidistant from one another, and as their orbits intermingle, individual stars can be thrown out of the multiple star system by chance encounters. Parenago and Ambartzumian estimated that this might happen very quickly, implying that the Trapezium is young. But since the Soviet astronomers could not be sure to what extent the stars of the Trapezium are separated along the line of sight, and could not measure their speeds accurately, they could not reliably predict the Trapezium's development.

The coming of large, fast computers has enabled us to say what *might* happen, even though astronomers are still no more certain of the relative distances of the Trapezium stars. Theoretical astronomers set up a Trapezium-like system in a computer and calculated what would happen under the influence of gravity.

Astronomers C. Allen and A. Poveda at the Mexican Institute for Astronomy computed 30 examples intended to embrace all possible states of the Orion Trapezium.

In the computer simulation, the stars were placed at random points and given an initial push in a random direction. Allen and Poveda ran the computer for the equivalent of a million years of the stars' lifetimes. In 19 cases, a Trapezium with at least four members still existed at the end of one million years, although quite often the lighter stars of the six had been thrown off. In three cases, four stars of the six had been thrown out of the Trapezium and only a binary system, consisting of the two heaviest stars, was left. In eight cases, the Trapezium had become a hierarchical multiple system.

Thus, according to the computer simulation, the most likely course of events is that the Trapezium will cease to exist as such after a few million years. This is not quite as quick as Parenago and Ambartzumian estimated. The Trapezium's age is however consistent with that of the Orion Cluster within whose boundary it lies. Presumably the Trapezium contains the few most massive stars in the cluster. The variable stars are the least massive ones and they are still forming. The Trapezium is indeed a young system, as, presumably, are the other 107 Trapezium-like multiple stars which Ambartzumian found.

Mu Columbae, AE Aurigae and 53 Arietis

Runaway stars

Among the normal slow-moving population of bright massive stars in our Galaxy, there are some so-called "runaway stars" with high speeds up to 200 km/second. Three of these are Mu Columbae, AE Aurigae and 53 Arietis. From their speed and direction of travel, Dutch astronomer Adriaan Blaauw calculated that all three left the constellation ORION possibly in a single event which occurred some three million years ago. Each has traveled across an intervening constellation to reach the present-day constellations of Columba, Auriga and Aries respectively.

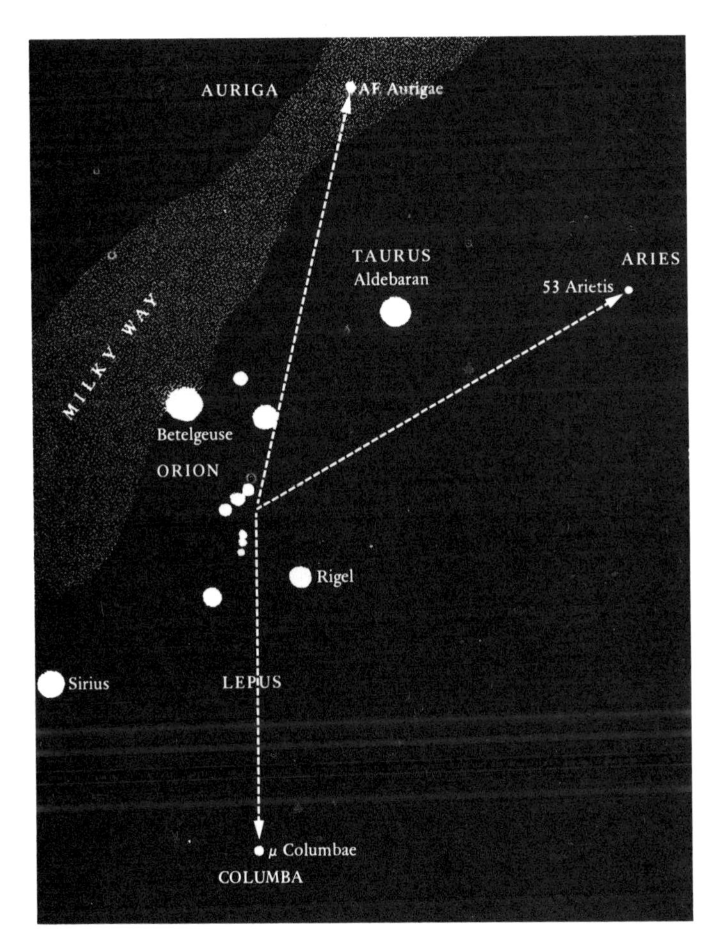

Runaway stars from Orion

Following a suggestion by CalTech astronomer Fritz Zwicky, Blaauw proposed that these stars were originally in a quadruple star system like the TRAPEZIUM, but that the heaviest member of the four exploded as a supernova. Having nothing left to orbit, the remaining three stars flew away, like shots from a sling, expelled far from their birthplace. An alternative explanation proposed by Blaauw's colleague, A. van Albada, is that the three were thrown from Orion by chance encounters with other more massive stars.

Whatever the truth of their origin, the destiny of the runaway stars is clearer. They are massive stars and ultimately will explode as supernovae. The stellar cinders which these explosions form—pulsars, black holes—will possibly continue their trajectory. Future astronomers will perceive them as fast-moving pulsars, flying high above the galactic plane. Some pulsars (like the CRAB PULSAR) are indeed known to be fast-moving. Some have speeds up to 300 km/second. At one time, these too may have been runaway stars like Mu Columbae, 53 Arietis and AE Aurigae.

Betelgeuse

The face of a star

Betelgeuse is also known as Alpha Orionis, the brightest star in the constellation Orion. It is a red supergiant, one of the largest stars known, and is relatively nearby, at 650 light years. Except for the Sun, it is the star with the largest apparent angular diameter, although to any ordinary telescope, its disk is still too small to be perceived.

Betelgeuse is a variable star and its spectrum shows evidence for a strong wind blowing from its surface indicating a low surface gravity. Theoretical

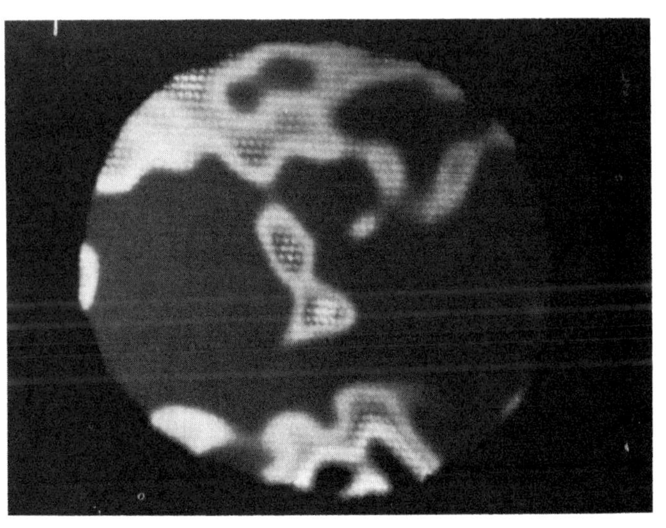

Betelgeuse KPNO

astronomers have given reasons for us to expect that enormous convection cells cause gas to rise and fall from Betelgeuse's surface. The combination of under-the-surface motion and weak gravity which could allow the uprising gas to break the surface layers gave rise to the expectation that blotches like large sunspots might be visible on its surface.

Kitt Peak astronomers have exploited new techniques to produce a picture of the surface of Betelgeuse which confirms this prediction. C. R. Lynds, S. P. Worden and J. W. Harvey used *speckle interferometry* to do this. If you look at a high-magnification image of a star in a telescope, the image seethes and writhes under the boiling action of the Earth's atmosphere, making it impossible to see the fine details which the telescope is capable of showing. High speed photographs show the existence of *speckles* in the image, each produced by individual cells of air in the atmosphere. Each speckle is the best image of the star that the telescope could produce and is near-perfect. If a way could be found to add up the speckles, the summed image would be the image of the star that would be seen by the telescope if it were outside the the atmosphere and free of terrestrial limitations.

The Kitt Peak astronomers found such a method, adding images from 1000 speckles seen on 20 photographs to produce the picture of Betelgeuse seen in Plate 5. Defacing the disk of the star are large spots. Betelgeuse's starspots are shown green and blue, cooler than the surrounding red surface of the star. Each of the spots is as large in diameter as the Earth's orbit round the Sun.

Antares

Slow occultation by the Moon

Antares is the brightest star in Scorpius. It is a first-magnitude, distinctly red star whose naked-eye appearance is like the planet Mars: hence its name, which means "Rival to Mars." It is a variable star, varying in an irregular manner by about one magnitude—a factor of 2.5—over an average period of 1733 days. At a distance of 3.4 arc seconds is a companion star, five magnitudes fainter. Antares itself is a binary star, since its velocity shows a cyclic shift with a period of 7.25 years. It is orbiting an unseen and near companion in an eccentric and lazy orbit.

Antares is more than 10,000 times as bright as the Sun, although its temperature is lower than the Sun's. The cooler an object, the lower its surface brightness—just as an electric fire glows dimmer as it cools down. Thus, if the surface brightness of Antares, area for area, is low, the star must be large to emit so much light. Its spectrum, which is that of a supergiant star, also points to a large dimension. Red supergiants are the largest stars known. However not even supergiants are large enough or near enough to Earth for us to perceive them as disks.

The actual size of Antares has been determined because of a lucky circumstance. Antares is in Scorpius, a sign of the zodiac—the path in the sky of the Sun, Moon and planets. Therefore, the Moon in its orbit may pass in front of the star and hide it from view. This phenomenon, a lunar occultation, happens many times each month to various random stars. In most cases the stars are rapidly extinguished: they are distant and small and the Moon's edge cuts across them very quickly. However, Antares takes an appreciable time to fade away because its disk is relatively large. The figure shows that the slow fade of light from Antares during an occultation takes about a quarter of a second.

The occultation measurements show that Antares is a virtually uniform disk 0.044 arc seconds in diameter. This is the same as a small coin (U.S. nickel or British penny) seen from 30 miles.

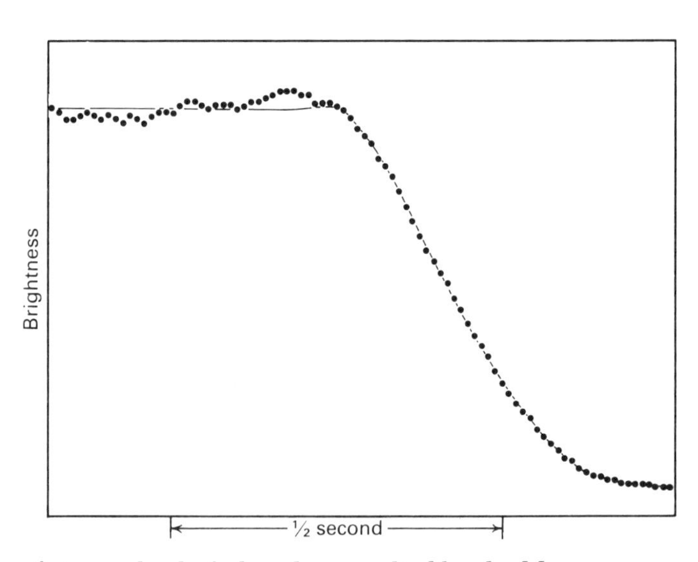

Antares slowly fades when occulted by the Moon

Algol

Eclipsing stars

Most stars in the sky are constant and do not change in the lifetime of a man or indeed in the lifetime of Man. Astronomers study constant stars, count and classify them, and make mathematical models of their atmospheres and interiors, but it is probably true to say that most astronomers are excited by those few stars which change in some way. The reason is clear. Changes do not occur in isolation. Changes in one property can precede, follow or coincide with changes in another property. The circumstances of such changes give

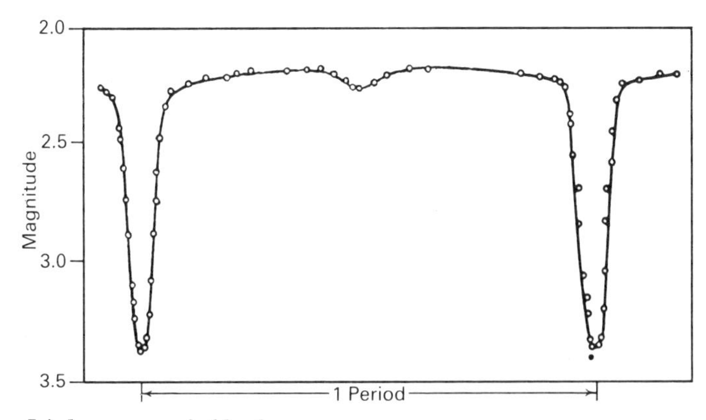

Light curve of Algol

clues to their nature and to the nature of the stars themselves.

One change which astronomers can see is the sudden death of a star. By studying the circumstances surrounding the death of the star, astronomers can better understand what brought the star to that state and thus how it had lived its life. Sometimes the changes are entirely superficial, like the interplay of a set of shadows on the wall. Yet even then the nature of the shadows depends on the dark star whose shadow is cast and the bright star which casts it.

In the sections which follow we describe different types of variable stars and give their light curves. It is necessary to explain just what the light curve represents. Astronomers have an awkward system for expressing the brightness of a star in terms of its "magnitude." Various historical accidents dating back 2000 years together with the strange nature of the human eye and brain have combined to produce the peculiar method of defining the magnitude of a star.

First, the brighter stars have smaller magnitudes. The light curves are plotted sensibly with the brighter stages of a variable star's cycle at the top of the graph, but seem to be upside down because large numbers appear at the bottom. Second, a star's magnitude is not its brightness, but the logarithm of its brightness expressed to the base 2.512. What this means is that if a star brightens by one magnitude, it has increased its brightness by a factor of 2.512. If it brightens by five magnitudes, it has increased its brightness by 2.512 raised to the fifth power, which is exactly 100 times.

It is not necessary to know these mathematical details to appreciate the following section. What is important is that an increase of one magnitude in the brightness of, say, a tenth-magnitude star, is not a small change of 10%: the star has more than halved its brightness.

Algol was one of the first variable stars to be discovered and it is the easiest to understand. Its recorded discovery was by Geminiano Montanari in 1667, but its Arabic name, which means Demon Star, and is cognate with the English word *ghoul*, suggests that its variability has been known since much earlier times.

Algol is usually the second brightest star in Perseus, and is therefore also known as Beta Persei following the system of giving Greek letters to the brighter stars of each constellation. Normally Algol is of magnitude 2.2. Every 2.87 days, though, its brightness dips to a minimum of magnitude 3.5, a drop of more than a factor of three in its light output. This change is easily seen with the unaided eye (which was how Montanari noticed it), and amateur astronomers have made regular observations of the minima for 200 years.

The explanation for this change was first given by John Goodricke, an English amateur astronomer, in 1782. He suggested that Algol consists of two stars of roughly equal size, circling each other with a period of 2.87 days. One star is hot and contributes most of the light. Because of this it is called the primary star. Another star, with only one-fifth the mass, much cooler and therefore much fainter, periodically passes between us and the primary, blocking off the latter's light. Only the fainter star is then visible and we see the star at its minimum. During the time the primary is behind the secondary, its light is eclipsed. If the eclipse were total, then the primary would be behind the secondary for some time and the light curve at the bottom of the minimum would be flat. Because in Algol the minimum is pinched to a point, we deduce that the primary is never fully hidden behind the secondary, as a result of the tilt of the orbit in the sky. When the primary comes out fully from behind the secondary, we see both stars again and the light from the pair of stars is at its maximum.

There is also a time when the primary star passes across the face of the secondary star, blocking off its light. Because the secondary is fainter than the primary, this does not make such a great difference to the light curve. This eclipse has a smaller effect and is known as the secondary minimum.

The mass of the hot bright primary of Algol is five times that of the Sun. The fainter, cooler secondary has the same mass as the Sun. Normally such a star would be Sun-sized and the primary would be about 3.5 times the diameter of the Sun.

Actually they are both 3.5 times the size of the Sun—information which is obtained from the time taken for the eclipses and the orbital speed, as measured by the doppler effect. The cooler star is a giant: it has expanded because it is running out of nuclear fuel. Yet the heavier star, which should in theory use up its nuclear fuel first, has not, and is still shining, apparently in contradiction of common sense. This is the so-called Algol paradox.

The paradox is explained by supposing that originally, the secondary *was* the more massive star. However, as it has expanded to become a giant, it had enlarged so much that its outer material has come within the gravitational influence of the other star. So in time, most of its material has dropped onto the other star, making that one the more massive and hence the brighter. In effect, the fuel of the present primary has been replenished by the unburned outer mantle of the present secondary. There is evidence in the spectrum of Algol that the transfer of material is still going on but that now the secondary is growing in mass at the expense of the primary. In addition, small period changes are probably due to the transfer of material from one star to the other.

In 1971 C.M. Wade and R.M. Hjellming of the National Radio Astronomy Observatory at Green Bank, West Virginia, discovered that Algol is a radio star. The star is erratically variable in its radio output, with strong flares up to 20 times its normal radio brightness. It is likely that the stream of gas onto the primary star is the cause of the radio emission.

Algol is a triple system, the close pair being orbited by a distant, third star with a period of 1.9 years. The third star swings the pair to and fro, sometimes closer to Earth, sometimes farther away. When the pair of stars is farther from the Earth, the light from them takes a little longer to reach us and eclipses are delayed five minutes behind schedule. They advance five minutes ahead of schedule as the pair swings closer to Earth.

Beta Lyrae

Ellipsoidal double star

Beta Lyrae is a double star in which the period of the orbit of one star around the other is nearly 13 days. Both stars are very disturbed by their rotation about each other. Instead of being a sphere, each star is an elongated ellipsoid. Each long axis points towards its companion and they circle each other with apexes nearly touching. When we see each star broadside-on, its presents a larger surface to us than when we see it end-on. The stars are therefore much brighter when seen side-on than end-on.

Beta Lyrae is what is known as an ellipsoidal variable star, because in this group the main cause of the variability is the rotation of ellipsoidal-shaped stars. In Beta Lyrae, the stars have about a 20% larger diameter as measured along the line joining them than across that line. Eclipses also occur, with one star totally eclipsing the other as seen from Earth.

This double star has a light curve which is complicated by many other changes occurring simultaneously with the rotation. Otto Struve interpreted the changes that he found as showing that the secondary star was surrounded by a disk of matter which had flowed from the primary.

Like ALGOL, Beta Lyrae is a source of radio emission which was discovered by Wade and Hjellming in 1971. This is further evidence of the transfer of material from one star to the other.

R Coronae Borealis

Sooty veils

R Coronae Borealis has been known as a variable star for more than a hundred years. It has the startling property of dropping from naked-eye brightness (magnitude 6.0) right out of sight, to as faint as the 13th magnitude.

The drops take place abruptly and unpredictably every five years, and last for one or two months, after which the star slowly recovers its original brightness, increasing in an irregular manner. Twenty-nine stars like R Coronae Borealis have been catalogued and they are called R Cor Bor stars for short.

The clue to the reason for this disappearing act is given by the spectra of R Cor Bor stars. Their atmospheres are rich in carbon gas—nearly two-thirds carbon according to one estimate. E. Loreta and J. O'Keefe have suggested that the fading occurs after some of the star's atmosphere is puffed off into space by some unknown mechanism. As the ejected gas recedes from the star, it grows colder. At some temperature the carbon atoms in it suddenly condense to make soot. When they do so, the soot powder obscures the light from the star and the drop in brightness occurs. Deep minima occur when the puff is ejected towards us; less deep ones when the puff is offcenter from the star. In any case, the soot cloud gradually expands into space, becoming more and more rarefied and letting through more and more light, until the soot is virtually transparent.

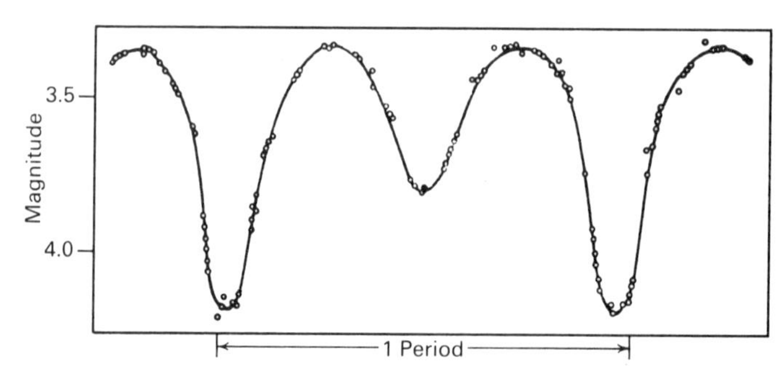

Light curve of Beta Lyrae

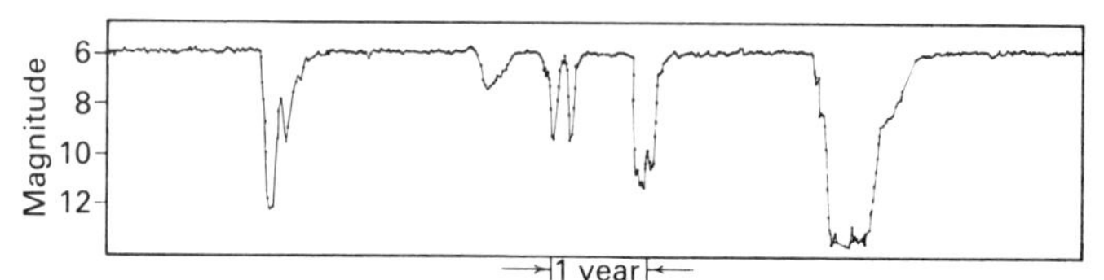

Light curve of R Coronae Borealis

How these smoky chimneys of stars get to that stage is not yet known. There is one clue which astronomers believe may be significant, although they have not yet worked out how. The carbon found in R Cor Bor stars is all carbon-12: each atom contains six protons and six neutrons. Most other carbon-rich stars contain carbon-13 as well, with one extra neutron. (The famous carbon-14, with yet one more neutron, used for archeological dating, decays too quickly after it has been made to be present in long-lived stars.) The carbon-12 clue must indicate something about the stage of nuclear burning which the star has reached, the step on the ladder of nuclear transformations where the R Cor Bor phenomenon appears. Judging by the small numbers of R Cor Bor stars in our Galaxy, the step is either brief or is found in peculiar and rare circumstances.

U Geminorum

Dwarf nova

U Geminorum is a star which is normally magnitude 14.3. Approximately every 100 days it suddenly brightens, usually by five or more magnitudes: that is, it increases its light output more than a hundredfold. Over the following one to three weeks, the star drops back to its original brightness, which, presumably, is its normal state. U Geminorum (abbreviated to U Gem) is representative of about 200 similar stars, all called dwarf novae, known in our Galaxy. Those that can be seen at all are easy to find since the change in their brightness is spectacular, but because they are relatively faint, only the local area of our Galaxy has been searched for them. Altogether there are ten to a hundred million of them in the vaster sea of the whole Galaxy. U Gem stars are, in fact, one of the more numerous kinds of stars in the Galaxy.

A clue to the nature of the U Gem stars lies in their spectra. In its normal state, the spectrum is that of a cool, faint dwarf star, similar to, but cooler than our Sun. In outburst particularly, the spectrum of the faint star is outshone by the hot, bright light from another nearby object. We know that there are two stars in U Gem because, whether it is bright or faint, it shows eclipses with a period of 4 hours 15 minutes. From the circumstances of the light curve, it is possible to say that the faint, cool star is the larger of the two and that the hot, bright light seen in outburst comes from a hot spot on the edge of a disk of gas encircling the other star. This star is very small, probably a white dwarf, and very little light comes from it. Most of the light comes from the hot spot and not from either of the two stars.

What makes the spot? A stream of gas pours from the cool star, as in the case of ALGOL and BETA LYRAE, attracted by the gravitational pull of the white dwarf. The gas hits the disk which encircles the white dwarf like a crash barrier. The hot spot is at the point of impact, heated by the energy released in the crash. In outburst the hot spot grows larger and more energy is released. Perhaps this represents the mayhem caused by the uncontrolled splash of the gas stream onto a smaller disk or even on the white dwarf itself. Perhaps the cool star has done something to the gas stream, as if opening a faucet to increase its flow.

Many of the older stars in the Galaxy have turned to white dwarfs and many of these will have companion stars because binary stars are by no means rare. Thus U Gem stars can be expected to be common.

Delta Cephei

Pulsating star

In 1784 John Goodricke, an English amateur astronomer, discovered the variability of Delta Cephei, which cycles from fourth to fifth magnitude and back in a period of 5.37 days. Its light curve is smooth with a steep rise and gentler fall, and it is the prototype of the so-called Cepheids whose periods lie in the range 1 to 50 days. Among the 700 known Cepheids is Polaris,

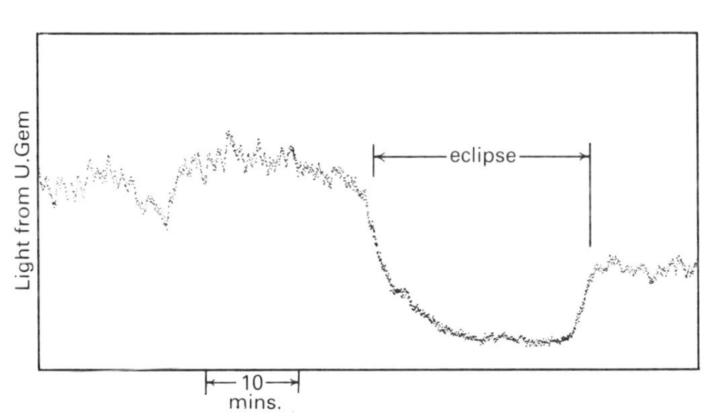

Light curve of U Geminorum

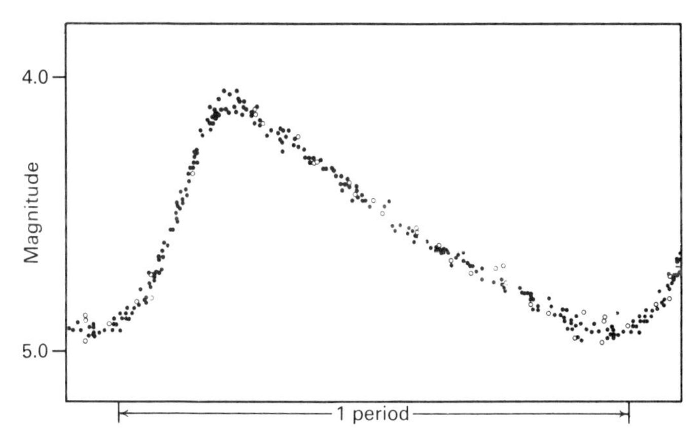

Light curve of Delta Cephei

the North Star. Cepheids are usually amazingly accurate clocks and their periods can be determined to within a small fraction of a second. They do, however, occasionally jump or even, like RW Camelopardis, stop altogether.

Cepheids played (and still play) an important role in astronomy because they can be used to measure distances. There are two reasons for this. First, because Cepheids are bright they can be seen at large distances. Second, a curious fact about these stars makes it possible to tell how bright a Cepheid is.

In 1912 Miss H.S. Leavitt noticed a relationship between the periods and average brightness of Cepheids in a nearby galaxy, the Small Magellanic Cloud. The brighter the Cepheid, the longer its period. Since the Cepheids which she was studying are all approximately the same distance from Earth, this meant that there was a relation between the intrinsic brightness of a Cepheid and its period. If an astronomer could identify a Cepheid in a galaxy beyond the Magellanic Cloud, he could measure its period, determine its intrinsic brightness and, by comparing this with its apparent brightness, deduce its distance and thus the distance of that galaxy. In the same way, if one could identify a faint glow in the night as a flashlight one would expect it to be much nearer than a similar faint glow known to be a lighthouse beam.

Harlow Shapley managed to calibrate this period-luminosity relationship and Edwin Hubble used it to determine the distance to the Andromeda Galaxy. Hubble found that the Cepheids in the Andromeda Galaxy were 4.6 magnitudes fainter, period for period, than the ones in the Small Magellanic Cloud. Therefore, its distance was 8.5 times that of the Small Magellanic Cloud. Thus Hubble estimated that the Andromeda Galaxy is almost a million light years from us. Later, Walter Baade realized that there were two kinds of Cepheids which had been mixed up in the period-luminosity relation. After disentangling them in 1952 Baade doubled the distance of the Andromeda Galaxy and therefore the rest of the Universe, whose scale had been determined by the Cepheids.

Cepheids are found in galactic clusters of stars and are used to determine their distance. In this way they outline the spiral arms of our Galaxy.

Why do Cepheids change in brightness? At first astronomers tried to explain them as eclipsing binary stars. It soon became clear that this was impossible. From the period-luminosity relation one might have guessed that the cause of the variation is intimately connected with the star itself, not just an accident. In fact, Cepheids are pulsating stars, throbbing like a regular heartbeat.

The cause of the oscillations in Cepheids lies in the structure of the stars. In the inner heated layers is a valve-type mechanism. When the valve is open, radiation passes through the star easily and the star shrinks. The valve then closes, traps radiation which pushes the star apart and makes it expand. The valve mechanism is the ionization of helium and hydrogen in the star.

The reason for the period-luminosity relation is that a pulsation is a wave, traveling in the star, whose period is the time the wave takes to move from the surface of the star to the ionization zone and back. Naturally, the bigger the star, the longer this takes, and waves also travel more slowly in bigger stars that are more rarefied. The bigger stars therefore have longer periods, just as bigger, hollow drums have deeper notes than smaller, solid chinese blocks. But because larger stars have more surface area and radiate more light, the longer period Cepheids are brighter than the others.

The analogy between musical instruments and pulsating stars like Cepheids can be pursued further. Some Cepheids show in their light curves the effect of a mixture of two periods at once. AI Velorum vibrates with periods 0.11157375 days and 0.08630767 days simultaneously. These are the fundamental and first overtone of the star's natural vibration period. Together they cause a rhythmic modulation of its fundamental oscillation just as the tone of an organ can be modified by allowing overtones into the fundamental vibration

RS Puppis

Cepheid in a nebula

It is of great importance to locate Cepheid variable stars whose distance can be determined. Because it illuminates a nebula, RS Puppis is one such important star. It is the only Cepheid variable known to lie in a nebula. The star RS Puppis lies in the middle of the overexposed part of our picture, at the center of the diffraction cross. (The latter is an imperfection of photos taken by reflecting telescopes and marks each bright star.) The nebula surrounds the burned-out image of the star.

The nebula which surrounds RS Puppis takes the form of a number of incomplete rings. This suggests that the dust in the nebula which is reflecting light from the star lies predominantly in a series of roughly concentric shells which, like soap bubbles, are seen only at their rims. Astronomers therefore believe that the nebula was generated by the star shedding small amounts of gas on four or five occasions. The mass of material in the nebula is estimated to be three times that of the Sun. This is too great to have been produced by RS Puppis alone. The expanding shells have grown by sweeping up gas that lay around the

Eta Carinae Nebula

North America Nebula

PLATE 10

Lagoon Nebula

Omega Nebula

PLATE 11

Pleiades

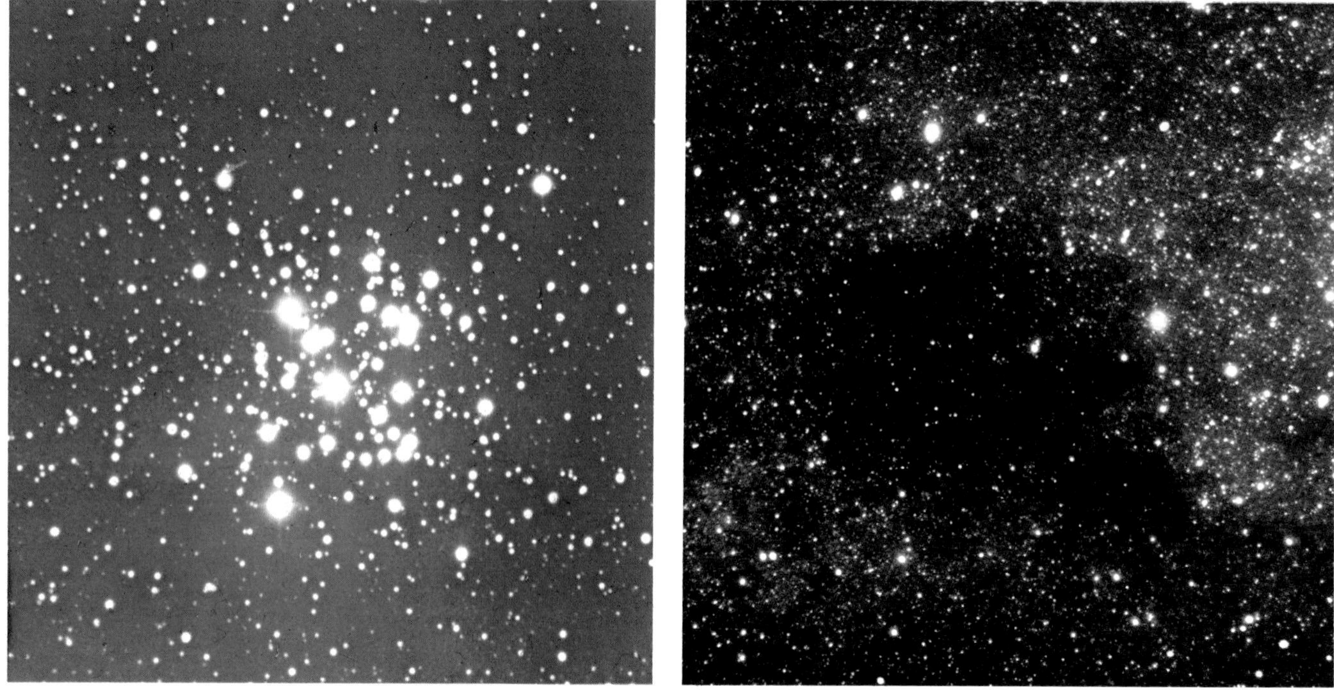

NGC 3324 *Coalsack*

PLATE 12

RS Puppis

UKSTU

star. But the amount of material swept up is also larger than would be expected, indicating that RS Puppis lies in a part of the galaxy dense with gas and dust.

A region of gas, dust and young stars lies in the same direction as RS Puppis. It is known as the Puppis III Association. Some of its member stars undoubtedly account for many of the scattering of stars of this photograph. The distance of the Puppis III Association is 6000 light years, so the distance one might guess for RS Puppis is much the same. It seems virtually certain that RS Puppis lies in, and is interacting with, the gas of the Puppis III Association; therefore astronomers can be confident that its distance is 6000 light years.

Cepheid variables are not known to eject gas shells. The shells around RS Puppis are therefore believed to have arisen at an earlier epoch, before the star became a Cepheid variable. Current models of stellar evolution suggest that some stars spend periods as red giants before becoming Cepheids. Many red giant stars are known to smoke off their outer layers, sometimes as distinct shells. The rings around RS Puppis afford an archaeological record of the star's former incarnation.

RR Lyrae

Measuring the Galaxy

RR Lyrae is a pulsating variable star like DELTA CEPHEI (p. 95). It throbs rhythmically with a period of 0.567 days and is typical of a group of similar stars whose periods lie generally between an hour and a day. They occur in large numbers. Over 4000 are known, but their distribution is not uniform and they occur only in the older parts of our own and other galaxies.

They are also found in globular clusters. M 3 holds the record, containing 183, and OMEGA CENTAURI (p. 135) is second with 140. Some globular clusters have none at all. In our Galaxy RR Lyrae stars are found in the center and in the old outer regions, the "halo."

Historically, RR Lyrae stars have played an important role in the study of the structure of the Galaxy. They are not as bright as Cepheids and therefore have not played as prominent a part in the study of external galaxies. The intrinsic brightness of all RR Lyrae stars is the same, to a good degree of approximation. This enabled Harlow Shapley to measure the distances of globular clusters containing RR Lyrae stars and to show that the clusters swarm around a point 30,000 light years away in the direction of the constellation Sagittarius, the center of our Galaxy.

Another determination of the distance of the center of the Galaxy was made by Walter Baade. He located the RR Lyrae stars visible towards Sagittarius along the length of a chance tunnel through the interstellar dust, like the one called M 24 (p. 76). After finding all

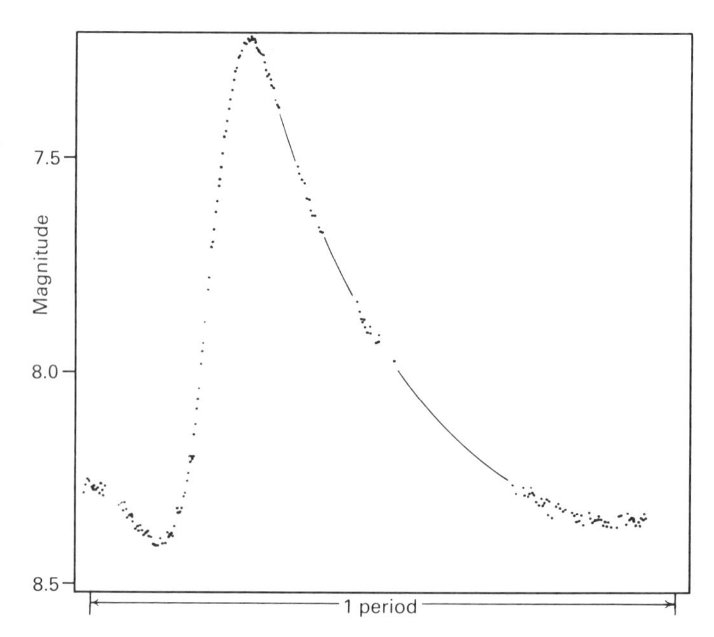

Light curve of RR Lyrae

the RR Lyrae stars Baade plotted their concentration along the tunnel and found that they peaked at a distance now known to be about 30,000 light years. He identified the peak in the distribution as the center of the Galaxy. The peak arises just because there are more old stars at the GALACTIC CENTER than elsewhere.

The fact that RR Lyraes all have about the same luminosity points to a remarkable sameness in their structure. This is the more remarkable since RR Lyrae stars have two kinds of light curve. RR Lyrae itself has a large change of brightness with a steep rise and gentle fall, and its period is longer than half a day. Others have a smaller change of brightness, flatter light curve and periods shorter than half a day. The two types of variation arise because the star can pulsate either in its fundamental mode or in its first overtone. This does not mean that the structure of the stars is radically different, but that the oscillation is produced somewhat differently. In the same way a skilled musician can produce either the fundamental or the harmonic from a trumpet by a particular method of blowing.

RR Lyrae stars are all giants of low mass. We can understand why they are found in the older parts of the Galaxy. Only the oldest low-mass stars have had time to evolve into giants. In fact, the RR Lyraes represent stars born throughout the first 90% of the life of the Galaxy. During all this time, more massive stars have been converting hydrogen to heavier elements such as calcium, then exploding. This mixed calcium into the interstellar medium from which subsequent generations of stars have been formed. Thus the RR Lyraes contain a spread in concentration of calcium. This became apparent when Mount Wilson astronomer George Preston realized that the strength of the calcium spectral line in the spectrum of the RR Lyraes showed remarkable differences from star to star in a manner which could best be explained as a deficiency or abundance of calcium. RR Lyrae itself has a large deficiency of calcium compared with the Sun and must have formed *very long ago* indeed, say six billion years, whereas other RR Lyrae stars like VZ Cancri have little or no deficiency and simply formed *long ago*, say one billion years ago.

Mira

Wonderful

Mira is the "wonderful" red star, the first true variable star whose discovery was recorded. In August 1596, German astronomer David Fabricius noticed a star in Cetus of second magnitude, and therefore easily visible to the unaided eye. The star faded from his view over the succeeding months and then sporadically reappeared. This amazing behavior earned the star the name *Mira*—Latin for "wonderful." In 1603 it was bright enough for the star-charter Bayer to see it and label it Omicron Ceti because he estimated it to be the 15th brightest star in the constellation Cetus (omicron is the fifteenth letter in the Greek alphabet). Evidently it was not at maximum brightness, else Bayer would have labeled it Beta or Gamma.

Holwarda suggested, from the observations of Mira made up to 1638, that it was periodic. The period of 331.62 days was established by 1660. The amplitude of its light curve is 8.1 magnitudes, which means that it changes its brightness by an astounding factor of 1700 times.

Mira is the type star for a large number of long period variable stars, all red, all bright, all with periods over 100 days. This is the most numerous single group of variable stars known, standing out from constant stars because of their large changes and their brightness. They are estimated to number almost 100,000 within our Galaxy alone. The Mira stars are periodic, but there does exist a similar group of bright, cool stars whose light changes are irregular or semiperiodic.

The changes in light output are not caused by any eclipse or shadow phenomenon, but are intrinsic to the star itself. The light output from a star depends on its area and the temperature at its surface. The larger and hotter the surface, the more light we see. In the case of the Cepheids and RR Lyrae stars there is a pronounced change in the surface area during the cycle of light output. In fact, by measuring the doppler shift of the spectral lines formed in the atmosphere of the star, astronomers can watch the surface rising and falling as the star expands and contracts.

The change in the size of the surface of Mira Ceti during its cycle is small, however. It would change its light output by only a small amount and not to the enormous extent of the change actually seen.

The change in light intensity is caused mainly by a change in temperature of the star, but even here the details of the story are complex. When Mira is faint its temperature is 1920 K (using the Kelvin scale of temperature, where 0 K is absolute zero), which is cool as stars go. Almost all of its radiation output is at infrared wavelengths and its visibility is low. At maximum, Mira's temperature is 2640 K. This increase in temperature causes much more of its radiation to be emitted at wavelengths visible to the human eye. The range of a factor 1700 in its light output from maximum to minimum arises mostly because of this wavelength shift. If we include the infrared radiation of Mira then its range in brightness is only a factor of four.

Mira is a double star. When it is faintest astronomers can see Mira's optical companion, a white dwarf star. Because Mira is double, astronomers can deter-

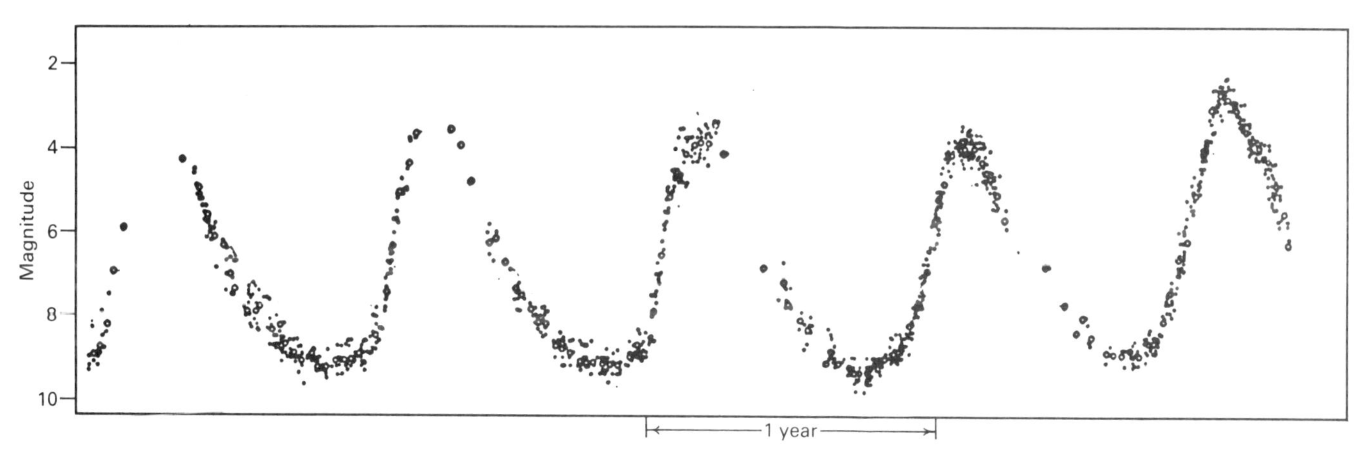

Light curve of Mira

mine its mass, which is found to be about the same as that of the Sun. Its brightness, however, is several hundred times that of the Sun. It is a red giant star and has exhausted the hydrogen in its interior as a nuclear fuel. It is now in a more advanced state of evolution, burning the helium which was produced from the hydrogen. When helium burning starts more energy is released. The extra outward flow of energy inflates the star to a giant.

Mira stars are a population of aging stars found all over the Galaxy. They are common because they are low mass stars and low mass stars are common.

Because they are 100 times the radius of the Sun but of similar mass, Mira stars have weak gravity at the surface, about 10,000 times less that that of the Sun, and hence a low escape velocity. Motion of the surface of Mira gives rise to ripples of gas flowing out through the atmosphere. The force of gravity is too weak to pull the gas back to the surface, and it escapes into interstellar space. Evidence of the motions of the gas is found in the spectrum of Mira.

Some of the gas which escapes from Mira is captured by the white dwarf companion. The gas which falls onto the white dwarf heats its surface, rendering it brighter than it otherwise would be. Its brightness seems to vary in synchronism with the pulsations of Mira, as waves of matter are thrown off one star and captured by the other.

The remaining gas cools as it expands away from the star. As it expands into cold interstellar space it reaches a temperature where solids can condense from the gas, just as crystals of salt will condense from a cooling solution of brine. Calculations show that grains of silicate minerals will form in preference to other materials. These calculations have been verified by infrared spectroscopy of Mira and other Mira-type variable stars. Their infrared spectra show the characteristic spectral features of silicates. It seems from our present knowledge that these minerals are formed only around Mira variables. The Earth is composed largely of silicate materials (sand, for instance) and astronomers believe that most of this was formed in the effluent gas streams of Mira variable stars.

In the spectrum of some stars similar to Mira, Paul Merrill in 1952 found four of the spectral lines of an element called technetium. This element does not occur naturally on Earth because it is radioactive and decays over a period of a few million years. This is short compared with the age of the Earth, so any technetium which was originally in the Earth has now disappeared. The properties of technetium are known only because the element has been made in nuclear laboratories. It is remarkable that this radioactive element has been found in stars that are longer lived than Earth, as any technetium which was originally present would have decayed. The only explanation must be that Mira-like stars have themselves recently manufactured technetium in their atmospheres. This was the first, and remains the clearest, observational proof that stars are celestial nuclear laboratories changing elements from one kind to another.

Nova Persei 1901

Superficial drama

Nova Persei 1901 was a star which shot from below magnitude 12 to magnitude 2.7 in 28 hours, when it was discovered by an Edinburgh amateur astronomer, Rev. T.D. Anderson on Feb 22, 1901. At maximum brightness a day and a half later, the nova reached magnitude 0.1. It faded rapidly at first, but then its decline in brightness slowed, and after oscillating between fourth and sixth magnitude for three months, it finally faded from naked-eye view in August 1901. It is now a star of magnitude 13.5. As it declined in brightness, European astronomers Flammarion, Antoniadi and Wolf were astonished to see expanding around the nova a shell of nebulosity which rapidly grew in size and intensity, although individual parts of the shell did not appear to move outwards. Kapteyn explained that the February pulse of light from the nova was propagating into a dusty nebula nearby. Light from the nova was then reflected towards Earth, reaching Earth months after the pulse of light, which traveled direct with no dusty detour. The shell was a briefly revealed reflection nebula which was four times the area of the Moon when it faded from view a year later.

To the surprise of most astronomers, save E. E. Barnard, who had early suspected that the image of the bright nova had appeared fuzzier than the other stars he saw through his telescope, a second nebula appeared on photographs of Nova Persei in 1916. This nebulosity was truly expanding outwards from the star with a speed of 1600 km/sec. Its present appearance, illustrated in our photograph, is of an incomplete circular nebula centered on the star.

The later expanding shell is a clue to what is happening in this kind of nova. There is an explosion just below the surface of a star, which ejects the surface at high speed into space. As it is thrown off, the surface expands larger and larger and so gives off more and more light. The star brightens spectacularly. But the thrown-off layers are spread thinner and thinner and become more and more transparent. The nova fades. Ultimately the ejected layers are transparent enough for us to see through to the star underneath. The ejected layers become clearly visible as a nebula when distant enough from the star that the star's light does not swamp the fainter light from the nebula.

Nova Persei *Hale*

After the effects of the nova explosion have died away, the underlying star can be studied. Always, when studied carefully enough, old novae turn out to be double stars consisting of a cool giant star, dribbling its atmosphere of hydrogen onto a white dwarf star. The white dwarf cannot tolerate more than a small shell of hydrogen on its outer surface and when too much hydrogen has built up the white dwarf explodes. Only about a millionth of the mass of the star is thrown off. For all its apparent drama, the explosion of a nova is very superficial.

Nova Ophiuchi 1977

X-ray nova

We include Nova Ophiuchi 1977 in this book partly because two of the authors were members of the team which discovered it. It was an x-ray nova, first perceived by the High-Energy Astronomical Observatory satellite as an intense source of x-rays suddenly appearing in the constellation of Ophiuchus. At that place in the sky a bright star became visible, as the pair of photos show. The star gradually faded away.

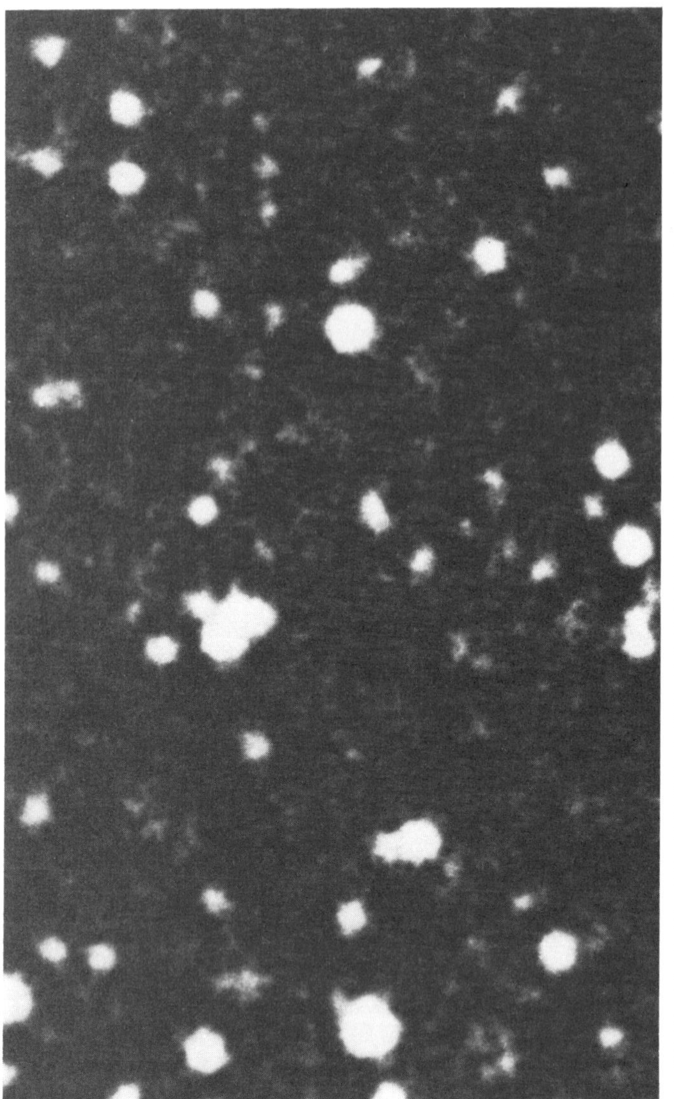

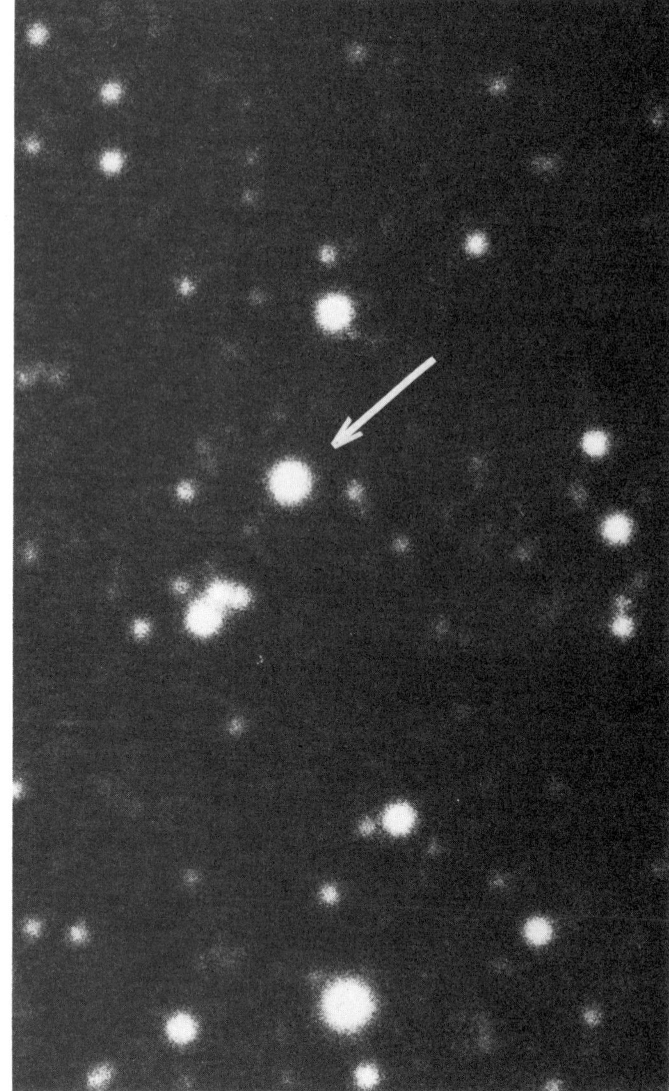

Nova Ophiuchi POSS ATT

Like ordinary novae, this kind of x-ray nova is probably a member of a binary star system. One member explodes, heating and brightening the surface of its companion star with x-rays.

Thus Nova Ophiuchi 1977 is in contrast to NOVA PERSEI 1901. In the latter, it was mainly the increase of surface area of the exploding star which caused it to brighten. In Nova Ophiuchi 1977 the increase in temperature of a nearby star made it become brighter.

Associations, Clusters and Nebulae

Orion Nebula

Brightest nebula visible from Earth

We think of galaxies as collections of stars, but irregular galaxies and spiral galaxies, like our own, contain gas as well. Some of the gas is in the form of hydrogen left over from the original formation of each galaxy, fresh from the Big Bang. Some has been recycled from dying stars back into space, by the persistent blowing of a stellar wind, by the sporadic puff of a quietly dying white dwarf or by the cataclysmic explosion of a massive star as it goes supernova. The recycled gas thus contains cosmic pollution, generated by the nuclear reactions inside the stars.

The interstellar gas is mostly in the form of individual atoms, approximately one atom per cubic centimeter in the Sun's neighborhood. One kind, the hydrogen atom, can be detected by its radio emission at a wavelength of 21 cm. Others, for instance calcium and sodium, absorb particular wavelengths from the light of distant stars, as the light traverses interstellar space.

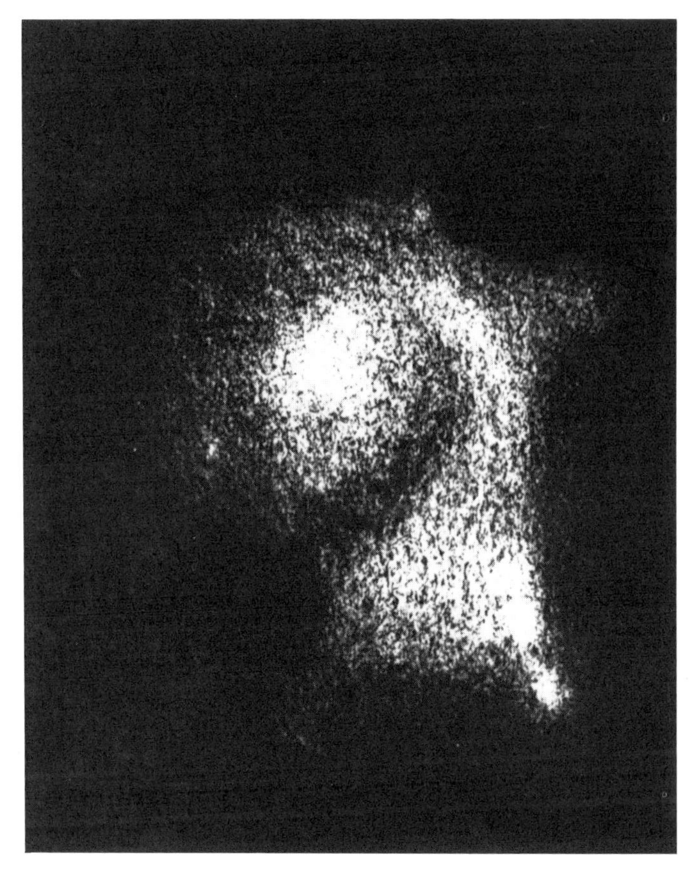

Infrared map of Orion DAA

In the cold of interstellar space molecules which would be destroyed by the heat of a star can survive. A remarkable variety has been detected by radio astronomers. The list of discoveries includes organic molecules of the type involved in the chemistry of life—water, ammonia, formaldehyde and ethanol are some of the more complicated molecules which have been detected. These discoveries have sparked new discussions on the origin of life, and give credence to those who claim that life is a universal phenomenon.

Interstellar space also contains dust. From its effects on the light of distant stars, astronomers know that a typical interstellar dust grain is of size one micron (one-25-thousandth of an inch), and that there is on average one grain in each 12 meter cube of space near the Sun.

The effects of the interstellar dust and gas are subtle, except where they lie in thick clouds near bright stars. This combination is in fact quite likely, since stars are even now forming from dust and gas clouds. The dust and gas left over from the formation of these newly born stars is illuminated and heated by them. To the astronomers of the 18th century the illuminated dust and gas had the form of clouds and they named them in Latin *nebulae*.

There are two kinds of bright nebula. Dusty nebulae merely reflect the light of any star which lies nearby. The emission nebulae, however, are those which generate their own light and do not simply reflect light from somewhere else. To shine as an emission nebula, gas must lie near particularly hot stars which emit plenty of ultraviolet light. The ultraviolet light ionizes hydrogen atoms in the nearby gas, splitting them into individual protons and electrons which come together again to reform their hydrogen atoms. In the process of recombining the hydrogen atoms predominantly emits the red hydrogen-alpha light which reveals the existence of the nebulae on photographs and gives them their generally red appearance.

The Orion Nebula is the nebula surrounding Theta[1] Orionis, the central star in the Sword of Orion, and in reality a cluster of four known as the TRAPEZIUM. The Orion Nebula was discovered by N. Peiresc in 1610 with a telescope given him by his friend Galileo. It is catalogued M 42; M 43 is a more northerly part of the same complex of dust and gas, and was discovered by de Mairan in 1731. Radio astronomers know the Orion Nebula as a strong radio source, Orion A.

The nebula can be seen with the keen unaided eye, and certainly with binoculars. Color photographs (Plate 6) show the red color of hydrogen light, while the eye is more sensitive to the green of the so-called nebulium line, in reality due to oxygen emissions. The Orion Nebula is the nebula of highest surface brightness visible from Earth, and is the best studied. It is a meeting ground of optical, radio, infrared and x-ray astronomers, and the combined efforts of all have served to reveal a large body of facts about it. This means that astronomers know the most about it of all nebulae, and therefore that they have more questions about it than any other.

The Orion Nebula's distance is 1300 light years. Apparently encircling it is a cluster of stars, known as the Orion cluster, faintly seen on red-filtered photographs. Actually the cluster lies a little behind the Orion Nebula, which seems to be the front region of a dusty cloud of gas and molecules, the largest dark cloud known in our Galaxy. Deep within the cloud many stars are forming. Perhaps 100,000 years ago the Trapezium stars were born. Being very hot, they are burning a hole in the dark cloud, ionizing the gas and blowing away the obscuring dust. They reside within an expanding bubble in the dark cloud. Quite recently, in astronomical terms, the bubble broke through our side of the dark cloud, enabling us to see into its hollow interior. Then the Trapezium stars became visible.

When we look at the photograph on Plate 6 we do not see the extent of the dark cloud, which in fact covers most of this picture. We see only the portion illuminated by the Trapezium. Some of the dark nebulosity still curls in front of the Trapezium's bubble, like the sinister fingers of a dark hand. The

Messier 42 UKSTU

sharper edges of the nebula are these remaining fingers of dust.

The dark cloud is opaque to visible light, allowing barely one-million-million-millionth of any incident light to pass. Nonetheless, we know much about its interior from infrared and radio observations, which are not impeded by dust. On p. 103 is an artist's impression of the appearance of Orion as we would see it if we had infrared sensitive eyes. At the bottom of the map near the center of the dark cloud is a particularly luminous star which has recently formed, probably within the last ten thousand years. It is known as the Becklin-Neugebauer object, after the Californian astronomers who discovered it in 1967. It, too, is evaporating its own bubble in the dark cloud, but this will not become visible to us for some millions of years.

Near the Becklin-Neugebauer object is another infrared source, called the Kleinmann-Low nebula. This represents the very heart of the dark cloud, where many stars are forming. The gas here comprises mostly hydrogen, helium and carbon monoxide; other abundant molecules include formaldehyde. All are at a temperature of some 80 K (-190°C). Further from the center the temperature falls to nearer 10 K.

The dust in the Orion Nebula is composed mostly of silicates, a mineral formed around cool stars such as MIRA and BETELGEUSE. This demonstrates that the entire nebula and dark cloud are new, having condensed from material processed at least once by cool stars.

Barnard's Loop

Cosmic bubble

In 1895 the American astronomer E.E. Barnard, of Yerkes Observatory, discovered a faint ring of gas surrounding the whole of the constellation of Orion. Revealed by his long-exposure photographs with a portrait camera, the ring is now called Barnard's Loop. It is in the form of an ellipse, approximately 14° by 10°, elongated in the same direction as the constellation. It is brightest in the east, to the south of the star Betelgeuse. The bright patch inside the Loop to the south is M 42.

The Loop is hot ionized hydrogen. The presence of large amounts of dust in Barnard's Loop was found in photographs taken by Gemini 11 astronauts Charles Conrad and Richard Gordon in 1966. The Earth's atmosphere absorbs ultraviolet light, but the astronauts were able to take photographs in ultraviolet as they orbited above the atmosphere in their spacecraft. The ultraviolet photographs showed Barnard's Loop as a 19° by 14° ellipse, enclosing the optical Loop.

C.R. O'Dell, D.G. York and K.G. Henize interpreted the Loop as a cosmic bubble blown in the interstellar dust by the stars of Orion. Pushing on this dust by the pressure of their radiation, the stars within the Loop have bulldozed the interstellar material into a region of above-normal density immediately outside the optical Loop. The three astronomers liken Barnard's Loop to the ROSETTE NEBULA. They calculate the time needed to blow a bubble the size of Barnard's Loop as three million years, which is close to the age of the older stars in Orion. The Rosette, much smaller than Barnard's Loop, is only 100,000 years old. Actually, the older stars in Orion are found in the Belt region and younger stars in the Sword region (including the TRAPEZIUM). This may explain why the Loop is elliptical. The push on the bubble is now coming from farther south than it once did.

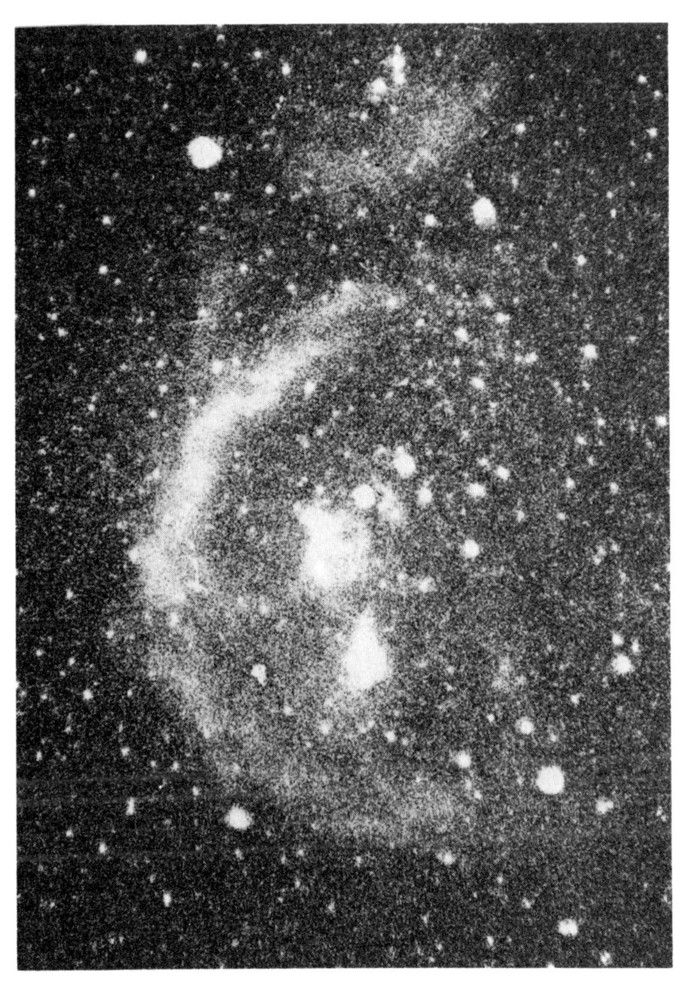

Barnard's Loop PGM

NGC 2244 and its Nebulae

Rosette Nebula

The Rosette is a beautiful nebula in the Milky Way constellation of Monoceros. Plate 7 reveals that it is spherical, and has a hollow center filled with bright stars. The bright parts of the nebula were discovered when visual observers scanned the area with telescopes, and are catalogued as NGC 2237, 2238, 2239 and 2246. Before the nebulae were found the central cluster of stars was discovered by John Flamsteed, Britain's first Astronomer Royal: it is called NGC 2244.

Although it bears a superficial resemblance to a planetary nebula like the RING NEBULA (p. 143), the Rosette Nebula is larger and more massive, and is associated with many stars rather than with just one. It is a zone of hot, ionized hydrogen embedded in a bath of cold, neutral hydrogen gas. The two brightest stars in the cluster are the ones mainly responsible for the ultraviolet light which ionizes the gas and heats it. However, the brightest star in the area, 12 Monocerotis, a fifth-magnitude star visible to the naked eye, is a cool, giant, foreground star, and is not connected with the nebula or the cluster, which are estimated to be 4500 light years away.

The stars in the cluster are very young indeed, perhaps less than 500,000 years old. The first explanation put forward for the hole in the center of the

Rosette Nebula *Hale*

Rosette was that it was a cavity about 12 light years across which had been generated by the formation of those young stars. The estimate was that about 100 solar masses of gas were "missing" from the hole, and most of this had formed the 20 or so brighter members of the cluster. It was calculated that the central hole would fill in after about 500,000 years as a result of the gas's random movements.

This last calculation brought about the demise of the theory that the gas in the center was used up in star formation. Similar central holes were found in other nebulae and it became clear that they are not uncommon. Evidently the central hole must last for an appreciable fraction of the lifetime of the whole nebula, that is for several million years, rather than half a million. Therefore the central holes must be supported from within.

A more acceptable explanation has been put forward by William Mathews. He suggests that radiation pressure from the bright central stars was the force which made the hole.

Embedded in the gas are cold dust grains. The existence of cold particles in a hot nebula came as a surprise to astronomers, but it is an observed fact. The dust grains are especially dense in certain parts of the nebula and show as black shadows on the enlargement. The grains are like little sails intercepting the light from the central stars. The energy which they catch and reflect gives them a push out from the center, a push called radiation pressure. The acceleration of each grain is limited by a drag produced by its passage through the gas. Each little particle meets a kind of wind resistance as it tries to move through the gas. The fastest speeds are usually limited to less than 10 km/second. The time such a grain would take to move six light years from a bright star, and thus create

Globules in the Rosette Nebula KPNO

a hole around it, would be somewhat more than 200,000 years.

The final link in the argument is the demonstration that the dust grains pull the gas through which they move, so that the brighter stars in NGC 2244 can create a hole in the gas as well as in the dust. Mathews assumed that each grain has a small electric charge equivalent to the loss of more than about 50 electrons. In a grain of graphite consisting of about a trillion atoms, even such a small charge as this is workable. Calculations indicate that it is likely an underestimate. The charged grains drag the charged ions of the hydrogen gas with them and thus create the hole.

The nebula itself is composed of hot gas and the hydrogen outside is cool. The pressure in the hot gas is many times that of the cold gas, and therefore the Rosette is expanding out into space. This crushes a shell of cold hydrogen at the periphery which has been detected by radio astronomer L. Raimond.

Thus the Rosette is a rapidly evolving ionized hydrogen nebula, changing completely over a time measured in just millions of years.

Messier 8

Lagoon Nebula

Situated in the constellation Sagittarius, towards the center of our Galaxy, the Lagoon Nebula M 8 (NGC 6523) is a complicated region of recently formed stars, gas and dark clouds, all seen through the peppering of stars which make up the Milky Way in this area. It lies at a distance of 4500 light years. Many of the more distant stars are blotted out by a curtain of dust which forms a backdrop against which we view the Lagoon Nebula (Plate 11). Near the middle of the nebula is the star cluster catalogued NGC 6530, discovered in 1680 by the English astronomer John Flamsteed, in the course of compiling his star catalogue at Greenwich. The nebula itself was first noticed in 1747 by the French astronomer Le Gentil. The brightest stars in the cluster, including 7 and 9 Sagittarii (both visible to the naked eye), are responsible for ionizing hydrogen atoms in the nearby gas.

The age of the star cluster is estimated at two million years: recent on the cosmic time scale. The nebula is the gas left over from the star formation within the cluster. The effects on the nebula of the bright stars in the young star cluster can be seen in the photograph. Like ripples of water spreading from a stone tossed into a calm pond, waves of hydrogen gas are spreading outwards from the cluster of stars, driven by the heat which they radiate. The outpouring gas collides with dark matter in the vicinity of the nebula, making bright rims of gas which wrap around the dark clouds at the nebula's edges.

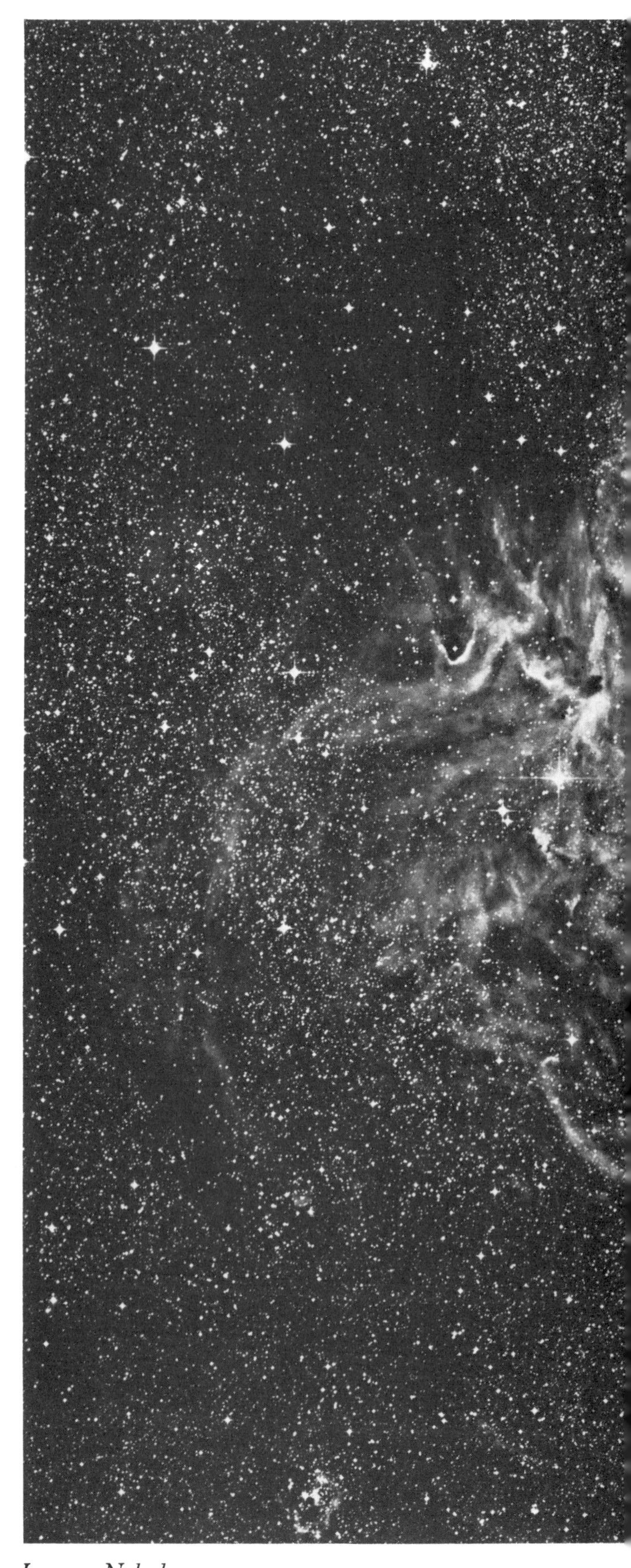

Lagoon Nebula

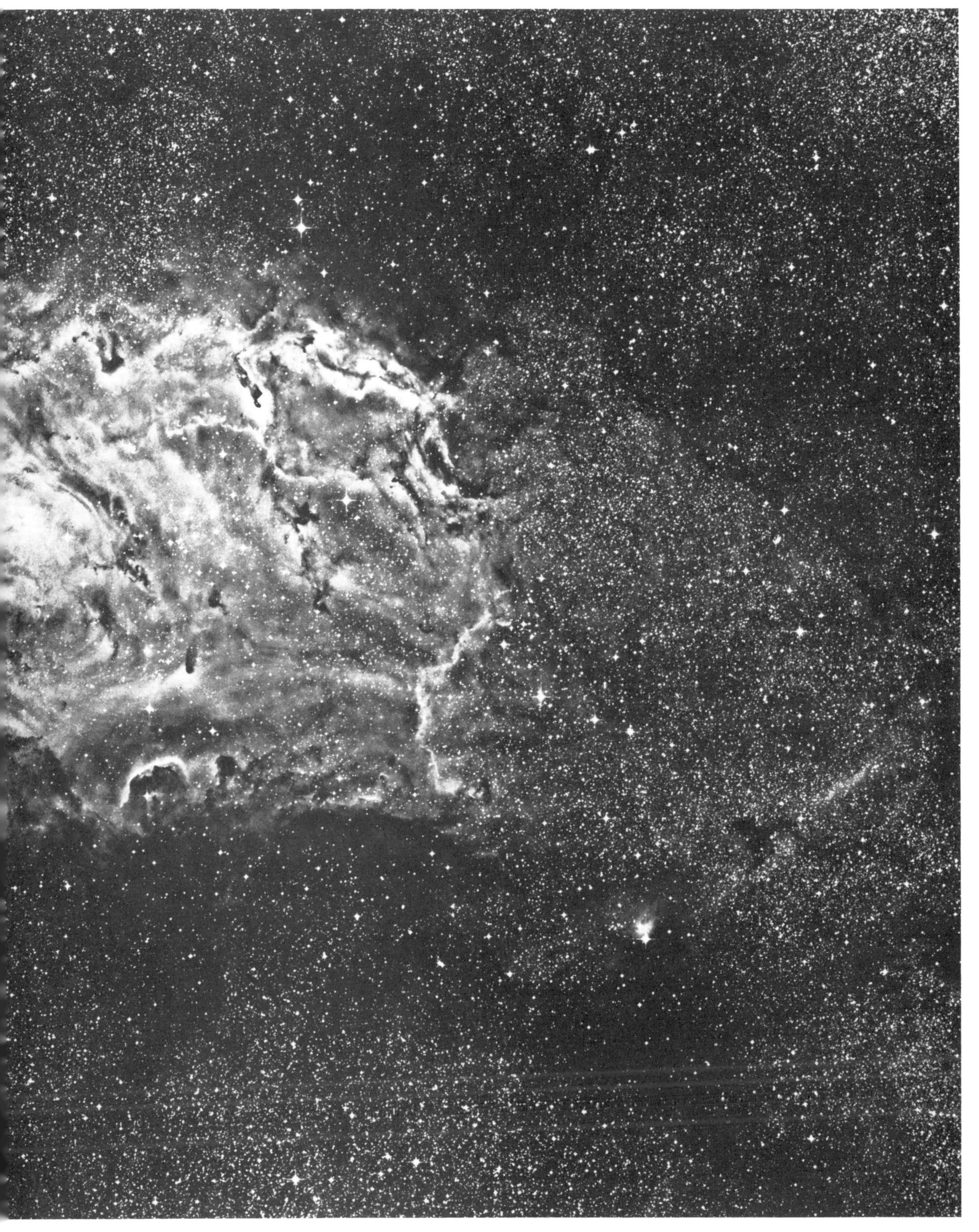

UKSTU

One such dark cloud cuts across the middle of the nebula (its appearance has given the nebula its name). It is embedded in the nebula and divides the stars of the NGC 6530 cluster from the brightest region of the nebula, which lies just to the right of center. The bright region contains the star Herschel 36 and a curious small very bright nebula known as the HOURGLASS. Herschel 36 is a very young star indeed, perhaps no more than 10,000 years old. Its birth has been induced by the activity stirred in the Lagoon by NGC 6530. Like a growing son, Herschel 36 is taking over control of the nebula from the parent stars of NGC 6530. It is probably Herschel 36 which will continue to ionize the nebula as the stars of NGC 6530 fade; it is already driving the flow of gas in the right hand half of the nebula, since the ripples of gas in this half seem directed away from it rather than from the stars of 6530.

The Lagoon Nebula illustrates the fact that although the Universe seems constant in the lifetime of a man, the processes of birth, death and change are in fact always altering its appearance.

Hourglass Nebula

Young nebula

The Hourglass Nebula was discovered within the Lagoon Nebula, M 8, by the great 19th-century astronomer Sir John Herschel, who noted a bright fuzzy patch which to his eye looked not quite starlike, near to a star which he numbered 36. The hourglass shape and star 36 are immediately apparent on our picture. Herschel mistakenly compared the Hourglass Nebula to the center of the Andromeda Galaxy, though the Hourglass is an emission nebula of glowing gas, heated by star 36, not the nucleus of a galaxy. The nebula was named the Hourglass in 1956 by A.D. Thackeray, who remarked that it was so bright that it could be seen in a telescope even when the Full Moon was nearby in the sky. The reason it is so bright is that it is a very dense nebula, and it is this property which in 1961 gave Nick Woolf the clue as to its age. Because it is dense, the pressure inside the Hourglass is very high and it will soon expand from its present half-light-year diameter and lose its identity in the rest of the nebula. Woolf calculates its age at no more than 10,000 years, by which time it should have dissipated. Thus it can be no longer than 10,000 years since Herschel 36 began to heat it. Herschel 36 must therefore be one of the youngest stars known.

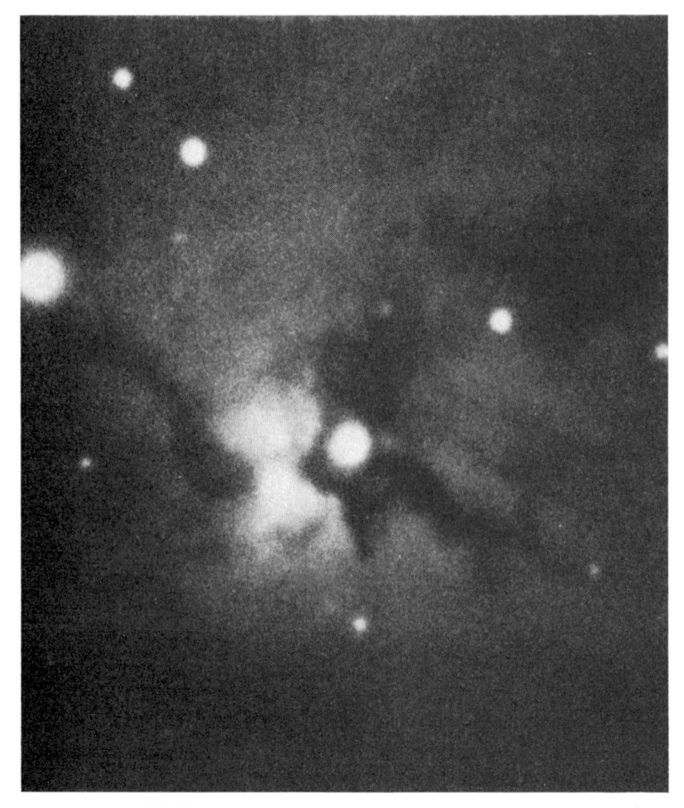

Hourglass Nebula AAT

Trifid Nebula

Caught in an interstellar net

The Trifid Nebula in the constellation Sagittarius is a region of hydrogen gas, glowing because it is excited by a central, bright, hot star, catalogue number HD 164492. Three radial dark lanes of dust divide the nebula into three sectors and give it its name (Latin for "cleft into three"). Because the light from excited hydrogen gas is predominantly the red H-alpha spectral line, the excited hydrogen appears pink in color photographs (Plate 8).

To the north of the Trifid is a blue nebula surrounding the star HD 164514. While bright, this star is not hot enough to excite its surrounding hydrogen and the light from the nebula is just starlight reflected by the dust mixed up with the cold gas which surrounds the Trifid. The nebula appears blue because the particles within it are of such a size that they selectively scatter blue light, just as sunlight scattered in Earth's sky is blue. Where not lit up, because no nearby stars illuminate it, dust in the cold gas surrounding the Trifid can nevertheless be detected because it obscures the faint background stars just outside the glowing nebula; there are fewer stars visible near the southern edge of the Trifid than farther away.

The three dark lanes could lie in front of the ionized hydrogen region. The ionized hydrogen could be expanding into a coarse lattice of stubborn lanes of dust which surround it like a fisherman's net. On the boundary of the excited hydrogen gas, especially to the

Trifid Nebula AAT

south, V-shaped luminous rims of gas point in the general direction of the central bright stars. The nebula has been heated by its central stars since they formed millions of years ago and the hot gas is expanding into the cold gas and dust surrounding it. Where the boundary of the outflowing sphere of hot gas encounters lumps in the cold gas, it divides round the sides of the lumps, leaving a V-shaped wake, like a river flowing past an island. At a later stage the nebula will have flowed on past the lumps so that they break off from their wakes and remain trapped inside the hot gas as dark globules and dust lanes, like the ones which give the Trifid its name. Some astronomers suggest that the globules will form new stars.

The star which excites the Trifid is one of a group of four like the TRAPEZIUM (p. 89), the four stars which excite the ORION NEBULA (p. 102). Two of the four are revealed by study of their spectral lines to be double stars. The distance of the cluster of stars is 4500 light years; the stars are young and were formed seven million years ago.

Eta Carinae's Homunculus Nebula

Brighter than a million Suns

Plate 8 shows the object called Eta Carinae. The star itself, which is one of the most luminous in the Galaxy, now lies shrouded within the Homunculus, a manikin-shaped nebula discovered by Innes 70 years ago which has been growing larger through the intervening decades. At the center of the Homunculus, unseen, is the star which has over the last century and a half given out as much energy as many stars emit in their entire lifetime.

When the first catalogue of southern stars was published, by Bayer in the early 17th century, Eta Carinae appeared to be a star easily seen by the unaided eye, probably around third magnitude. As more astronomers studied the southern skies, Eta Carinae was recognized to be a variable star, ranging irregularly between second and fourth magnitude. Early in the 19th century, however, the object began to brighten considerably. By 1843 it was the second brightest star in the sky, at magnitude –1. Then Eta Carinae faded slowly by a factor of 1000, and for several decades it has remained fairly steady between sixth and seventh magnitude.

Innes was an assiduous observer of double stars, and when he discovered the nebula, he thought he saw several companion stars within it. These have all moved radially outwards from the central bright knot, and have become more diffuse. They are no more than denser portions of a nebula which is expanding at a rate of 500 km/sec. Today the Homunculus Nebula measures about 12 by 8 arc seconds, has a rich orange center and a yellow-white frothy surround.

In 1948, while testing the then newly installed 74-inch Radcliffe Telescope in South Africa, David Thackeray discovered a faint outer ellipse of nebulosity. This too is expanding and is now very faint and difficult to photograph. It seems that the central object has ejected more than one shell in its lifetime. Careful measurements of the Homunculus show that various

Homunculus *Ed Ney*

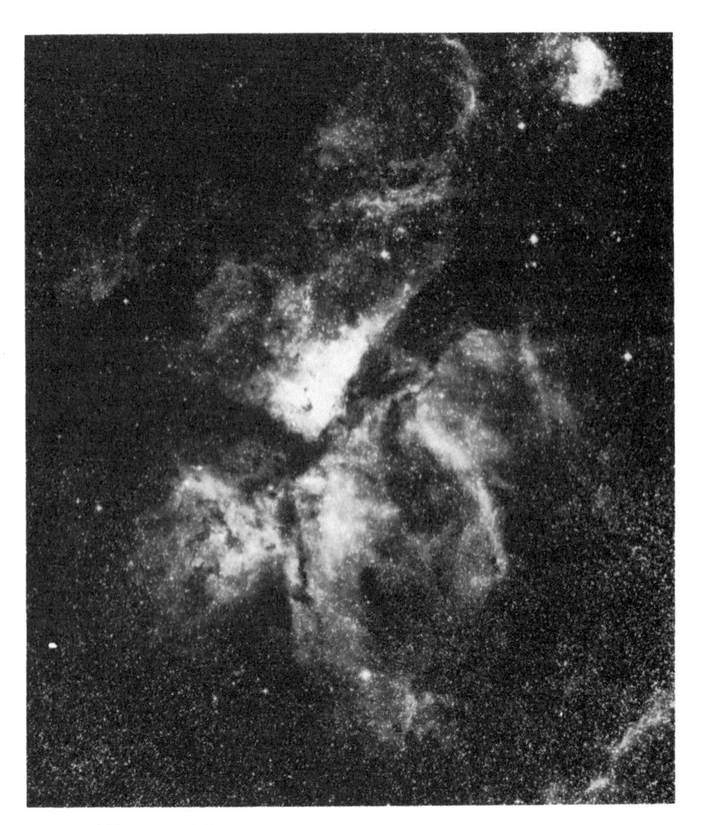

Eta Carinae Nebula UKSTU

parts of it were thrown out from one central star at different times in the mid 1840's, around the time of the star's maximum brilliance.

Unfortunately, no spectra were taken of Eta Carinae before 1892. In that year the object had the spectrum of a normal star a little hotter than the Sun, around 7500°C. In 1895 a dramatic change was seen to have occurred, for the spectrum then, as now, comprised strong emission lines. The emission is unlike that of most nebulae in two ways. First, the gas is excited by a much cooler star than those which excite gas in most ionized hydrogen nebulae. Second, there is a striking absence of lines due to oxygen. Normally, oxygen lines are the strongest of all emission features in the spectra of nebulae, in part because oxygen is the third most abundant element in our Galaxy, after hydrogen and helium.

Since 1895 the spectrum has changed hardly at all. Astronomers explain this behavior, which is also displayed by novae, quite simply. The shell of gas which formed the Homunculus was small and opaque in 1892. The light which reached astronomers in that year came only from the outer surfaces of the shell, which were at a temperature of 7500°C. As the shell expanded, it eventually became transparent. The transition from opaque to transparent can occur very quickly; in many novae it requires only a few days, and in Eta Carinae it happened between 1892 and 1895. Ever since the latter date we have been seeing through the shell to the inner nebula. This inner nebula is still opaque at its center, for we do not yet see the star through it.

Eta Carinae almost certainly lies within the ETA CARINAE NEBULA (above). Various estimates of the distance of this nebula have been made, and the most reliable currently seem to agree on 6800 light years. This is far greater than the distance of any of the brightest stars in the sky. Sirius, the only star to outshine Eta Carinae in 1843, is a mere nine light years distant. When we correct for the distance of Eta Carinae, we find that in 1843 it shone with the light of six million Suns. This exceeds by almost a factor of ten the luminosity of the next brightest star known!

In 1968 and 1969 California astronomers Gerry Neugebauer and Jim Westphal made infrared observations of Eta Carinae. They discovered that at wavelengths 20 times that of visible light, Eta Carinae is the brightest star in the sky. The reason for this great infrared brightness is the presence of dust grains in the inner regions of the Homunculus. The dust grains absorb the central star's light. In so doing they become hot, and radiate in the infrared to cool themselves. Most of the dust in Eta Carinae is at about freezing point. The dust grains cannot be radiating more energy than they receive from the central star. Thus in assessing the present luminosity of Eta Carinae we must include the infrared component. When they did so, Neugebauer and Westphal found that Eta Carinae is still radiating several million times as much energy as the Sun. In all probability the central star has faded very little in the last 140 years. Only the formation of dust grains in the Homunculus caused its visible light to be dimmed.

The chemical composition of the dust includes a considerable proportion of oxygen, bound up in silicate minerals such as form around MIRA variable stars (p. 99). This may in part explain the absence of oxygen in the spectrum of the Homunculus nebula.

From the evidence of the ionization of the nebula, the temperature of the central star is not particularly high. Most astronomers agree that a value around 25,000°C. is about right. Knowing the temperature and luminosity, astronomers can estimate the size of Eta Carinae, and they find it to be ten times that of the Sun. This is not unduly large, for many stars are known with diameters several hundred times that of the Sun.

In all probability, it is the mass of Eta Carinae which is extreme. Astronomers believe that stars heavier than 100 solar masses cannot form, and that at masses near this value stars are unstable. Eta Carinae is probably a star of about 100 solar masses. Its

Horsehead Nebula

NGC 6302

PLATE 13

Ring Nebula

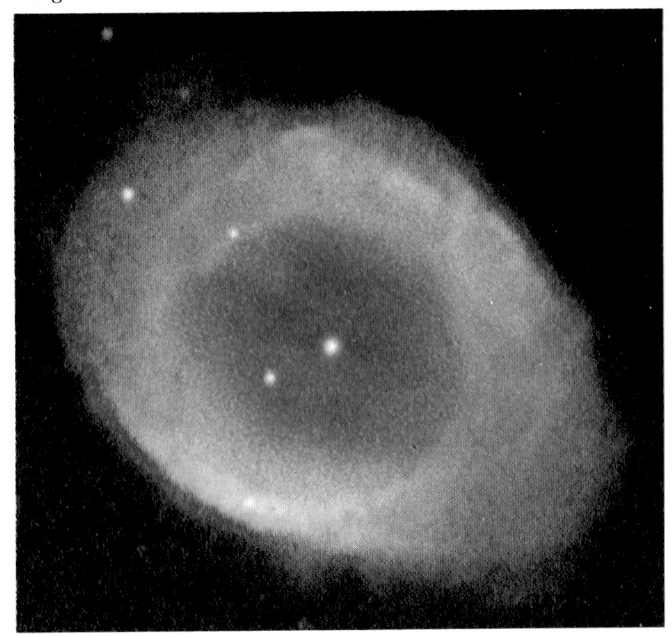

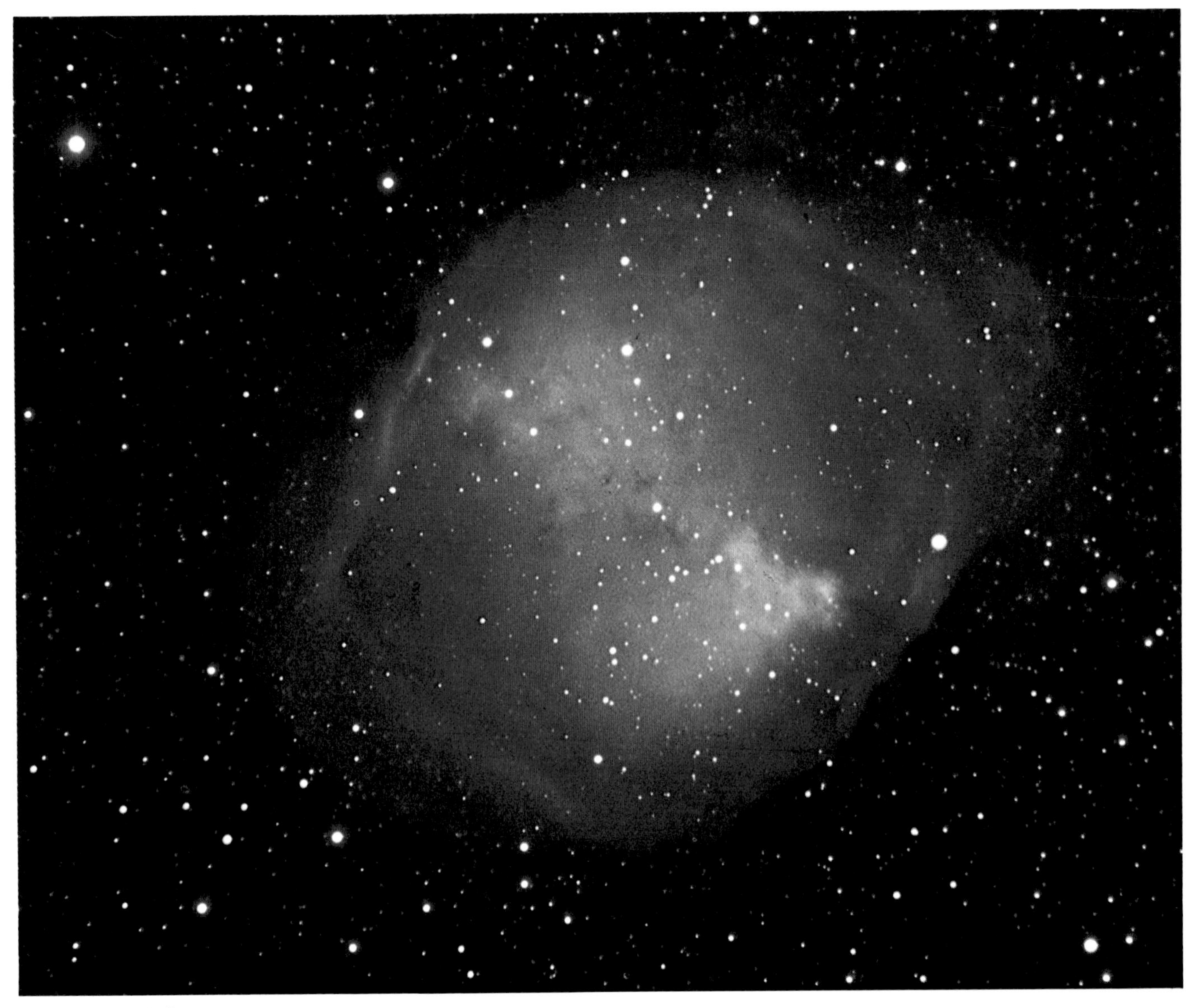

Dumbbell Nebula

PLATE 14

Crab Nebula

PLATE 15

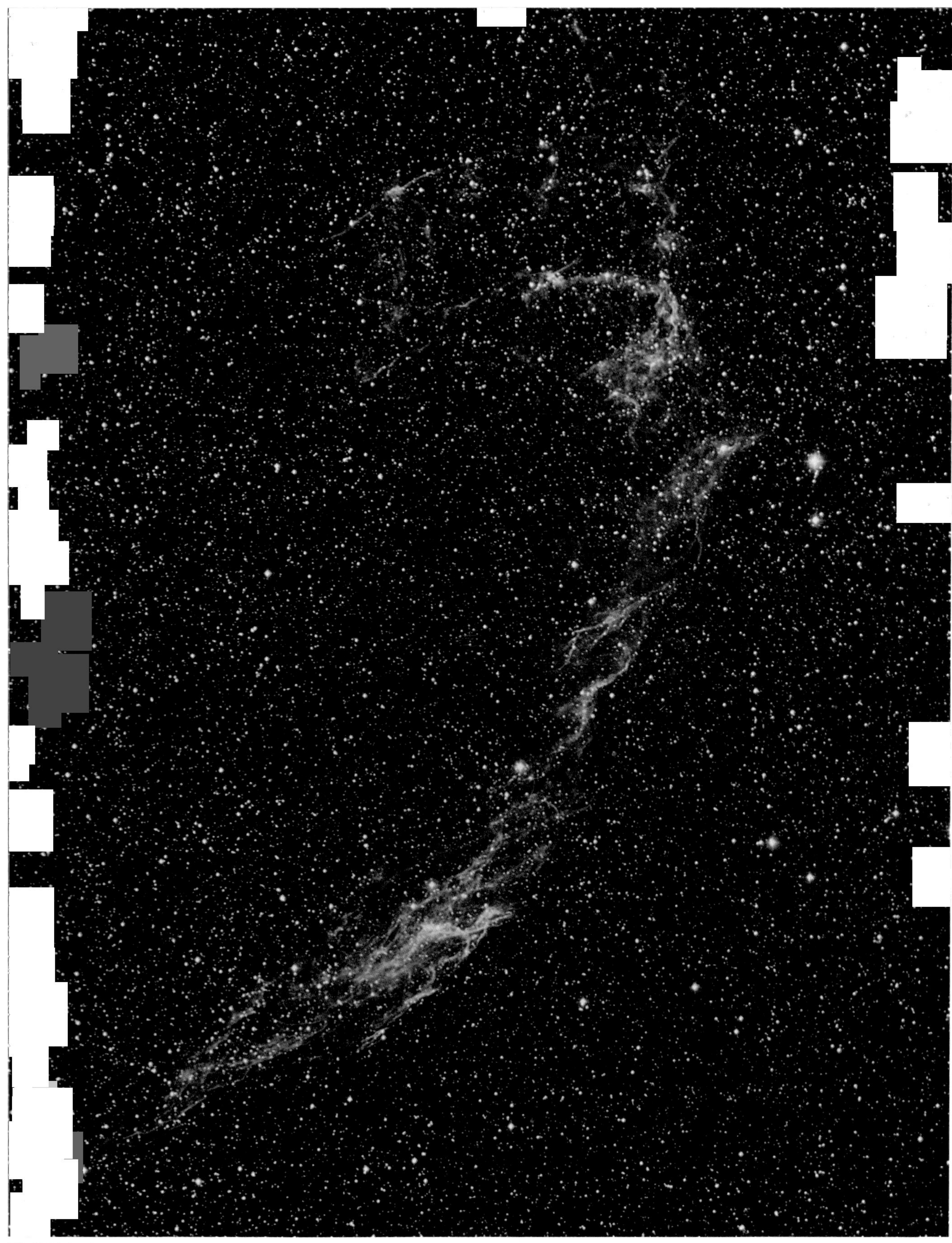

Veil Nebula

PLATE 16

instabilities have resulted in the ejection of two or more shells of gas, each of which represents an attempt by the star to regain stability. Such attempts serve no more purpose than to delay the inevitable. Canadian astronomer Nolan Walborn speculates that Eta Carinae may be nearing the time of its life when it can no longer stabilize itself, and that it will soon erupt into a supernova. If Walborn is right, Eta Carinae is destined to become one of the greatest explosions mankind will ever witness.

Eta Carinae Nebula

And Keyhole Nebula

At the center of the Carina star cloud in the southern Milky Way lies the bright patch of the Eta Carinae Nebula, NGC 3372, a region of glowing interstellar gas more than 1° across (four times the area of the Full Moon). Discovered by the Abbé Lacaille in 1751, the first accurate map was drawn by Sir John Herschel during his visit to Cape Town (1834-1838) in which he catalogued the nebulae visible from the southern hemisphere.

The Eta Carinae Nebula shines because there are hot, bright stars within it, notably Eta Carinae itself. They irradiate it with ultraviolet light which ionizes the hydrogen atoms in the nearby gas. Oxygen atoms, too, are split by the light of embedded stars and on recombining emit strong blue and green light. These and other colors produce the purple and red tints of the nebula (Plate 9).

The intrinsically brightest and hottest star in the Eta Carinae Nebula is HD 93129A, which is one of the most luminous and most massive star in the Galaxy. It is the brightest member of the cluster of stars to the northwest of Eta Carinae, known as Trumpler 14. The cluster is above and to the right of center of Plate 9. The cluster and therefore Eta Carinae lie at a distance of 6800 light years.

The nebula is a whirling patchwork of gas and dust. Striking dark lanes cut across the glowing gas behind. Some dust intrudes into the nebula itself, like the central Keyhole Nebula embedded in the brightest part of the nebula (center in our illustration, Plate 9). The Keyhole was verbosely described upon its discovery by Herschel as "a singular lemniscate-oval vacuity." When originally drawn by him, the keyhole shape of this nebula was complete, but now is not so clear, because the eastern edge has faded. This edge of the Keyhole shines not by its own light, but by reflecting the light of the star Eta Carinae. This is proved because the spectrum of light from the faded eastern edge of the Keyhole is the same as the spectrum of Eta and is not at all typical of the spectrum of light from the rest of the nebula. Eta itself has faded considerably, being at its brightest at the time Herschel made his sketch. At that time, in fact, it was temporarily one of the brightest stars in the sky; it faded throughout the 19th century and is now around sixth magnitude.

Radio and optical observations of the Keyhole Nebula show that the "eye" of the keyhole is a bursting bubble, known to radio astronomers as Carina II. It is expanding overall with a speed of about 40 km/second and was produced by an unknown violent event (a supernova?) about 25,000 years ago. Another dramatic event in this violent region of the Galaxy was the supernova which produced a supernova remnant and x-ray source 4U 1043–59, somewhat to the east of Eta Carinae on the near side of the nebula. The reason for this coincidence of explosions is that the Eta Carinae Nebula is the scene of recent formation of massive stars, which evolve quickly and explode within a relatively short time.

Lambda Orionis Nebula

Mistiness in Orion's head

The Greek astronomer Ptolemy (AD 144) first recognized the "mistiness in Orion's head" as a small star cluster. Although the nebula itself is too faint to be seen with the naked eye, on a clear night the eye is attracted to it on account of the presence of numerous faint stars.

Long-exposure photographs show that the cluster's brightest star, Lambda Orionis, irradiates its sur-

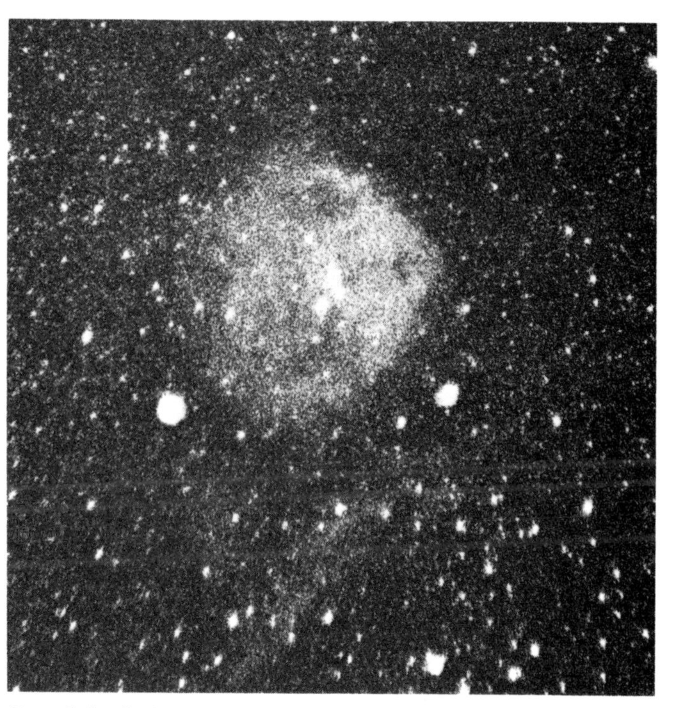

Lambda Orionis Nebula PGM

rounding gas. The nebula which it has heated is an almost perfect sphere 8° across centered on Lambda. The gas in the nebula is hotter than the cold interstellar hydrogen surrounding it and thus has a pressure some 1000 times greater. As it expands into space, it compresses cold gas ahead of its edge like a piston. Gas and dust, swept up ahead of the exploding sphere of hot gas, is visible on small-scale photographs as a dense shell of absorbing clouds, hiding the light of stars behind the shell. The shell is also detected by radio astronomers since the cold hydrogen emits radio radiation with a wavelength of 21 cm, and they can deduce that the shell is expanding at 8 km/sec.

American astronomer A. Joy and Soviet counterpart G. Markova have found young stars in the dark shell. The collapse of their parent gas clouds was apparently triggered by the compression. Thus the formation of Lambda Orionis was a spark to an explosion which ignited further star formation, in a cycle that perhaps will repeat itself until no more dust and gas is left in the area.

The Lambda Orionis Nebula lies at a distance of 1300 light years, much the same as the other stars of Orion. The diameter of the shell is 180 light years, and its age is two million years.

Omega Nebula

The Horseshoe and the Swan

Discovered in 1746, the Omega Nebula M 17, also NGC 6618, is a bright nebula excited by a group of at least five hot, intrinsically bright stars in Sagittarius. John Herschel's 1833 description of it was "very remarkable, consisting of two loops like a capital Greek 'omega', the one bright and the other exceedingly faint." It is also known as the Horseshoe and the Swan. It is shown in Plate 11.

The Omega Nebula was the first to be photographed with red-sensitive film to record the red hydrogen-alpha line of hot hydrogen. The photograph, by P. C. Keenan in 1936, began the recognition of this kind of nebula. The western (right) edge of the excited gas abuts a dark, dusty nebula containing at least two bright stars whose light is all but extinguished by the dust, but whose infrared radiation penetrates the dust more easily. The stars heat the dust; it therefore emits infrared radiation itself. Colder dust is situated to the east and is weak in infrared radiation. A tide of ultraviolet radiation from the group of five stars penetrates the dust all around. The hot edges of the dust clouds are "boiling" off and moving in waves back to the center of the nebula.

The Omega Nebula is 4800 light years from Earth, and about 27 light years in diameter. It contains as much as 4000 solar masses excited by the stars.

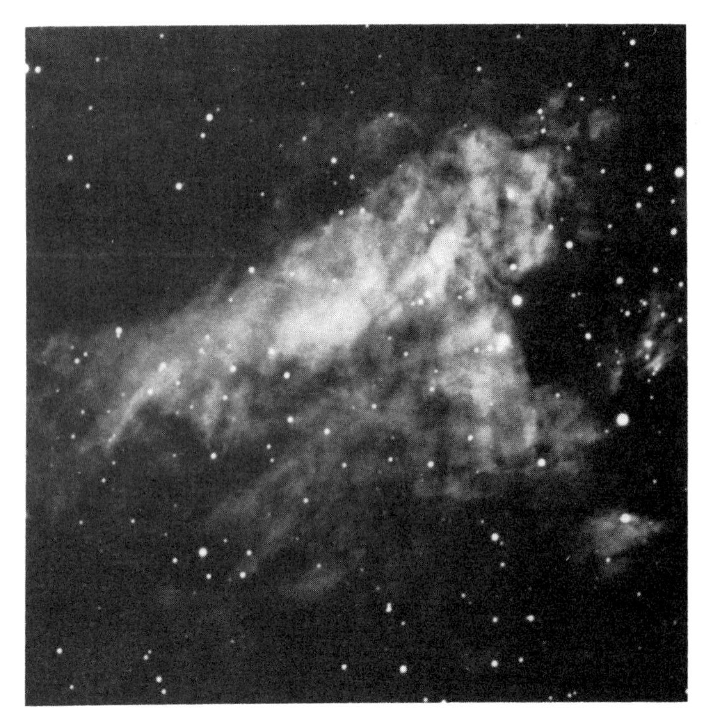

Omega Nebula KPNO

As a mixture of cold dust, warm dust, hot gas, cold gas and stars, the Omega Nebula well illustrates the aphorism that the interstellar medium in our Galaxy is more chaotic than random.

North America Nebula

Interstellar chaos

The North America Nebula is also a chaotic area of gas and stars. It is formed of reflection nebulae, dark nebulae, and excited gaseous nebulae. In Cygnus, it comprises the nebulae NGC 7000, IC 5067, 5068 and 5070. Originally discovered by William Herschel in 1786, it was first photographed by Max Wolf in 1890. On his pictures it was revealed with the shape of the North American continent (Plate 10).

The nebula is estimated to be 2300 light years distant. Since the line of sight to the nebula passes along a spiral arm which lies in this direction, unrelated objects appear one behind the other. Thus disentangling the three-dimensional structure of the nebula is very difficult. The star HD 199579, also known as BD +44° 3639, appears to be the hottest star in the region. It produces the ultraviolet radiation which ionizes the excited gas. Light from other, cooler stars, including Deneb (Alpha Cygni) is reflected from the dust as well.

The North America Nebula is about 1° across and may be glimpsed with the unaided eye in very good skies.

North America Nebula *Hale*

California Nebula

Chance encounter

The California Nebula, NGC 1499, is the brightest nebula in a region of the constellation Perseus which is rich in nebulosity and dark matter. The California Nebula is 1000 light years from Earth.

The California Nebula is excited and heated by the star Xi Persei. Since Xi Persei does not lie inside the cloud of gas, the nebula lacks the usual roughly circular outline. The long, thin shape which inspired its name results because it forms the rim of a dark nebula of dust and gas which is illuminated by Xi Persei.

Xi Persei is one of 17 stars recognized as the II Perseus association of stars. An association of stars is a loose group thought to have originated together but now merging into the general Milky Way collection of stars. Associations of stars are young because they cannot be recognized for long as a group and II Perseus is estimated by Adriaan Blaauw to be 1,300,000 years old.

Another member of the II Perseus association is the star X Persei. It is a weak x-ray source, one of the weakest known. It consists of a pulsating neutron star of period 13.9 minutes, orbiting X Persei itself every 22.4 hours.

Xi Persei is running away from the rest of the stars in the II Perseus association. Its speed is about 70 km/sec, in contrast to 10 km/sec which would be the normal speed for stars of the same kind. As described in the case of MU COLUMBAE (p. 91), there are two rival theories to explain such "runaway stars." They may have once been part of a binary system which has since been disrupted, or they may have been thrown out of a small stellar cluster by a chance acceleration. Be that as it may, the fact is that Xi Persei is moving through space at high speed and is by chance in the area of NGC 1499. Unlike most bright nebulae, in which stars and gas are intimately connected by a family relationship, the California Nebula has thus been made visible by a chance encounter.

California Nebula *Lick*

Gum Nebula

Fossil nebula

While most emission nebulae glow because a bright, hot star shines ultraviolet light at nearby gas, providing the power for the gas to convert to visible radiation, the Gum Nebula may have been powered by a flash from a long-past supernova. The nebula, the largest in the sky, was discovered by the late

Australian astronomer Colin Gum. For his doctoral thesis he had photographed the southern Milky Way to discover new nebulae. Gum made a mosaic of the photographs which he took of the constellations Vela and Puppis and the nebula which bears his name became apparent. It has a diameter of 30° on the sky.

Within the nebula lie the stars Zeta Puppis and Gamma Velorum. Gum thought that they powered the nebula. Recent calculations by John Brandt, Theodore Stecher, Steve Maran and David Crawford, however, indicate that the two stars are not powerful enough. These astronomers proposed instead that the hydrogen atoms in the nebula had been split apart not by the steady radiation from the stars within but by a burst of radiation from the explosion of the VELA SUPERNOVA some 12,000 years ago.

Eventually all the gas in the nebula will recombine to make neutral hydrogen atoms and the nebula will cease to shine. But the recombination process takes many hundreds of thousands of years because the distance between the electrons and protons of the gas is so vast that a wandering electron seldom passes close enough to a proton to be grabbed in order to make a hydrogen atom.

According to the supernova burst explanation the Gum Nebula is not a nebula in the usual sense since the stars present within are not powerful enough to keep the nebula constantly ionized. It is a transient nebula, formed in a flash and not being renewed, a fossil left by an event long past.

Herbig-Haro Objects

Early phase of stellar evolution

In 1946 and 1947 the American astronomer George Herbig took a series of photographs of NGC 1999, a small, diffuse nebula which lies due south of Orion's Sword. Three peculiar nebulous objects turned up on these photographs. A Mexican astronomer, Guillamero Haro also noted them, and they came to be known as Herbig-Haro objects numbers 1, 2 and 3.

More than 100 further examples have since been found. Each is composed of one or more diffuse, starlike objects. H-H 2, for example, consists of at least ten individual bright spots, each quite variable by a few magnitudes. Two have increased in brightness by three magnitudes since around 1952.

These little nebulae have distinctive emission-line spectra which show that gas molecules in them are crashing together. The energy of the crash heats the gas and causes it to emit light.

In his discovery paper George Herbig remarked that his objects "define another type in the growing list of peculiar objects that occur where stars and nebular material are intimately associated." Herbig had in

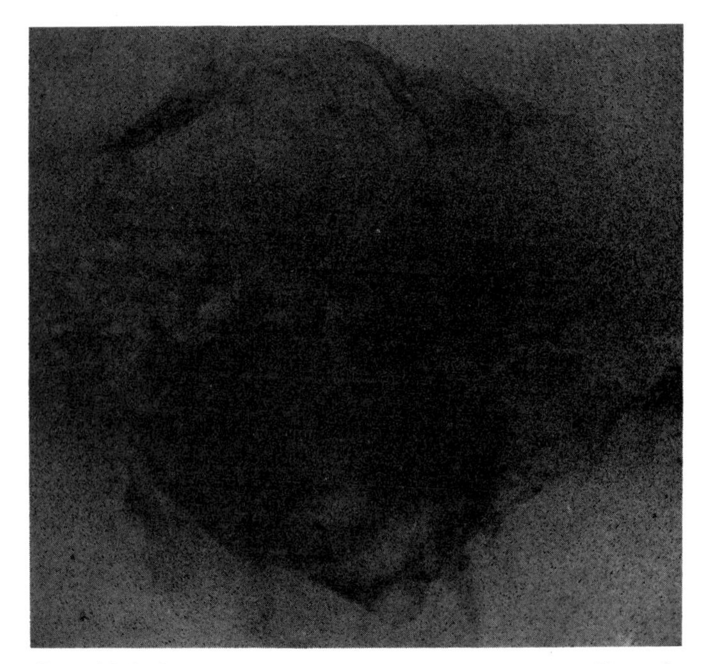

Gum Nebula *Brandt*

mind the T Tauri stars, which are irregular variable stars found only in very young clusters called T-associations. T Tauri itself is embedded in a nebula (Hind's Nebula) which, observed on its own, would look like a Herbig-Haro object.

A typical T Tauri star has an unsettled atmosphere, extending far out from the surface of the star and showing evidence of infalling and outflowing gas. Astronomers believe that T Tauri stars and other stars like them—RW Aurigae stars, YY Orionis stars, T Orionis stars—represent stars still contracting from the interstellar gas, newborn stars still not fully formed. Perhaps Herbig-Haro objects, which at present are wholly nebulous and contain no visible stars, will eventually produce T Tauri stars. A further indication that this may happen is that H-H objects themselves do not show strong infrared radiation as T Tauri stars do, though single infrared sources do often lie nearby. The usual explanation for excess infrared emission from objects like T Tauri is that it is produced by dust heated to a few hundred degrees by a star embedded within. No star—no heat—no excess radiation; this seems to be the logical chain with H-H objects.

An alternative view of H-H objects is that they are mere reflection nebulae giving off the light of a very young star which lies within. Only infrared radiation from the star could penetrate the dusty cloud and reach Earth directly, and this accounts for the nearby infrared object. (Long wavelength infrared radiation penetrates obscuration more easily than short wavelength blue light. Thus infrared photographs of distant landscapes reveal far-off details more crisply.) Some light from the star may shine through gaps in

Herbig-Haro Objects *Lick*

the dust onto particular places in the surrounding nebula, giving what we see as the Herbig-Haro object. As the dusty clouds shift to and fro the gaps open and close and the reflections come and go. Thus they appear as individual variable stars.

This hypothesis, according to Steve and Karen Strom, renders the Herbig-Haro phenomenon less exotic than it seemed before. "If accepted," they say, "it suggests an uncharacteristic kindness on nature's part, in that we are permitted, albeit by reflected light, to observe a very early phase of stellar evolution." Perhaps in H-H objects we see the first 100,000 years of a star's life, or an even earlier period.

One surprise is that Herbig-Haro objects show an outflow of material as judged from their spectral lines, rather than the infall which the contracting dust cloud forming the star might lead us to expect. Astronomers have mixed feelings about this. On the one hand, it is difficult to see how an outflow of material can help a star to get together. On the other hand, a gas cloud which begins by rotating slowly spins up to rotate very fast indeed when it collapses, in order to conserve its momentum. The same effect results in the rapid spin rates of objects such as the CRAB PULSAR (p. 153). But stars are observed to rotate slowly—the Sun rotates once per month for example. The excess speed of H-H objects' spin could be carried off by an outflow of dusty material from a newly-formed star. Paradoxically, the outflow of material observed in the stars' spectra may indicate that they are contracting.

At the least, the astronomer who studies H-H objects can say with Robert Frost

"Some dust thrown in my eyes
Will keep my talk from getting overwise."

MWC 349

Proto-solar-system

MWC 349 is a faint star in Cygnus, surrounded by gas, whose identifying spectrum of bright lines earned it a place in the Mt. Wilson Catalogue of such stars. Nearly 50 years after its original discovery, new observations were made with infrared sensitive tele-

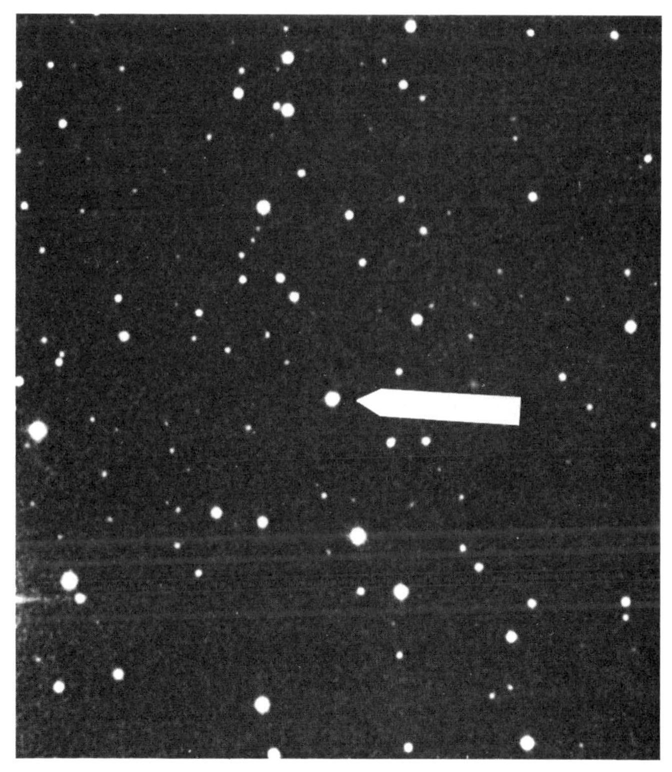

MWC 349 POSS

scopes aboard the Kuiper Airborne Observatory, a NASA-owned C-141 jet containing a 36-inch telescope and able to fly as high as 45,000 feet, well above most of the Earth's atmosphere. These observations showed that MWC 349 is a massive, young star surrounded by a disk of luminous gas about the size of the orbit of Earth. Colder, invisible gas extends out to a distance greater than the size of our solar system.

Steward Observatory astronomer Rodger Thomson speculates that the disk is like the so-called solar nebula, envisaged as orbiting the Sun at its birth and from which the planets of our solar system condensed. MWC 349 may be a system of planets in the act of formation.

NGC 6888

Flung out by a weight-watching star

NGC 6888 is a wispy, filamentary nebula in the constellation Cygnus. At its center is the associated star which rejoices under the unglamorous title of HD 192163.

But HD 192163 is one of a rare breed of objects, the Wolf-Rayet stars. Only about 150 such stars are known in our Galaxy, the first having been discovered in 1867 by the Paris astronomers Wolf and Rayet. These stars are distinguished spectroscopically: unlikely the narrow, dark Fraunhofer lines found in most stars, their spectra contain only broad, bright features. These

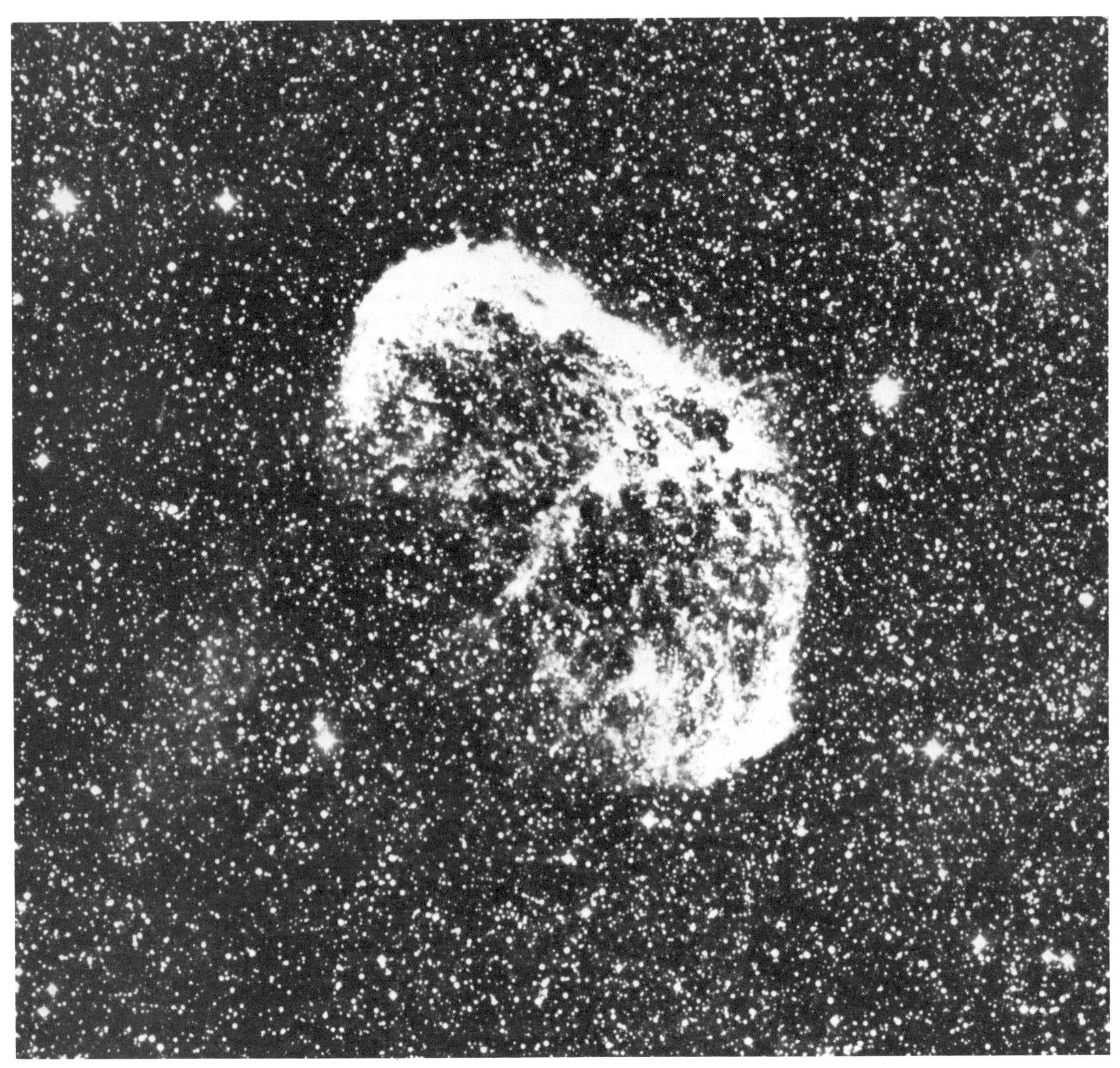

NGC 6888 POSS

spectral features divide the stars into two groups. In one, the strongest features are due to carbon; in the other, nitrogen predominates. There are no carbon features in a nitrogen Wolf-Rayet star, and vice versa. This clear division probably, but not certainly, indicates a difference in chemical composition of the two types.

HD 192163 belongs to the nitrogen sequence. It is one of the brightest members, both because of its large surface area and because it is hot. Astronomers estimate its temperature at near 40,000°C, and its dimensions are about seven times greater than the Sun, giving it 50 times the area to radiate from. Only the great distance of HD 192163, nearly 5000 light years, makes the star rather inconspicuous in our skies.

Wolf-Rayet stars are continually shedding their outer layers. Gas is streaming away at speeds up to 3000 km/sec. This is as fast as gas moves in those stellar explosions we call novae (p. 100), yet novae erupt only once or twice in a star's lifetime, whereas Wolf-Rayet stars undergo continuous activity. Astronomers estimate that most Wolf-Rayet stars would lose their own mass in a few million years. Thus longevity cannot be one of their attributes. It seems that Wolf-Rayet stars are very young, and may be a phase that all large stars go through at a time when the laws of physics dictate that they must lose weight.

The nebula, NGC 6888, probably represents material lost by HD 192163. This is not entirely certain, since it may have been gas left over after the star formed. In either case, it is being blown away by the Wolf-Rayet star, and is still traveling supersonically despite having plowed into considerable quantities of stationary gas. The present dimensions of NGC 6888 are 16 by 24 light years, and astronomers estimate that it contains one or two solar masses of gas.

Hyades

Cosmological foundation stone

Our Galaxy contains ten trillion stars scattered over wide distances. They were born at various stages of the Galaxy's life, out of interstellar gas polluted with a range of materials. It would probably have been impossible to make sense of the properties of stars and determine how they evolve if some stars had not clumped together in clusters. There are several reasons why this clustering makes the job of classifying and understanding stars easier.

Since all the stars in a cluster are at the same distance from us, a cluster star which appears four times as bright to us as another is indeed four times brighter and not half the distance. In studying a cluster therefore astronomers can safely rank the stars in it from brightest to faintest.

A further simplification to be found in clusters compared to the general run of stars in the Galaxy is that astronomers have been able to assume (without contradiction until recent fine interpretation of subtle evidence) that all the stars in a cluster were born at the same time and are all, therefore, of the same age.

Astronomers make the further assumption that the cloud of gas from which each cluster of stars formed was well stirred. Therefore all the stars in it are of the same composition. If astronomers can determine the ages of many clusters of stars they can find out how the composition of interstellar gas changed in the lifetime of the Galaxy.

Just as a series of panoramas in a museum might illustrate Earth's geological history by showing the forms of prehistoric life against a backdrop of the scenery inferred from the rock strata of each geological era, so astronomers regard star clusters as a gallery of stars of various ages set in a changing Galaxy. Unfortunately star clusters are not arrayed in a sequence around us and it has been hard work to set them in the right order.

Star clusters are of two kinds. Populous, old, spherical clusters are called globular clusters. The more sparsely populated, relatively young clusters are called open clusters (or galactic clusters because they concentrate to the Milky Way).

In the next few pages we show a panorama of open clusters. Our first example is the Hyades.

The Hyades is a prominent star cluster near the star Aldebaran in Taurus. At 150 light years distance it is the nearest star cluster to the Sun and appears the largest to the eye, spreading over an apparent diameter of 8°. The Sun in fact lies closer to the center of the Hyades cluster than do some of its outlying members. Because it is so close that the individual stars in it are well separated from each other neither Messier nor Dreyer thought to give it an M or NGC number in their catalogues of clusters and nebulae. Our photograph has (inset) a chart naming the individual members of the Hyades which can be seen with the unaided eye.

Its closeness has made the Hyades the foundation stone on which Allan Sandage and Gustav Tammann based the measurement of the distances to other star clusters and other galaxies and ultimately the size of the Universe itself. Because the Hyades cluster is so close its distance can be measured by a unique method relying on geometry alone, with very few assumptions built into it. Astronomers find the distance of the Hyades by the moving cluster method.

The method depends on knowing that all the stars in the cluster move in a star stream of equal, parallel space-velocities, like a military band marching in step

Hyades cluster (inset: key to star names) POSS

straight across a parade ground. In fact the Hyades stars are all moving at 45 km/sec relative to the Sun. Their individual motions with respect to each other are less than 0.75 km/sec, so if they were indeed bandsmen we would say that they were well disciplined with little propensity to step out of line. Because they are moving parallel to one another, the Hyades stars appear to stream towards a single point in space, just as parallel railroad tracks converge to a point on the horizon and meteor showers like the LEONIDS (p. 242) diverge from a single radiant point.

The Hyades stars' destination is called their convergent point and it lies just east of the star Betelgeuse. The nearer a star is to the convergent point, the more its speed is wholly *along* the line of sight. A cluster member right at the convergent point would be moving directly away from us. The farther a star is from the convergent point, the more its speed appears as proper motion, or angular speed *across* the line of sight. The trade off between radial velocity and proper motion for a star in the cluster depends on the angle between the convergent point and the line of sight to the star. It also depends on the star's distance. Once the convergent point of the Hyades star cluster has been identified, therefore, the distance of each star in the cluster can be found.

The distances to other clusters of stars like the Hyades are found by matching the stars within them to similar Hyades members and assuming that stars which have the same properties (same spectra and so on) are intrinsically the same brightness. Comparing the apparent brightness with known brightness makes it possible to estimate the distance of the cluster.

Thus from the foundation stone of the Hyades it is possible to step on the ladder of astronomical distances to other star clusters in our own Galaxy, and from these, eventually, to galaxies beyond.

As well as this, the Hyades cluster is used as a

reference point in the study of the chemical composition of star clusters. Astronomers often refer to other star clusters as more metal rich or metal poor than the Hyades. This is because, being so close, and the stars in it therefore so bright, the Hyades has been well studied from the point of view of stellar evolution. (When astronomers speak of "metals" in a star's composition, they are referring to all elements other than hydrogen or helium.)

The Hyades was in fact one of the first two star clusters studied by Ejnar Hertzsprung at the turn of the century. Its properties made him realize that the stars in a cluster came only in certain combinations of brightness and color, and that this was a property of the structure of stars. The other star cluster which he studied was the PLEIADES. Hertzsprung remarked upon the fact that the Pleiades do not contain any bright yellow stars (yellow giants), while the Hyades contain four. The reason soon became clear. The Hyades stars are ten times older than the Pleiades and some of the stars have exhausted their hydrogen. They are now burning helium. Indeed, from the fact that at least six white dwarfs have been found in the Hyades astronomers deduce that some of its stars have completely exhausted all their available nuclear fuel.

The Pleiades

Seven stars—or six

The individual stars of the Pleiades in the constellation Taurus were so well known in antiquity that the Pleiades has never been recorded as other than a star cluster. In this it is unlike any other star cluster, all of which were first recorded as nebulae before more powerful telescopes separated their stars.

Most people (including Galileo) can see only the six brightest Pleiades stars with the naked eye. In classical literature however there are always seven sisters or seven doves mentioned in the legends connected with the Pleiades, although nine of the stars have names. There has been speculation that the seventh "lost" Pleiad, presumably the star Pleione, has faded since ancient times. This is not certain, however, because the keensighted can see more than six stars there: in 1579 Kepler's teacher Mästlin drew a chart showing 11 of the stars in the Pleiades in their correct places. This was 30 years before the invention of the telescope with which Galileo counted 36 member stars. It is now estimated that there are three to five hundred members, of which 262 were catalogued in 1958. They are clustered 400 light years from Earth and most are contained within a sphere 30 light years across.

Each bright star of the Pleiades cluster is embedded in a blue streaky nebulosity (Plate 12). Unlike those nebulae which shine by emitting their own light, the Pleiades nebulosity is cold gas mixed with dust which reflects the light of the nearby stars. There are two reasons for the blue color of the reflection nebulae. First, the light of the Pleiades stars is itself blue, as can be seen with the eye when the Pleiades is inspected with a moderate-sized telescope. Secondly, starlight reflected off interstellar dust has an intensified blue color, just as sunlight becomes sky-blue when scattered in Earth's atmosphere. The dust in the Pleiades is lined up, presumably by magnetic fields in the region of the star cluster, and so the Pleiades reflection nebulae have a streaky look.

The stars of the Pleiades range from bright blue stars to fainter red ones. This progression in color and brightness simultaneously is called the main sequence on which most stars lie. After the discovery of the same effect in the stars of the solar neighborhood by the American astronomer Henry Norris Russell, the plot of star brightness against star color became known as the Hertzsprung-Russell diagram. It was later recognized that the main sequence of the H-R diagram represents the status of all stars during the majority of their active lifetime while they are converting hydrogen to helium in nuclear reactions in their centers.

The bright blue stars of the Pleiades are so young (50 million years old) they only just show traces of having consumed much of their nuclear fuel, hydrogen. American astronomer Allan Sandage predicted that the Pleiades probably contained two brighter bluer stars in the distant past which have run out of hydrogen fuel and are now white dwarfs; one has indeed now been found.

All the bright Pleiades stars save the one named Maia spin very fast. Typically each star rotates about its axis in two Earth days; the speed of its equator is typically 200 km/second. The centrifugal force at the

Pleiades cluster *Hale*

equator of such a rapidly rotating star is a substantial fraction of the force of gravity there. The effect is to raise the equator of the star and flatten the poles, and if we could view in close-up we would find them noticeably squashed compared to spheres. Indeed, the lowering of the effective force of gravity at the equators of the rapidly rotating Pleiades stars has allowed some of their atmosphere to leak off in a disk of gas reminiscent of the shape of Saturn's rings. Evidence of the disk which surrounds four of them can be seen in their spectra.

Pleione, the seventh "lost" Pleiad, which is often just invisible to the naked eye, threw off shells of gas in 1938 and 1970, probably repeating an episode in its life which was incompletely observed earlier, in 1888. Probably all heavy stars like Pleione go through a series of hiccups as they begin to run out of hydrogen fuel, and eject puffs of gas into space. Heavy stars end their lives as white dwarfs even though white dwarfs cannot be much more massive than the Sun, and throwing off shells may be one way in which heavy stars get rid of their excess matter. The ejected matter mixes back into the gas from which the stars originally formed. The late Otto Struve once stated that more photographs have been taken of the Pleiades than of any other stellar object. Considering their nearness and their youth, poised at a crucial moment of the stellar lifetime, it is not surprising that they are at the focus of observational astronomy.

h and Chi Persei

Perseus double cluster

The stars h and Chi Persei are the two brightest in a striking pair of star clusters in Perseus. The clusters, named for the two stars, are also known as the Double Cluster, visible to the naked eye as hazy patches

h and Chi Persei Double Cluster *Lick*

between Perseus and Cassiopeia. Their catalogue names are NGC 869 and 884.

The two clusters are less than 1° apart and of similar brightness and appearance. One would expect them to be associated with each other, and indeed they are. They both contain the same kinds of stars, the brighter being blue and red giants. Their ages are about the same, though Chi Persei could be younger than h Persei—10 million years old as against 20 million years. They are at the same distance, about 7100 light years, are separated by only 50 light years or so, and have very similar motions in space. h Persei is slightly the larger and contains 3900 solar masses of stars while Chi Persei contains 3300 solar masses.

Dust in the area of the Double Cluster is patchy, and concentrated near the brighter stars in the central regions of the clusters. Curiously, there is little trace of a nebula in this area, although bright, hot stars exist to excite any gas present and make it visible, and the clusters are considerably younger than the nebulous Pleiades. Presumably there is not much hydrogen near to h and Chi Persei. First of all, this is because they are about 500 light years below the galactic plane. Most hydrogen lies in a layer about 600 light years thick centered on the galactic plane. The second reason is that the clusters lie in a region on the extreme edge of a spiral arm. Outside spiral arms the hydrogen density is low.

Double clusters of stars are not uncommon among the younger galactic clusters. This must simply be caused by the way in which a contracting gas cloud can break up into two or more fragments, each generating a young star cluster. Other manifestations of the same phenomenon are BARRED SPIRAL GALAXIES and double stars.

The h and Chi Persei Double Cluster will not survive as a pair for long. They will journey around the Galaxy in 100 million years or so, passing near the gravitational tug of spiral arms and large gas clouds. The effect of the gravitational force on each cluster will be different, and they will drift apart as they age. Significantly, globular clusters (page 135), the oldest star clusters in our Galaxy, are never found as pairs. If pairs had ever formed, they would have been separated by the tidal disruption of the Galaxy as they passed round it, which most have done dozens of times in their 10 billion-year lifetime.

Thus only the youngest clusters, say of age less than 100 million years, can be found in pairs.

Praesepe

The Hyades' twin

Praesepe is a bright, open, star cluster in the constellation Cancer. It is bright enough to be clearly

Beehive cluster POSS

visible without a telescope, and the keen-eyed have from ancient times recognized it as a cloudy patch. Hence its macabre Chinese name, meaning "The exhalation of piled-up corpses" and its friendlier western names of the Beehive, and the Crib. In 1609 Galileo Galilei, the first scientist to record the sky as seen through a telescope, found to his astonishment that Praesepe contained "a multitude of more than 40 small stars." More than 200 members of Praesepe are now known, isolated from the chance stars in the same direction by the fact that all the cluster's stars are moving together across the sky.

Praesepe is a cluster very like the HYADES star cluster across the Milky Way in Taurus. Its brightest stars are only of sixth magnitude, considerably fainter than the brightest stars of the Hyades which are of third magnitude. This is because Praesepe, at 500 light years, is four times farther away than the Hyades. For the same reason Praesepe appears only one-quarter of the diameter of the Hyades. Praesepe stars are found up to 0.75° from the cluster center, while Hyades' members swarm out to 3° before it becomes too difficult to pick them from the background stellar multitude. Both clusters are old enough for all their hot blue stars to have exhausted all their nuclear fuel and to have become white dwarfs.

About 20 white dwarfs are expected in Praesepe, having evolved from the same number of massive stars, the brightest of which, when it existed at the birth of Praesepe, would have outshone any star in our present sky. One small difference between the two clusters, for which there is no generally accepted explanation, is that Praesepe contains fewer double stars than the Hyades. Only 20% of the Praesepe stars are double, whereas as many as 70% of the Hyades stars have companions.

The two clusters are so alike that one must suspect

that they shared a common origin. The member stars of both the Praesepe and Hyades clusters are all moving through space at the same speed (47 km/second) and in the same direction (towards a point near Betelgeuse, the brightest star in the constellation of Orion). Both clusters "remember" the motion of their parent interstellar gas cloud, from which they were born as twins, 40 million years ago. Like h AND CHI PERSEI Praesepe and the Hyades form a double cluster.

NGC 3324

Another Jewel Box

Discovered by the Abbé Nicolas-Louis de Lacaille in 1751 on his astronomical expedition to the Cape of Good Hope, the Jewel Box is NGC 4755, an open cluster of stars 7800 light years from the Sun. It was named the Jewel Box by Sir John Herschel because of the startling contrast in color between its brightest star, Kappa Crucis, and the remainder. The color of Kappa, easily seen by eye with a modest telescope, is brilliant red; the remainder, which give various impressions of color, are all more or less blue when measured. We show in Plate 12 another jewel-box-like cluster, NGC 3324, which shows the same distinction of color between the brightest star and the remainder.

The reason for this distinction is that each cluster is a young cluster of stars, somewhat younger than the PLEIADES and a little older than the DOUBLE CLUSTER in Perseus. Each is at a stage when all its brighter, hotter stars (the blue ones) have just exhausted their central supply of hydrogen. For the past 30 million years the hotter stars in each cluster have been converting the hydrogen in their centers to helium, and now they have little of this nuclear fuel left. Less hydrogen is available to sustain the rate of energy production needed to hold each star up against its own gravity. The stars have become unbalanced. Their centers are releasing energy by contracting, but their outer parts are expanding.

Paradoxically, although they are reaching the end of their supply of fuel, the contraction causes them to shine brighter than ever before. They become so much brighter that astronomers call them giants. At this stage they are still hot and so shine with a blue light; all save one of the bright stars in each cluster are blue giants. The very brightest star has passed this point. It has precociously expanded beyond the blue giant stage, so much that it has cooled, dropping in temperature from some 40,000° to 3000°. However, the contraction of its central core still maintains its total power output to some extent, and the star has remained as bright as the other blue giants. Indeed, since the blue giants radiate so much of their energy as invisible ultraviolet light, it seems brighter than any of the other stars in the cluster. It therefore remains the brightest ruby in a jewel box of sapphires and diamonds. Such bright red stars, known as red supergiants because they are so much bigger and brighter than other red stars, are comparatively rare, because the transition stage to red supergiant and beyond is a brief passage in the lifetime of a star.

NGC 3324 AAT

Dusty Nebulae

Hubble's Variable Nebula

Shadow play

During the 18th and 19th centuries, the task of many astronomers was to scan the skies in ever-increasing detail, charting the objects they found. Occasionally they would come across what some of them imagined were "holes in the heavens"—apparently starless fields in the Milky Way, which they imagined were caused by fast-moving star clusters tearing the fabric of the Milky Way.

It was the American E.E. Barnard and the German Max Wolf who proposed the real explanation. The starless fields are places in the sky where dark clouds were silhouetted against the stars beyond.

These dark clouds are composed of dust. Its existence was deduced from the light of the stars which it obscured, but occasionally it is itself well-enough illuminated by close stars that it reflects light, just as individual motes of dust can be seen in sunlight shafting through a window. Indeed, dark nebulae and reflection nebulae have long been known to occur together. Sir William Herschel noted the dark areas which he saw while "star gaging," sweeping his telescope across the sky. He warned his sister, Caroline, to be prepared to note down bright nebulae when he saw these "holes" because he had learned from experience that bright nebulae often appeared in association with them.

One nebula which Herschel discovered, NGC 2261, has become known as Hubble's Variable Nebula. It has the distinction of having been the first object photographed by the 200-inch Palomar Telescope, and is a triangular reflection nebula pointing at the star R Monocerotis (abbreviated to R Mon). The "star" is actually the bright tip of the nebular triangle, and the whole has a generally comet-like appearance. R Mon appears perceptibly non-stellar viewed through large telescopes at moments when the Earth's atmosphere is particularly steady and the celestial images are crisp. Evidently the real star is hidden deep within a dusty cocoon which we see as a small, bright nebula.

However, on most occasions, R Mon has the appearance of an 11th-magnitude star, and that is how it appeared to Schmidt, an astronomer in Athens who discovered in 1861 that the star was variable. It varies by four magnitudes in an irregular way. Most changes are drops in brightness reminiscent of R CORONAE BOREALIS, but some are flashes of light as if more starlight were briefly showing through a veil.

R Mon has a distinctive spectrum (including some emission lines), very similar to that of the surrounding nebula. Astronomers infer from this similarity that the nebula is a reflection nebula, composed of dust, which simply reflects the light from R Mon, rather than absorbing light and reemitting it at a different wavelength.

The rare property of the nebula NGC 2261 is that it too is variable. Edwin Hubble discovered this from photographs taken at several observatories between 1900 and 1916. He saw that the nebula varied in shape and brightness, and since then it has been known as Hubble's Variable Nebula. C.O. Lampland assembled a collection of almost 1000 photographs of it, the last one taken only the month before his death in 1951. According to Lampland the entire nebula has at one time or other undergone changes ranging from mild dimming to complete obliteration of detail.

Hubble had thought that the changes were caused by the motion of lumps of nebula; Lampland disagreed, remarking on the correspondence of delicate details on different photographs. He was convinced that the photographs showed a permanent framework swept by waves of luminosity and obscuration.

What then is the cause of the changes? Because R Mon itself is variable, the light reflected from R Mon off the nebula should vary too. The two do not, however, seem to vary together. Over the 30 years to 1966 R Mon gradually increased in brightness, while, if anything, the nebula faded. Of course the reflected light from the nebula has had to travel from R Mon to the dust before setting out on its journey to Earth. Thus the star and nebula would inevitably be out of step by the length of time that light takes to cross the nebula. It is, however, unlikely, that the nebula is as large as 30 light years in diameter.

Hubble's Variable Nebula KPNO

More probably, the dense clouds of dust in the vicinity of R Mon pass across in front of the star causing the light which we see coming from it to dim and brighten. The same clouds, or others associated with them, jostle in orbit round R Mon, also passing between R Mon and the nebula and causing shadows to play across it like a slide show.

The nature of the projector in the slide show, R Mon itself, is a separate mystery in the large enigma of NGC 2261. It appears to be similar to RW Aurigae variable stars. These are often found near nebulae and dust and are thought to be similar to T Tauri stars in that they have recently formed from the interstellar material. R Mon itself is cocooned in dust which it heats and which then emits copiously in the infrared part of the spectrum.

Is the leftover dusty gas still falling onto the recently formed star? Or has the star overdone its collapse so that it now needs to rid itself of matter and is spewing the excess back off into space?

Possibly both of these ideas contain a germ of truth. The emission-line spectrum of R Mon hints at the ejection of gas in a disk around the star—an embryo solar system, like MWC 349. The dusty clouds which shadow the nebula could be left over after the collapse of the star from the interstellar gas.

Evidently, through the dusty shadows, we dimly see a star recently formed here on the edge of NGC 2261.

The Footprint Nebula

Donut nebula

The late 1970's witnessed the discovery of about a dozen very small nebulae of a type scarcely recognized before. They have become known as bipolar nebulae. The Footprint Nebula, pictured here, is a typical example of these objects.

All bipolar nebulae share well-defined characteristics. They are small pairs of nebulae, frequently less than 10 seconds of arc in total extent. Most are mistaken for stars on Schmidt photographic survey plates, and their true nature becomes apparent only when examined closely. All shine by reflected light, and as most are bright, there must be a very bright star illuminating them. In no cases is that star seen. Intense infrared radiation is a third characteristic of bipolar nebulae.

The Footprint Nebula was the first of its kind to be discovered, in 1946 by Mount Wilson astronomer Rudolph Minkowski. Its name was given by George Herbig, who was the first person to account for its shape.

Herbig's explanation of the Footprint Nebula applies equally to all other examples. The illuminating star lies at the very center, between the two bright blobs. It is hidden from our view by a particularly dense cloud of dust and gas. This cloud forms a donut-shaped ring around the star. There are two directions from which light can escape—the two poles of the donut, for in these directions the nebula is more tenuous. Light is funneled along the polar directions and illuminates the more tenuous portion to generate the two nebulous blobs. In the same way a lighthouse beam illuminates only a small portion of a fog into which it is shining. The dust of the donut which serves to hide the star from us becomes heated by the starlight. Temperatures of a few hundred degrees are the norm. Dust at those temperatures emits all its radiation in the infrared, which accounts for the third characteristic of the bipolar nebulae.

Because the illuminating star is hidden, astronomers cannot easily study the bipolar nebulae. Thus their distances are not well known. Estimates of the distances of the Footprint Nebula range around a few thousand light years.

HD 130079

Yardstick to a dark globule

Patches of obscuration occur in all shapes and sizes, but are always recognizable when something luminous lies behind. In this case the background is formed by distant stars, and the dark material occupies an elliptical area through which few of them are seen.

Isolated dark nebulae such as this are often found on photographs of star fields. It is difficult to learn much about them because their distances are generally unknown. In this example, however, the distance can be determined. At the eastern (left-hand) end the globule is illuminated by a bright star, and forms a reflection nebula to that star. Although included in catalogues as HD 130079, prior to the discovery of its

Footprint Nebula *Lick*

HD 130079 and a globule ESO

association with the dark nebula, this star had received little attention. Indeed, it is one of the faintest stars in the Henry Draper catalogue. The spectrograph reveals that it is a star very like SIRIUS. Foreground dust slightly dims HD 130079, but its faintness is mostly an effect of distance. Since it appears 23,000 times fainter than Sirius, it must be 150 times as distant (the brightness varies with the square of the distance), and thus about 1420 light years away. This, then, is the distance of the dark nebula. From the apparent size of the nebula, eight by five arc minutes, its linear dimensions can be calculated as three by two light years, or about 1000 times the diameter of the solar system.

Counts of the star density both inside and outside the dark nebula give an estimate of its obscuration. At its densest, this nebula dims the background stars by at least a factor of 400. Making plausible assumptions, this can be converted to a mass of dust producing the obscuration: the result is one-eighth the mass of the Sun. Since there is usually about one or two hundred times as much gas as dust in nebulae, the whole complex probably weighs 10 or 20 times as much as the sun. Its density is 10^{20} times less than that of an average star.

Isolated dark nebulae such as this are often called Bok globules after the astronomer Bart J. Bok, who first drew attention to them. The Bok globules associated with H II regions have similar sizes and densities to this isolated example. They are not stable, but must slowly contract, typically taking a million years. Such a collapse would be expected to produce one or a few stars, possibly a little heavier than the Sun, and initially shrouded in a reflection nebula. It seems very unlikely that HD 130079 has formed within so tenuous a globule. Rather, astronomers believe that the globule and the star have followed courses around the Galaxy which, by chance, brought them close enough together at the present time for the star to act as a convenient yardstick to the globule.

HD 155578 UKSTU

Giant Reflection Nebula

Reflecting the Milky Way

Reflection nebulae, by their very nature, tend to be rather small objects and to huddle close to relatively bright stars. It is thus unusual to find a reflection nebula covering fully 10° of sky and associated with no bright stars. This photograph shows part of such a nebula.

That this object is a nebula is undeniable. That it is not an H II region is shown by its absence on photographs taken in the principal hydrogen emission line, H-alpha, and by spectroscopy of the brightest portions, which also reveals no emission lines. Reflection of light remains the only credible explanation for the nebula's visibility.

The star HD 155578 (arrowed) is near a locally bright portion of the nebula, so it is highly likely that the nebula lies at the distance of this star. From the brightness and spectral type of HD 155578 we get a distance of 250 light years, which means that the nebula lies about 100 light years below the general plane of our Galaxy. This places it well within the spread of stars which form the disk of the Galaxy. HD 155578, and other stars at 100 light years from the galactic plane, are quite inadequate to illuminate this nebula. Instead it reflects the light of the whole Galaxy, and in particular of the stars in the plane immediately above it.

An exactly parallel analogy occurs on Earth. From the outskirts of a large city one often sees bright clouds at night. These lie above the city and are illuminated by all its lights. No single light is responsible for the brightness of the clouds.

The dust which reflects light back to us also obscures the stars beyond. By counting the number of stars seen through the nebula and the number in adjacent, clear portions of sky, astronomers can estimate how much dirt lies in the nebula. This amounts to a few times the mass of the Sun. If the normal quantity of gas coexists with the dust, the total mass of the nebula is several hundred times that of the Sun, spread over an area of about 120 square light years.

The Thumbprint Nebula

Bright dark globule

The Thumbprint Nebula, discovered by M.P. Fitzgerald, got its name from its ambiguous appearance on his photograph. It is an unusual combination of a bright and dark nebula. Actually, the distinction between a bright and a dark nebula is a fine one. A cloud of dust which is dark when viewed from one side may appear, like for instance THE CALIFORNIA NEBULA, to be a prominent reflection nebula to an observer on the other side. Thus bright and dark nebulae are frequently found side by side. But individual small nebulae, such as the Thumbprint, which so clearly show both characteristics, are rare.

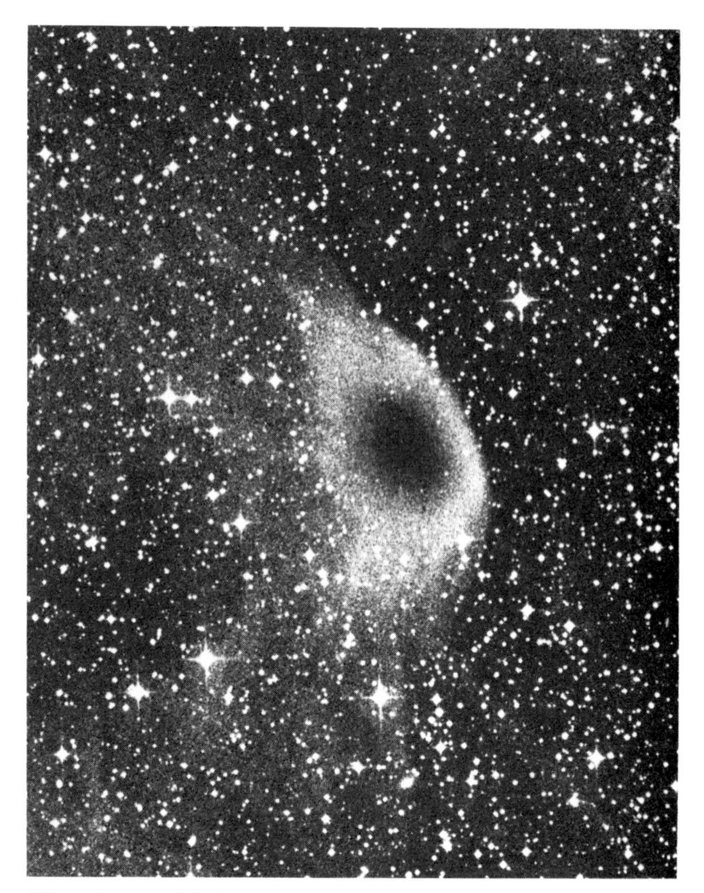

Thumbprint Nebula UKSTU

The center of the Thumbprint looks extremely dark. In part this is a contrast effect, for measurements show it to be a few percent brighter than the surrounding sky. The Thumbprint lies well below the plane of our Galaxy, and must therefore be illuminated by light from all the stars above it in the galactic plane. It is one prominent feature in an extensive network of bright and dark nebulae, most of which are so illuminated, and represents a locally dense portion of a giant agglomeration of gas and dust.

An estimate of the distance of this complex can be made by observing stars at various distances from the Sun. The distance of a star can be estimated by comparing its brightness with that expected for its spectral type. The amount of intervening dust can be determined by the reddening effect it has on the star's light. Thus stars act as probes of dust, and by examining a large number of stars within a few degrees of the Thumbprint it can be shown that the dust complex begins about 1300 light years away. It is assumed by astronomers that the Thumbprint lies at that distance, in which case it is about 450 light years below the galactic plane.

By measuring the brightness of different portions of the Thumbprint, astronomers can determine many of its properties. The dust grains which produce both the reflection and the absorption are found to become much more numerous towards its center, and this suggests that it is collapsing under its own gravity and may one day form a small group of stars. Its present dimensions are 1.6 by 2.6 light years, and the mass of gas and dust it contains is about ten times that of the Sun. It is therefore comparable in size to the Bok globules, and is distinguished from other examples principally by the illumination provided by the stars of our galaxy above it.

Terrestrial clouds sometimes exhibit the same appearance, though their geometry differs because they are illuminated by a single bright source, the Sun (or Moon). When the Sun lies behind a dense cloud it gives the cloud bright edges because light can diffuse through the thin, outer portions. The Thumbprint is the celestial analogy of a cloud with a silver lining.

IC 2220

Toby Jug Nebula

This photograph shows the reflection nebula IC 2220. An observer of this nebula might be struck by its similarity in shape to a tankard of a type called in Britain a Toby Jug, but it has yet received no nickname in the literature.

IC 2220 is distinctive not only for its shape. Almost hidden at its center, but visible on this photograph because of the four diffraction spikes which project from it, is the illuminating star. This is known by its unglamorous catalogue designation HD 65750, and it lies at a distance of about 1300 light years.

The nebula is unusually bright to be reflecting HD 65750 alone, but there is no other source of illumination. This suggests that the dust which gives rise to the reflection may be of a different kind from that in most nebulae. This possibility is made more credible by the unusual nature of HD 65750. The central stars of reflection nebulae are usually young, hot and blue; HD 65750 is old, cool and red. It is the type of red star which is losing gas from its outer layers in a steady trickle. Dust grains rich in silicate minerals are condensing from the gas as it cools, and it seems possible that most or all of the dust and gas of the nebula have been produced from material lost by the star. The process is called mass loss.

There are large quantities of dust swirling around the star. As thicker portions pass in front, HD 65750 is sometimes dimmed by a factor of up to two. The irregular nature of this variation is what indicates that it is not the star itself which is varying.

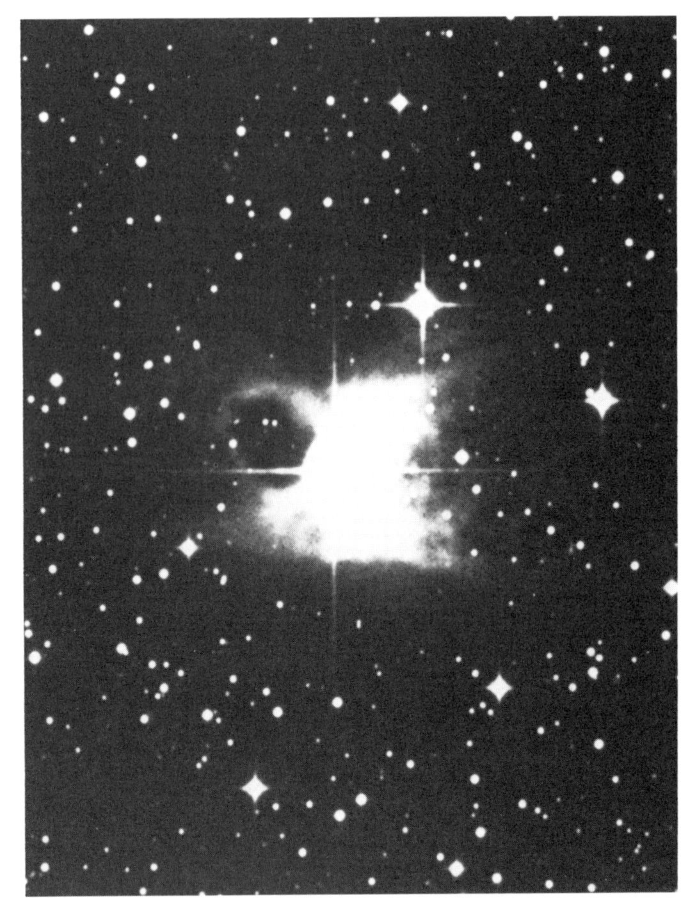

Toby Jug Nebula UKSTU

The nebula measures 1 by 0.3 light years. The gas travels at about 10 km/sec, and therefore requires about 100,000 years to cover the distances involved. The mass of IC 2220 is about three-quarters that of the Sun. Red stars like HD 65750 are known to lose material at rates near one solar mass in 100,000 years, which is consistent with the other estimate for the age of IC 2220. However, such stars are not normally expected to lose so much mass. HD 65750 and IC 2220 appear to represent an extreme example of mass loss.

Horsehead Nebula

Dark dusty droplet

The Horsehead Nebula is a dark dust cloud in the shape of a horse's head and neck which protrudes into the bright emission nebula known as IC 434 in the constellation Orion. It is virtually opaque, as there appear to be no stars visible through it. IC 434 emits light because of the energy input into it by the nearby star Sigma Orionis. The color picture (Plate 13) shows that IC 434 emits red H-alpha light, in contrast to the blue reflection nebula in the dark half of the picture.

Zeta Orionis (easternmost star of Orion's Belt) is brighter than Sigma and nearer the nebula, but is less effective in emitting the ultraviolet light which causes nebulae to shine. IC 434 is expanding and colliding with a large cloud of dust which fills the eastern quadrant of the constellation of Orion (dark half of picture). The presence of the dark cloud can be seen in the way it cuts down the number of stars visible behind it (compare the two halves of picture). The boundary of the cosmic collision between dark cloud and nebula can be seen as the bright rim of IC 434 which abuts onto the dark cloud (center, across picture). In the collision, droplets of the dense dark cloud are penetrating the less dense, lighter nebula, like oil poured into vinegar salad dressing. The Horsehead is the largest known of these elephants' trunks of dark material. At a later stage the droplets detach from the main body of the dark cloud, when they are called globules. Some astronomers think that this may be the first stage in the formation of some stars, though others believe that the globules will eventually disperse into interstellar space.

The American astronomers Henry Curtis and John Duncan used the Horsehead as the first proof that the dark nebulae were not due simply to the absence of stars. Said Curtis in 1918: "It is impossible to look [at the Horsehead] and not be convinced that there is something dark between us and [the bright nebula]."

The Coalsack

Apparently black

Ingenuous astronomy students have been known to mistake the 4° diameter circular patch in the constellation of the Southern Cross for an isolated cloud silhouetted black against the Milky Way, for that is the appearance of the Coalsack. It is, in fact, a dark nebula of dust absorbing the light of stars beyond. Its striking appearance arises because of its relative nearness, at 550 light years, and because all of it lies in front of a

Horsehead Nebula UKSTU *(color plate 13: Hale)*

The Coalsack DFM

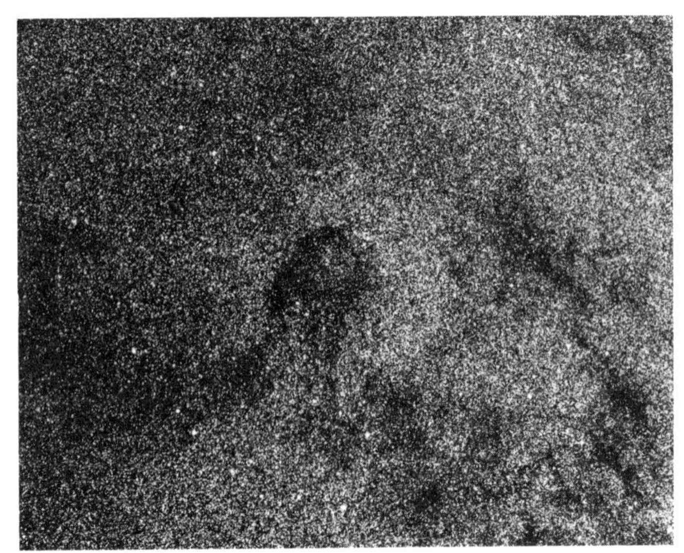

Parrot's Head POSS

bright area of the Milky Way (Plate 12).

According to recent studies, the Coalsack is actually a complex of small, dark clouds in close proximity, including some particularly small, dense globules. It also contains some COMETARY GLOBULES (p. 132).

In 1970 K. Mattila of Heidelberg used the Coalsack to prove what E.E. Barnard suspected from visual observations since 1919, that the so-called dark nebulae are not in fact dark clouds but "are a dull, feebly luminous mass." They appear black only because they extinguish the light of stars behind.

Some of the starlight falling on the Coalsack is absorbed by the dust grains in the cloud. However, nearly all the incident starlight is reflected by the dust grains, although it no longer comes directly from the stars. Thus the stars behind the Coalsack are dimmed, while the Coalsack itself shines with a feeble glow. According to Mattila, the Coalsack itself has about 10% of the brightness of the Milky Way. Nevertheless, the contrast between the Coalsack and the surrounding star clouds gives the impression that it is darker than any other part of the sky.

Barnard 87

The Parrot's Head

E.E. Barnard found 349 dark nebulae in the course of photographing the Milky Way. He nicknamed number 87, in Sagittarius, the Parrot's Head. The beak points left and the bright patch of stars, including the one catalogued CD−32° 13679, represents the parrot's eye.

B 87 is the most prominent dark nebula in the area but it is just one of a chain which have a generally ropelike appearance.

Many nebulae appear like this, forming crescents, S-shapes and arcs. The ropelike nebulae in a given area have a tendency to run parallel with each other, forming patterns much like iron filings placed near a magnet. Some astronomers believe that the dark nebulae in fact align with the magnetic field of the Galaxy.

In the Sun's neighborhood, the galactic magnetic field runs towards the constellation Aquila, roughly parallel to the spiral arms. This magnetic field manifests itself in several rather subtle ways. For instance, radio astronomers see a kind of Milky Way of radio waves in which individual objects like nebulae or supernova remnants are seen against a background of radio emission. This radio Milky Way is apparently caused by stray electrons orbiting in the general galactic magnetic field. In another manifestation of the field, starlight is polarized. Your view of the Galaxy while wearing polarizing sunglasses is fractionally different when you tilt your head, but in practice the difference requires careful measurement to detect it. The polarization is caused by the passage of the starlight through needlelike dust grains which align themselves in the Galaxy's magnetic field, spinning like paper windmills on a piece of string over a used-car dealer's stock.

These independent lines of evidence suggest the presence of a magnetic field, and agree on its direction. The long nebulae which Barnard found seem generally to run parallel to it. There is no satisfactory explanation for this.

Cometary Globules

Tail of a dark globule

Sixteen of these so-called cometary globules are presently known, nearly all on the outskirts of the GUM

NEBULA (p. 115). Most were discovered quite recently by Edinburgh astronomers Peter Brand and Tim Hawarden on photographs taken by the U.K. Schmidt Telescope. Each is an elongated, comet-like object with a small dusty head which is almost or completely opaque (white on this negative picture, means emitting no light). A faint luminous tail up to a degree in length trails from one side of the head. The other side of the head is bright, presumably heated by a star within the Gum Nebula (Gamma² Velorum). The tails of all the cometary globules point away from the center of the Gum Nebula; each represents material dragged off the head by a tide of gas flowing past it.

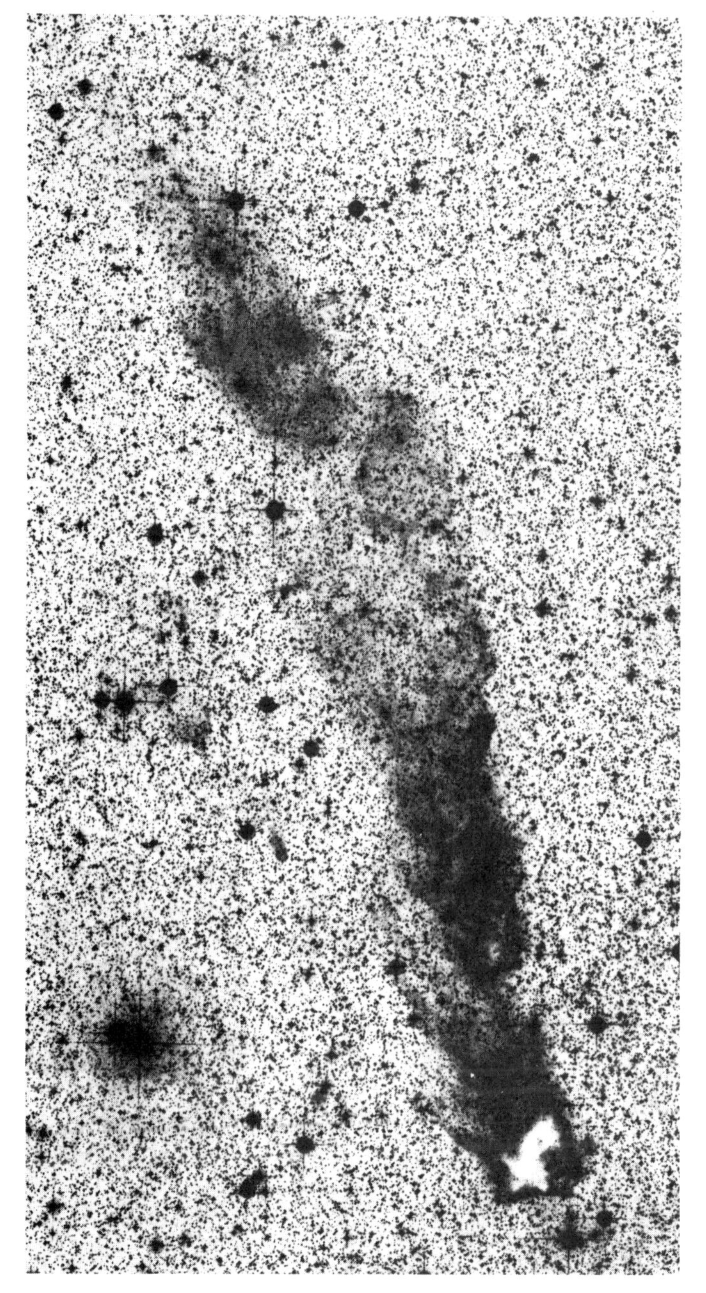

Cometary globule UKSTU

Rho Ophiuchi Dark Cloud

Dense dust

Near the star Rho Ophiuchi lies the densest part of an opaque dark cloud which blots out all the stars behind. The cloud is amazingly large—its effects can be traced over an area of 1000 square degrees, which is 2% of the whole sky! It is a mixture of reflection nebulae, dark clouds and bright nebulae. The star to the south of Rho Ophiuchi, HD 147889, is embedded in a reflection nebula, while Sigma Scorpii, farther to the southwest (lower right), is hot enough to heat the gas and make it glow. The reflection nebulae show the same streaky appearance as the PLEIADES nebulosity (p. 120).

The distance of the Rho Ophiuchi dark cloud is 700 light years. This is relatively close to us, as might be expected from the large size of the cloud. Most of the bright, blue stars in the area within the constellations Scorpius and Ophiuchus lie in the same region of space and are part of the same stellar association, which itself is part of GOULD'S BELT (p. 79) on our interstellar front porch.

The dust is amazingly patchy. In parts it is virtually opaque, while just to one side, objects like the globular cluster MESSIER 4 can easily be seen in the distant background (5700 light years). The densest region lies near to the star HD 147889. According to optical studies, the light of the fainter stars visible through the murk is diminished by a factor of 10,000. However, these studies were carried out in 1955, before the arrival of infrared astronomy. Optical astronomers can only talk about the faintest stars they can see, while observations in the infrared can detect stars whose visible light is totally absorbed by the intervening dust. Infrared emission penetrates dust more easily than light can, and over 40 very heavily absorbed stars were detected in a Kitt Peak Observatory survey made by Frederick Vrba, Karen and Stephen Strom and Gary Grasdalen. These stars appear to form a cluster. Their record-breaker was called VS-17 and is an intrinsically bright star whose light is attenuated by an astounding factor of 10,000 trillion! The age of the cluster is not great, perhaps several million years, and presumably it will drift out of the cloud eventually, perhaps after a galactic rotation of 100 million years, or after a supernova has purged the dust. Meanwhile, if life has evolved quickly on a planet orbiting a star within the cloud, the astronomers of such a race will have an unobstructed view of their own sun and only an attenuated view of the nearby stars of the cluster. Stars outside the cloud will be invisible except by infrared techniques. But, given little stimulus by the nearly empty sky, no one might think to apply these techniques to astronomy.

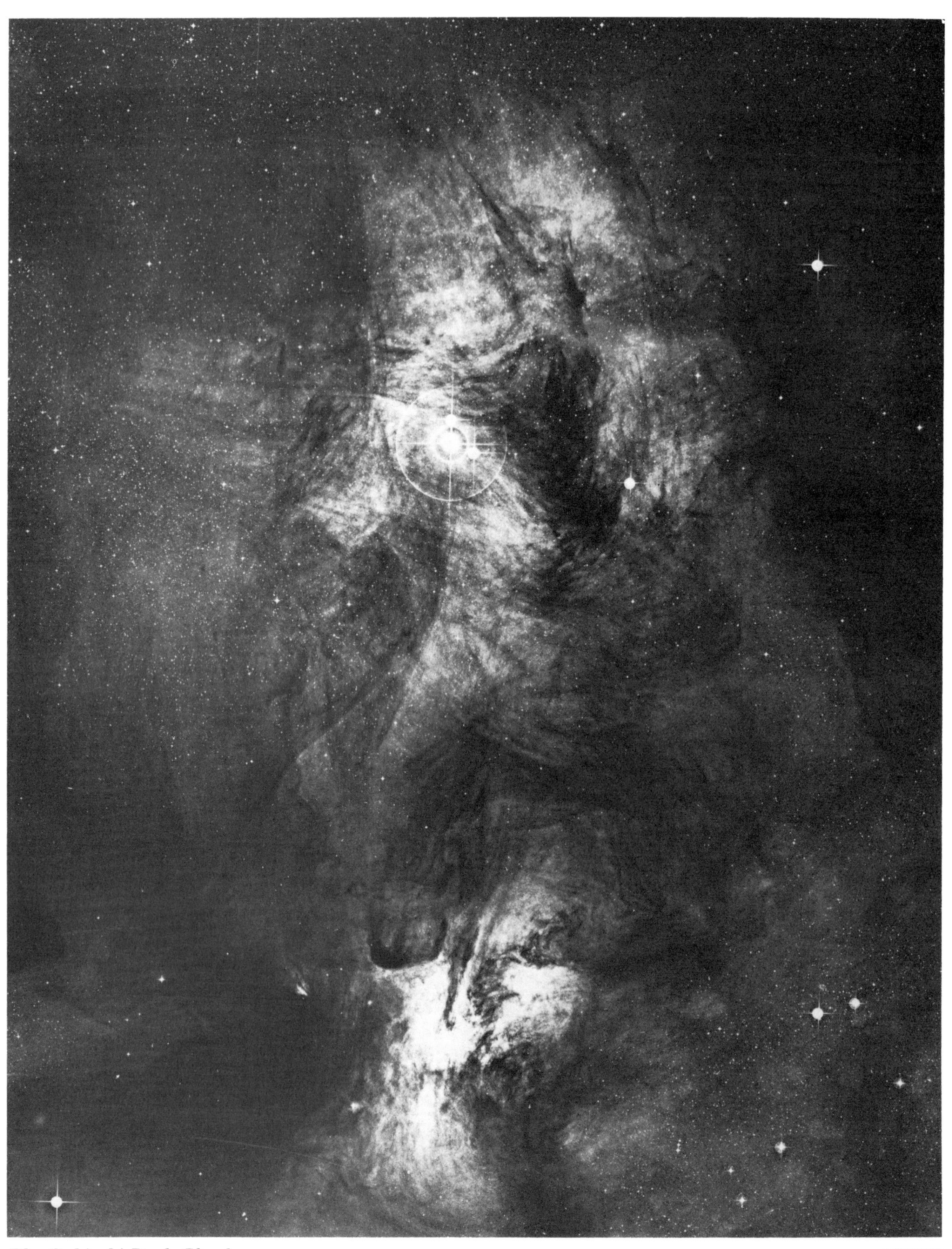

Rho Ophiuchi Dark Cloud UKSTU

Endpoints of Star Evolution

Omega Centauri

Largest diameter globular cluster

Messier's catalogue lists 26 clusters having a characteristic globular shape. This is an astonishingly high fraction (20%) of the known globular clusters.

The reasons why 18th-century observers were so successful in picking out the globular clusters are as follows. In the first place the globular clusters consist each of hundreds of thousands, or maybe literally millions of stars packed so close that they appear to touch when viewed in a small telescope. The clusters are therefore bright and easily recognizable. Secondly, globular clusters contrast with open clusters which concentrate to the Milky Way and can be difficult to pick out from the background and foreground stars. Rather, many globular clusters are seen out of the Milky Way band and are conspicuous against the emptier, dark, intergalactic sky.

Globular clusters are remarkably similar to one another. This is because they are so old. Any differences which they might have had at their formation have been smoothed out in the course of time. Yet behind their similar faces, some have distinct and individual personalities which we show in this representative selection. We start with the grand example of Omega Centauri.

Omega Centauri sounds like the name of a naked-eye star, but it is actually one of the two brightest globular clusters, along with 47 Tucanae. It is in the southern sky and does not rise above the horizon when seen from the latitude of northern Europe. It rises only about 10° at the latitude of the Mediterranean or the southern U.S. Thus Ptolemy and other northern hemisphere astronomers perceived it, dimly and blurrily, as a fifth-magnitude star above their southern horizon. But to the resident of Australia, South Africa or South America who sees the bright center of the globular cluster, Omega Centauri is not point-like, but has a perceptible shape. The young Edmond Halley first recorded its nonstellar appearance during an expedition to St Helena in 1676. Its number in Dreyer's *New General Catalogue* is 5139.

With a large telescope, stars belonging to the cluster can be seen extending out to more than 1°. The total magnitude of the globular cluster is 3.6, and its distance is 16,500 light years. It could be the most massive globular cluster in the Galaxy: it certainly has the largest diameter of any globular cluster known, 620 light years.

Apart from its size, Omega Centauri is an unusual, even unique globular cluster in other ways. It is distinctly nonspherical. Overall, it is an ellipse with its axes in the ratio of about 5:4. Astronomers believe this shows that Omega Centauri is rotating about its axis more rapidly than is usual for globular clusters.

Omega Centauri is one of the oldest globular clusters known, being 15 billion years old. The Universe has doubled in size during the lifetime of this cluster. Astronomers believe that it is because of its age that Omega Centauri's stars have such featureless spectra compared with most stars in our Galaxy. The elements which cause spectral lines had not been made in any abundance when Omega Centauri formed out of a primordial cloud of gas. Measurements show that generally its stars contain only a few percent of the "metal pollution" present in a star like the Sun.

The brightest stars of Omega Centauri are red giants of 11th magnitude. When examined with a telescope as large as the 150-inch Anglo-Australian Telescope, such stars have the same apparent brightness as Venus seen with the naked eye, and their red color is immediately apparent.

Sprinkled in among the mass of red stars are white ones which, in globular clusters, are known as "horizontal branch stars." In Omega Centauri, the horizontal branch stars are unusually varied in their spectra. Apparently, some of them are churning deeply into their nuclear reactor interiors, and are bringing

Omega Centauri AAT

new elements up to their surfaces where they show up in the spectra. Perhaps the stars which do this are especially rapidly rotating. Mt Stromlo Observatory astronomers Mike Bessell and John Norris speculate that if this is so, the proportion of rapid rotators in Omega Centauri is connected with the rapid rotation of the cluster as a whole.

One problem in studying Omega Centauri is that it is not far from the Milky Way and therefore is seen through large numbers of confusing stars unrelated to the cluster. These so-called field stars have been weeded out from the stars which belong to Omega Centauri by using the fact that all stars in the cluster are moving together whereas the field stars are moving at random. Lists of genuine cluster members have been prepared at the Royal Greenwich Observatory by a large team, which included almost a generation of British student astronomers at one time or another. A by-product of the weeding process was that the team measured the proper motion of Omega Centauri more accurately than that of any other globular cluster. It was then possible for Richard Woolley and Mike Candy to calculate its orbit around the Galaxy. This turned out to be a retrograde motion, that is, it moved against the general rotation of the spiral arms of the Galaxy. The cluster loops in a three-petalled roseate pattern with perigalacton (closest approach to the Galactic Center) being 6200 light years. Apagalacton (farthest point from the Galactic Center) is 21,000 light years.

This orbit is typical of globular cluster orbits in that they jaywalk across the orderly flow of most galactic traffic. This accounts for the most significant difference between globular clusters and galactic clusters: globulars have no gas. Any gas that they had originally or that their stars have ejected has been swept out by the rest of the Galaxy as they have repeatedly looped through it. They have not been able to produce new stars for a billion years. In effect, they are progressively sterilized on each pass through the Galaxy.

47 Tucanae

PIE in the sky

In theory, 47 Tucanae should be the name of a star, but 47 Tucanae is a globular cluster: like OMEGA CENTAURI, it was thought by early observers to be a star of the fourth magnitude until Lacaille, during a southern expedition in 1751, saw that it was non-stellar. It is catalogued NGC 104. Being in the far southern skies it is never seen from northern observatories. Australian, South African and Chilean observatories have studied it well, however, as it is a particularly clearly visible globular cluster.

47 Tucanae UKSTU

For a globular cluster, 47 Tucanae is relatively nearby at 13,000 light years. Its position, well away from the Milky Way band, has two useful consequences for astronomers. The first is that not much interstellar dust lies in front of it, because dust concentrates in the Milky Way. The second is that the number of unrelated stars seen in projection against it is small, (compare the field stars on our picture with the number on that of Omega Centauri), and therefore there is little confusion.

47 Tucanae has turned out to be an unusual globular cluster in several ways. It has at most three RR Lyrae stars in it, in contrast to Omega Centauri which has more than 150. Secondly, there is evidence that it contains more galactic metal pollution than typical globular clusters like Omega Centauri. Metal pollution is measured in terms of the abundance of metals in the HYADES star cluster. Omega Centauri's metal pollution is only a few percent of that of the Hyades, while 47 Tucanae contains about a third of the quantity of metals that the Hyades have.

If 47 Tucanae were a particularly young globular cluster, astronomers could understand its high metal pollution. It would mean that 47 Tucanae formed out of interstellar gas which had been enriched with metals from previous supernova explosions, while other clusters formed out of relatively pure hydrogen left over from the Big Bang. 47 Tucanae does indeed show some evidence that it is one of the younger globular clusters—a mere 10 to 13 billion years old! This may be about half the age of Omega Centauri, although astronomers are not too confident of their ability to judge the age of objects as old as this.

It is still surprising that the metal pollution in the Galaxy could increase from zero to a third of its present value so quickly. However, there are other indications that this did occur and astronomers refer to the theory that it did as Prompt Initial Enrichment, or PIE, in the sky.

Messier 13

Great Globular Cluster

M 13 is NGC 6205, the Great Globular Cluster in Hercules, the largest globular cluster in the northern sky. It was discovered in 1714 by Edmond Halley, who noted that it could be seen with the naked eye. The cluster contains perhaps half a million stars.

M 13 shows the slightest traces of departure from the generally spherical shape from which the globular clusters get their name. Why globular clusters are usually so spherical was until recently a puzzle. Most collections of stars in the Universe are rotating and are distorted into ellipsoids which are either oblate (flattened at the poles like a tangerine, or like the Earth), or prolate (pointed along the polar axis like a cigar). In ELLIPTICAL GALAXIES (p. 14) the ratio of the short diameter to the long reaches 3.3 to 1.

In S0 galaxies (p. 17) and in SPIRAL GALAXIES (p. 16), which are extremely oblate, the long diameters are up to 20 times the short diameter, the latter being the rotation axis of the galaxy. The most flattened globular cluster, M 19, has long and short diameters in the ratio 5:3. Most, however, are barely distinguishable from spheres.

Why then do galaxies show elliptical forms while the rather smaller globular clusters do not? Theoretical astrophysicists sketch out a scenario for the formation of a group of stars in which the stars are made from a contracting gas cloud. Even if the gas cloud rotates very slowly when it is large, it will rotate

Messier 13 POSS

quickly when it contracts as a result of the need to conserve its angular momentum. This will tend to flatten the star system into a disk. Some stars in the group will be wider-ranging than others and may be thrown out of the cluster. Random encounters between stars will cause others to escape. These stars carry off the angular momentum or rotation speed of the group. In time the group will then, in the astrophysicists' term, relax into a more spherical shape. The more closely packed the stars in a group are, the more quickly the relaxation will occur.

In globular clusters there may be a density of about one star per cubic light year. The time needed for a globular cluster to relax is calculated at 100 million to a billion years. As the globular clusters are billions of years old, there has been plenty of time for any clusters which were elliptical to change to their present globular shape. However, the time for a galaxy to relax is longer than the age of the Universe because galaxies are less densely packed with stars than globular clusters, with perhaps one star every five cubic light years. Thus galaxies have generally not relaxed and they show the variety of elliptical and disklike shapes which we have seen.

NGC 6624

X-ray globular cluster

NGC 6624 is a dense cluster of stars of roughly spherical or globular shape. It contains an unusual x-ray emitting star, catalogued 3U 1820–30, which was discovered by x-ray-sensitive telescopes aboard the satellite *Uhuru* (3U in its catalogue number refers to the third catalogue of x-ray stars found by *Uhuru*.)

As a typical globular cluster, NGC 6624 has more than 100,000 stars within a sphere some 20 light years in diameter. The same sphere in the neighborhood of the Sun would on average contain just about 100 stars. However, the middle of NGC 6624 is especially crowded—perhaps 2000 stars are packed into the central one light-year diameter core. Princeton astronomers Jeremiah Ostriker and John Bahcall have speculated that the x-ray object in NGC 6624 is a massive black hole, perhaps formed in the collapse of the central region under its own force of gravity. They attribute unusual brief recurring "bursts" of x-rays from 3U 1820–30 to the effects of a neutron star in orbit around the black hole. Periodically, they suggest the neutron star crashes through a disk of gas around the black hole like the rings of Saturn.

Seven globular clusters have x-ray sources in their centers. Considering the number of stars in globular clusters in fact, x-ray stars are hundreds of times more common in globular clusters than in the Milky Way galaxy as a whole.

NGC 6624 POSS

Messier 15

Globular with planetary nebula

M 15 in Pegasus is considered to be one of the four great globular clusters in the northern sky, along with M 3, M 13 and M 92. M 15 was discovered by J.D. Maraldi in 1746. Observing a few years later with his relatively poor telescope, Charles Messier could not distinguish individual stars in the crowded cluster, but it was resolved into stars by Herschel. In the *New General Catalogue* it has the number 7078. One feature in particular distinguishes M 15: it contains a planetary nebula (p. 140) the only one found in a globular cluster in spite of the hundreds of thousands of giant stars in the 125 globular clusters known. This demonstrates how briefly planetary nebulae last compared with the lifetime of stars.

Like NGC 6624, M 15 is an x-ray source and the SAS-3 satellite team have located the x-ray object right in the core of the globular cluster. According to Mount Stromlo Observatory astronomers Barry Newell, Garry Da Costa and John Norris, M 15 contains a bright, almost starlike nucleus reminiscent of the nucleus of the ANDROMEDA GALAXY, or even our own GALACTIC CENTER. They estimate the mass of the starlike object at 800 times that of the Sun, and suggest that it is a massive black hole.

NGC 5694

Escaping from the Galaxy

NGC 5694 is a globular cluster of stars which is moving so fast that it is destined to escape from our Milky Way galaxy into intergalactic space. In general all the 125 known globular clusters can be assumed to be permanent members of our Galaxy. They speed through it in elongated orbits around its center like comets around the Sun. They often swing large distances away from the center of the Galaxy, but they always fall back from the edge.

NGC 5694 is the only clear exception. Its exact orbit is not known since astronomers cannot measure its velocity in all three dimensions of space, but only in the one: radial velocity along the line of sight. This speed is 180 km/sec towards the Earth. The Earth and Sun are themselves in orbit around the Galaxy and correcting for this gives at least 273 km/sec for the velocity of NGC 5694 compared with a stationary point at the Sun's galactic position. The question is: is this fast enough for the cluster to leave the Galaxy? If its speed is larger than a certain value, the so-called escape velocity, it will leave and never return. 190 km/sec is reckoned to be the escape velocity from the Galaxy for an object at the position of NGC 5694, which is 85,000 light years from the Galactic Center on the far side from us.

Thus Cerro Tololo astronomers William E. Harris and James E. Hesser believe that NGC 5694 possesses more than the escape velocity and that it must indeed be leaving the Galaxy for ever. They discount the alternative possibility that the Galaxy has at least twice the currently accepted mass, and so can bend NGC 5694's orbit back from escape.

Messier 15 POSS

The two scientists deduce that NGC 5694 could be an intergalactic tramp, simply passing through on a once-only visit, or that it is a normal globular cluster that has somehow been pushed into a higher energy orbit. Because NGC 5694 is so similar to other, typical, globular clusters, they favor the latter idea. They identify the Magellanic Clouds as the only obvious candidates at the right distance and with sufficient mass to throw a globular cluster into space. However, they speculate that NGC 5694 may have belonged originally among the Clouds' globular cluster family and was never part of our Galaxy. Although its past is uncertain, its future is to course for ever in the void between galaxies.

NGC 5694 POSS

Messier 4

Nearest globular cluster

M 4 is the nearest globular cluster, at a distance of 5700 light years. It is not the brightest globular cluster, however, mainly because light from it is absorbed by the RHO OPHIUCHI DARK CLOUD (p. 133). If the cloud were not there M 4 would be easily visible to the unaided eye even in the northern hemisphere. Although M 4 is south of the celestial equator, it is visible to European and American amateur astronomers who envy their southern hemisphere colleagues their two naked-eye globular clusters, 47 TUCANAE and OMEGA CENTAURI.

M 4 is not, however, a very dense cluster. It contains an estimated 60,000 solar masses of stars, in contrast to the seven million of a more distant globular cluster, M 22. A stellar population like that of M 22 would put M 4 in a position to rival Sirius in the night sky.

Messier 4 POSS

Shapley 1

Fine-ring nebula

Although modern astronomers know a great deal about the nature of all kinds of objects, it was not always so. When the 18th- and 19th-century astronomers studied the nebulae, they had no notion of what sort of object they were observing. All they could do was to classify objects according to their appearance.

Some nebulae appeared in their telescopes as disks, reminiscent of the visual appearance of the distant planets Uranus and Neptune. This led to the name "planetary" nebulae for such objects.

Astronomers believe that the production of planetary nebulae is a feature of the evolution of many, indeed perhaps all lightweight stars of, say, twice the mass of the Sun. In their dotage such stars appear to give a hiccup, or perhaps a series of hiccups, which puff off part of their outer layers. They eject about a tenth of their mass, which is a small but not insignificant proportion. The core left behind is hotter than a normal star, and consequently emits a great deal of ultraviolet radiation. It is this radiation which heats and excites the ejected gas and enables us to see it.

The star cools down because the nuclear burning process in its interior is no longer producing significant quantities of radiation. It becomes a white dwarf and in the end, a black dwarf. The nebula meanwhile fades from view, merging into the general background of interstellar gas. The white dwarf passes on its journey around the Galaxy leaving its cosmic litter, which will perhaps contribute towards the formation of new stars at some later time.

Shapley 1 is a typical planetary nebula. It was discovered by Harlow Shapley. It is a fine-ring nebula, nearly circular in appearance, nearly uniform, and it has a bright central star. Planetary nebulae are popularly supposed to consist of hollow shells of gas, appearing ringlike only because the line of sight passes through a greater depth of material at the edges than at the center. Instead, Shapley 1 has more the appearance of a genuine ring, a toroid or donut-shaped object.

Planetary nebulae have a great variety of shapes, and do not always look so much like planets. This variety leads astronomers to suspect that some of their shapes, such as that of Shapley 1, arise because a three-dimensional object such as a toroid can have different appearances when seen from different angles. Shapley 1 is a case of a true ring which we are seeing face-on.

Abell 36

Evolved planetary nebula

Abell 36 is an intrinsically faint and large planetary nebula discovered by American astronomer George Abell in 1955. Abell estimates its distance at 1360 light years and its diameter at three light years. Because it is so large, he classifies it as an old planetary nebula.

The thought behind this is simple. The planetary nebula is thrown off the star and expands into space. Therefore the larger it is, the older it must be. Moreover, even assuming that the overall light emitted by the gas, atom for atom, does not decrease, its surface brightness will decrease as nebula grows and

Shapley 1 AAT

Abell 36 UKSTU

the gas becomes thinner and thinner.

In practice the total intensity of light will fall with time, since this depends on the excitation from the ultraviolet radiation of the central star, which diminishes as the star cools. Thus aged planetary nebulae are both larger and fainter than young, compact specimens.

NGC 7293

The Helix Nebula

The Helix Nebula, NGC 7293, is the planetary nebula with the largest angular size—about half a degree overall, the size of the Moon's disk! It is nevertheless a bright object, which suggests that it is the nearest to us: its great angular size owes as much to its large diameter as to its proximity.

Only on deep photographs does the helix shape become apparent. The visual appearance of the nebula in a telescope is of a ring with a star at its center—the star which ejected the nebula. This star can be singled out from the numerous stars which lie by chance in much the same direction for several reasons. First, it lies exactly at the center of the nebula. Second, the inner edge of the ring has a streaky appearance, like petals (hence its other name, the Sunflower Nebula) and the streaks, traced backwards, point to the suspect star. Finally, the spectrum of the star shows it to be unlike the vast majority of Milky Way stars which are relatively cool. Instead it is very hot, like other central stars in planetary nebulae.

The streaks and the general appearance of the ring give the impression that the planetary nebula is expanding away from the central star as the puff of a cosmic smoke ring. This is indeed the case. Measurements of the speed at which the nebula is expand-

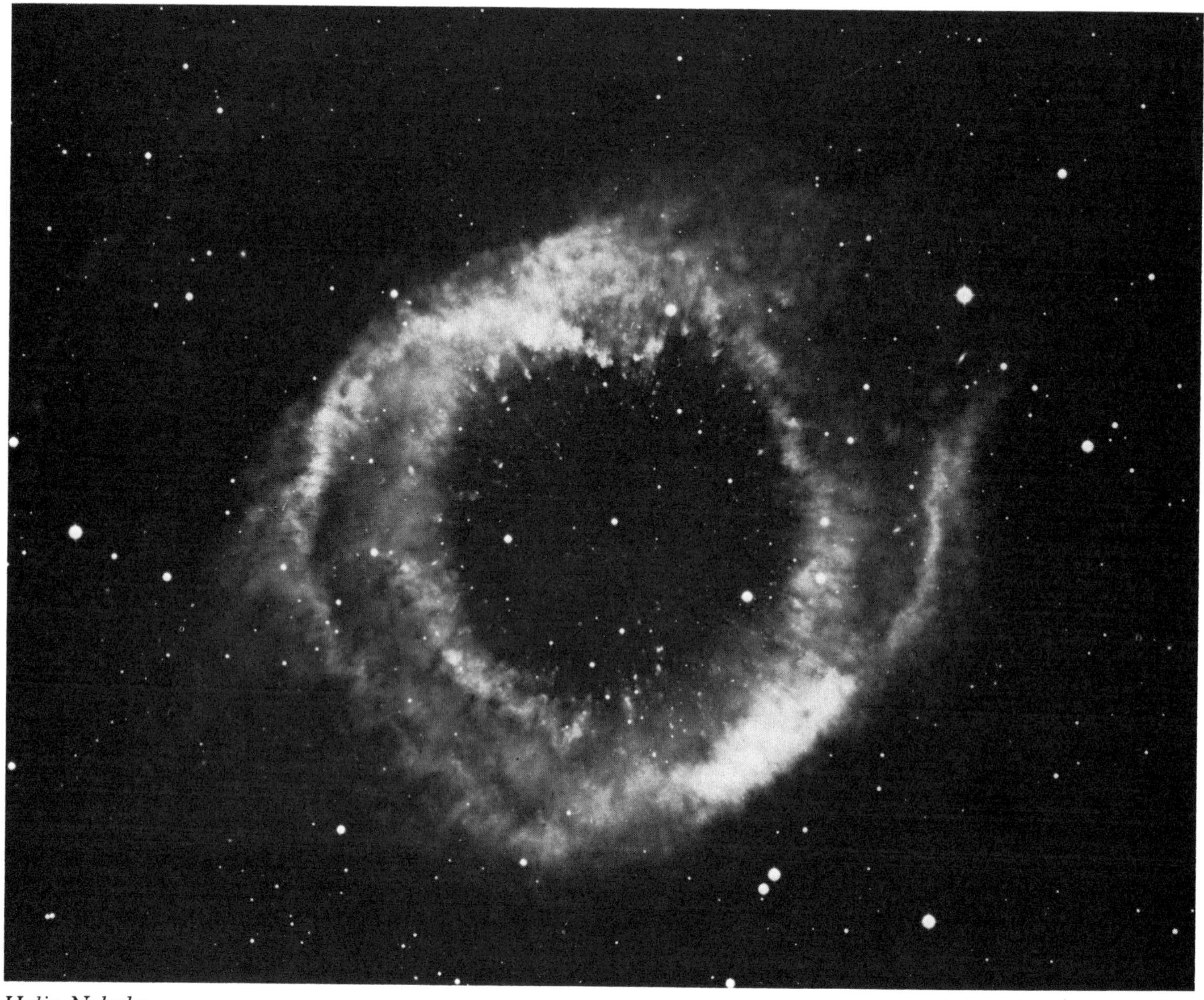

Helix Nebula AAT

ing, made by observing the doppler shift of the spectral lines of the gas, put the expansion speed at between 20 and 40 km/sec.

The nebula has the appearance of two overlapping circles or a helix—the shape of a spiral staircase. This has been connected with the possibility that the "smoke ring" was thrown off in the orbital plane of another star circling the central star. There is no evidence which suggests that the central star of the Helix is double, but the central stars of some other planetaries are indeed known to be binaries. The material in a typical planetary nebula expands at a rate of a few tens of km/second—a speed comparable with the orbital velocity of a star round a companion. So if material is ejected over even a fairly short period of time (hours rather than days) the motion of the star through space can change appreciably. The orbital velocity of the star is added to the ejection speed of the material in different directions, so after a few thousand years, when we observe the nebula, the material ejected at different times is separated by a visible amount. If the material is not ejected exactly in the orbital plane of the stars, there will be an apparent twist in the ring of ejected material.

The filaments in the Helix Nebula could be produced by some subsequent minor explosions on the central star. An alternative explanation is that they are caused by a steady wind of gas blowing outwards from the central star, overtaking the gas in the nebula and breaking it into droplets.

The Helix hints that the ejection and detailed shape of a planetary nebula is affected by many of the circumstances of its parent star—its rotation, the presence of a binary companion or a magnetic field, for example.

Messier 57

The Ring Nebula

The Ring Nebula is a fine planetary nebula in the constellation Lyra, and a great favorite with amateur astronomers as it is bright and easy to observe even with fairly small telescopes. It was discovered by Antoine Darguier in 1779. In his description he said that it was "as large as Jupiter and resembling a fading planet." This was the first time that such a nebula had been compared with a planet, although it was Sir William Herschel who gave currency to the description "planetary nebula."

The nebula, catalogued as M 57 or NGC 6720, is more than 2000 light years from us, according to modern determinations. The elliptical shape indicates that it is a donut or toroid seen at a somewhat inclined angle. The ellipse is centered on a star which is one of the hottest to lie within a planetary nebula. Another star of similar apparent brightness lies within the boundaries of the planetary nebula, but this is a chance projection into the line of sight. The spectrum of the central star of the Ring is reminiscent of the spectra of many white dwarf stars. Astronomers believe this suggests that many, perhaps all, white dwarfs go through a planetary nebula stage.

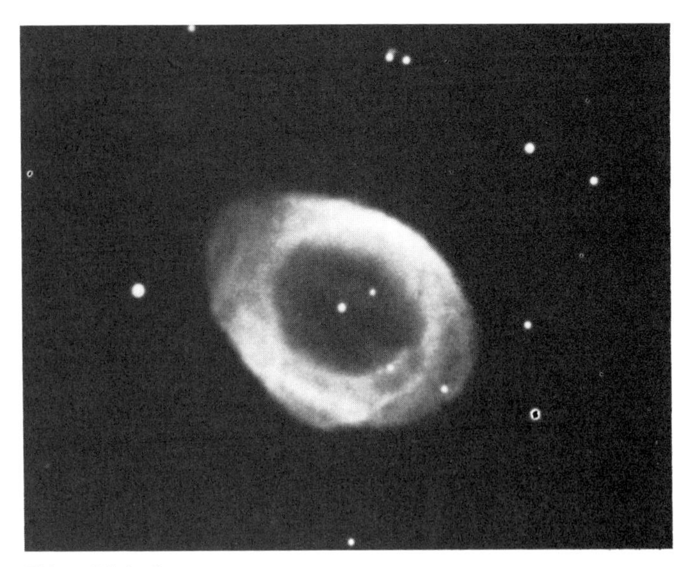

Ring Nebula KPNO

The Ring Nebula is not as uniform as many other planetary nebulae of large apparent diameter: within it there are filaments and large lumps. Its outer edge is cleanly defined, however, which has made possible the measurement of its growth in size over the 40 years in which large modern telescopes have been taking photographs of it. It is expanding at 19 km/second, or 60-millionths of the speed of light. Since its radius is one-third of a light year, the time which the nebula would take to get to its present size is about 5500 years. This figure gives the true age of the nebula if it has expanded at a more or less uniform rate. Uniform expansion is not wholly likely, but is sufficiently plausible to make the calculation interesting.

A striking effect seen in color pictures (Plate 14) of the Ring Nebula is the gradation of color from the green center to the edge where it is red. The reason lies in the amount of energy required to make different gases glow. The central star's ultraviolet radiation excites nearby gas molecules, causing them to glow their characteristic colors. The gas in the nebula is of relatively uniform composition, but the atoms of each element are affected differently according to their proximity to the central star. Hydrogen is readily excited, but oxygen and nitrogen require greater amounts of ultraviolet. Near the center, where ultraviolet light is abundant, the green color of oxygen and nitrogen dominates. Farther out, however, the radiation from the star is weakened by distance and is of longer wavelength as a result of absorption closer in. At the rim, it is only strong enough to excite the predominantly red color of hydrogen.

Messier 97

The Owl Nebula

A round "face" and two large "eyes" distinguish the Owl Nebula, which is a planetary nebula in the constellation Ursa Major. It is one of the largest planetary nebula catalogued and was discovered by Pierre Mechain in 1781. In Messier's Catalogue it is M 97; it is also known as NGC 3587. Sir William Herschel was of the mistaken opinion that it was a star cluster just beyond the resolution of his telescope, and it was Lord Rosse's drawing in 1848, which suggested the likeness to an owl. Herschel's son, Sir John Herschel, noted that the light from the Owl was very uniform and smooth, apart from the eyes, without the increase in brightness seen towards the center of an unresolved globular cluster. Thus he concluded that the Owl was not a star cluster but a nebula.

As with other planetaries, there is a central star, from which the nebula was ejected. The nebula is evidently remarkably symmetric and its shape suggests that it is a hollow cylinder, seen obliquely so that the "eyes" are formed by the open ends.

The Owl is 1600 light years from us and its diameter is about 1.5 light years. Its density is remarkably low, 1/10 to 1/100 of the normal planetary nebula value. Generally if the same mass had been thrown off in every planetary nebula we would expect the density to decrease as the nebula grew so that the larger nebulae would indeed be the most rarefied. This low density suggests that the Owl is a rather old planetary nebula, since the larger a nebula grows, the more rarefied its gas becomes. Perhaps this also explains why the Owl has such a generally smooth appearance, as though the lumps have had time to smooth out.

Messier 27

The Dumbbell Nebula

Messier discovered this planetary nebula, known as M 27 or NGC 6853, in 1764 while sweeping his telescope across the constellation Vulpecula. Lord Rosse, the 19th-century observer who built a giant 72-inch reflector, discovered its hourglass or dumbbell shape and was misled by a chance sprinkling of Milky Way stars into thinking that it was two clusters of stars embedded in a nebula, like many galactic star clusters.

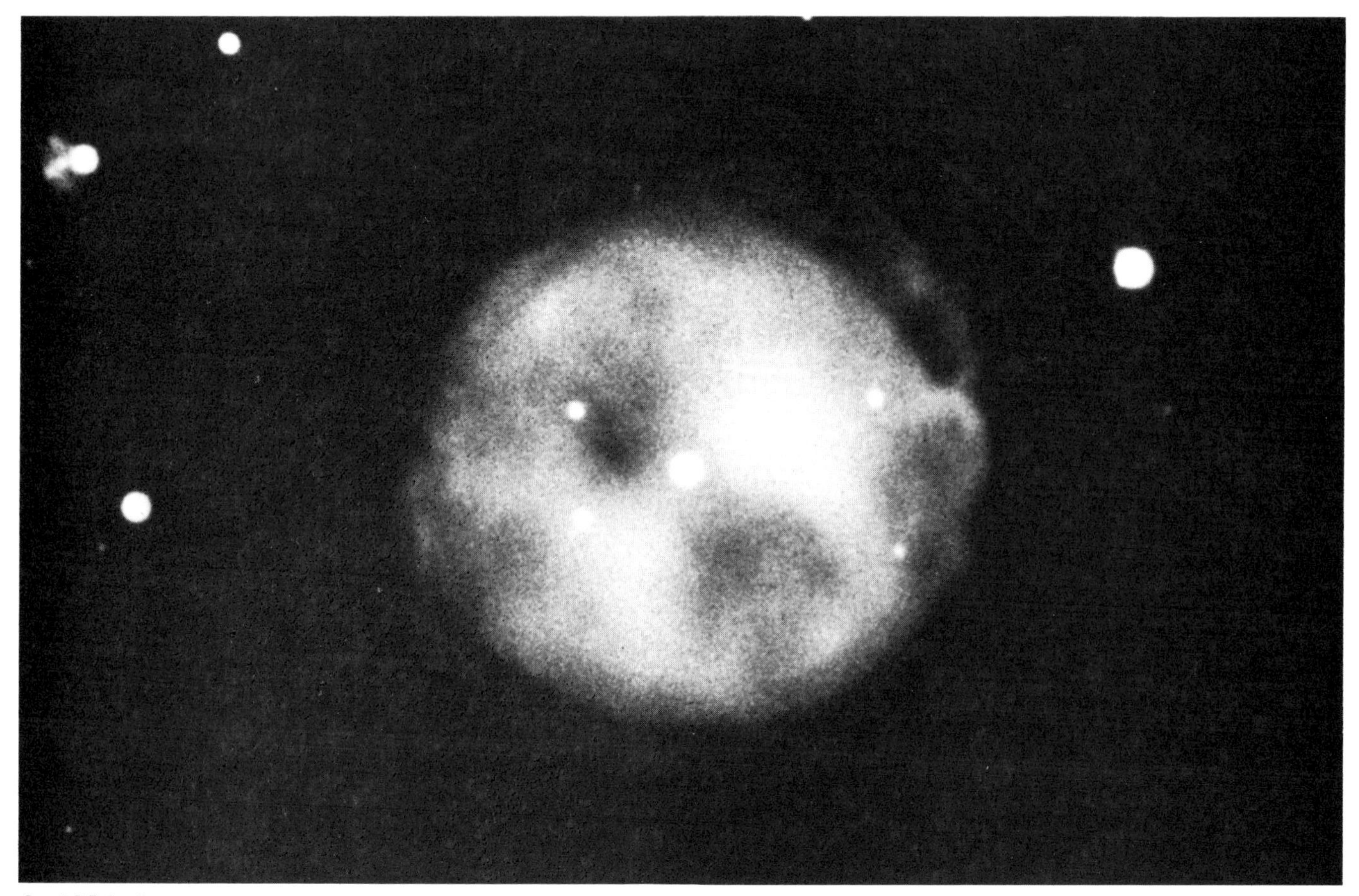

Owl Nebula *Hale*

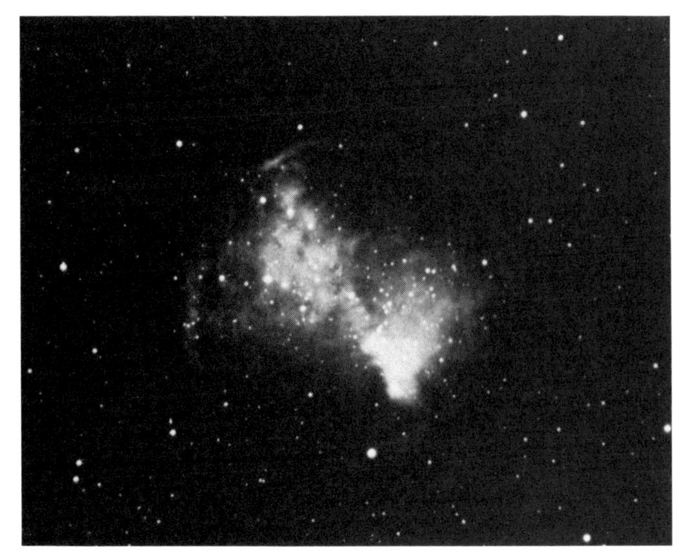

Dumbbell Nebula *Hale*

The Dumbbell has a hot central star which shows blue on color photographs (Plate 14). Red and blue nebulosity is separated as in the RING NEBULA. On long-exposure photographs, the dumbbell shape appears filled in to something more like a sphere.

The large apparent size of the Dumbbell (among the brighter nebulae it is second only to the HELIX) suggests that it is nearby. This is supported by measurements made by astronomers at Pulkovo Observatory in the U.S.S.R. They detected a growth in its angular size of 0.068 arc seconds per year; this is the same annual growth rate as that of a tree whose rings grow a quarter of an inch a year, as seen from a distance of 12 miles!

NGC 2932

The Eskimo Nebula

Despite its unique appearance, the Eskimo Nebula, NGC 2932, is a planetary nebula 3500 light years away from us. It has a "face" with "eyes," "mouth," and "nose" caused by a chance condensation in the nebula, or by some projection effect of a complicated three-dimensional figure. The "face" is surrounded by a "fringe" which has the appearance of fur around the hood of an eskimo.

There are several planetary nebulae which are surrounded by fainter haloes of this kind. Although the haloes are faint, they can contain a great deal of mass if they are large. The faint haloes may represent matter expelled from the parent star while it was a red giant, before it passed to the stage at which it hiccupped and produced the planetary nebula.

In the case of the Eskimo, however, Charles O'Dell has suggested that the fringe represents the dissipating remains of a previous planetary nebula shell, thrown

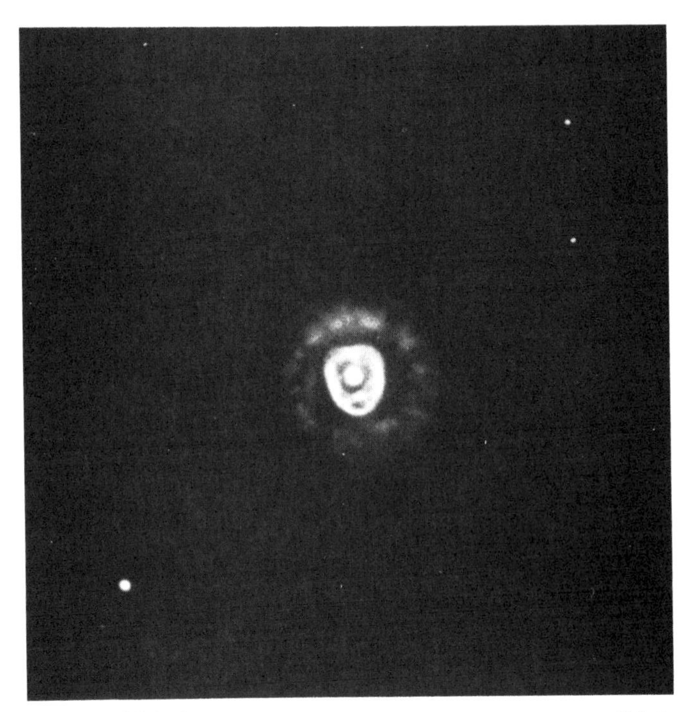

Eskimo Nebula *Lick*

off before the bright inner one. Perhaps the parent star went through a series of hiccups before some metaphorical drink of water calmed the spasms and it began to cool more gently to a white dwarf star.

NGC 7009

Saturn Nebula

The Saturn Nebula, NGC 7009, in Aquarius, is a planetary nebula of curious shape. Superimposed on the inner, brighter, elliptical ring of the nebula is an outer, faint ring with *ansae* (Latin for "handles") projecting from the long diameter. It resembles the planet SATURN with its rings almost closed; hence its name. The double rings of the nebula may be different shells thrown off the central star.

FG Sagittae

Rosetta Stone for astrophysics

FG Sagittae is a remarkable variable star whose brightness rose steadily from magnitude 13.6 in 1894 to magnitude 8.9 in 1970, and has remained steady since. G.H. Herbig and A.A. Boyarchuk noticed in 1960 that photographs taken with the Lick 120-inch reflector showed that the image of the star was surrounded by a nearly circular nebula with radius 18 arc seconds. At the distance of FG Sagittae, 8000 light years, this corresponds to a diameter of 1.5 light years.

In shape the nebula is reminiscent of a planetary

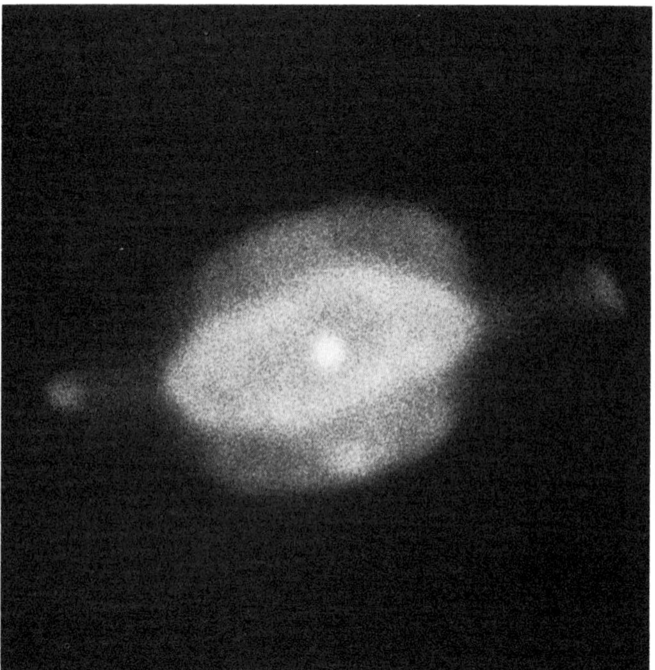

Saturn Nebula KPNO

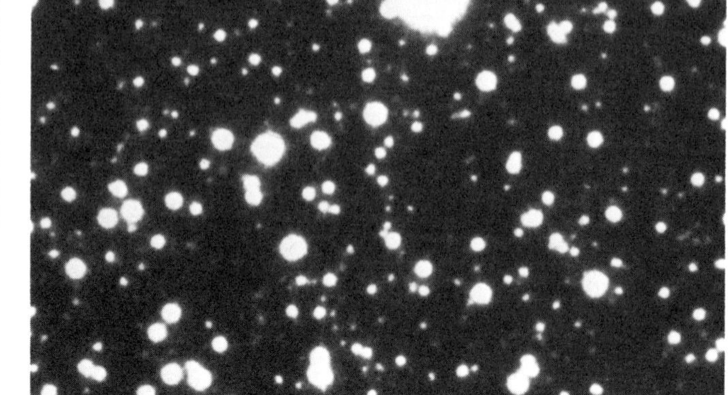

FG Sagittae POSS

nebula, and indeed its spectrum was shown to be a kind similar to some planetary nebulae. The nebula around FG Sagittae could have been present since 1894, since old photographs showing the star have an unusually fuzzy appearance. Perhaps it simply remained unnoticed until the better quality modern photographs.

In the 1960s Herbig and Boyarchuk found evidence in FG Sagittae's spectrum that its atmosphere was expanding. They thought it was ejecting a further shell destined to become another planetary nebula around the star. This shell, however, turned out to have a most unusual composition. Besides the normal spectral lines of the common chemical elements, in 1969 there appeared the spectral lines of such unusual chemical elements as yttrium, zirconium, cerium, lanthanum, neodymium, samarium, and praseodymium. These elements are significant in that they are all formed by the so-called *s*-process. In this, neutrons are added *slowly* (hence *s*-process) and in a progression to the nuclei of various middleweight elements to make new kinds of atoms.

Astronomers believe that all chemical elements heavier than hydrogen and helium have been made in stars, by a total of about half a dozen different processes of which the *s*-process is one. Nearly always the element factory lies within a star and the results of the element production below its surface are hidden from astronomers' view. Only if the elements produced rise to the star's surface can they be detected.

In FG Sagittae the newly produced *s*-process elements were churned to its surface and its expanding atmosphere in 1969. It is the only star we know of in which contamination of its atmosphere by newly made elements is proceeding so quickly as to occur before astronomers' very eyes. According to G.E. Langer, Robert Kraft and Kurt Anderson, FG Sagittae is a Rosetta Stone for nuclear astrophysics. It is still not clear, however, how the curious production of s-process elements is connected with the birth of FG Sagittae's planetary nebula. Can it be associated with the recent discovery that FG Sagittae is a binary star?

NGC 6302

Bipolar nebula

NGC 6302 is a remarkable nebula discovered by E.E. Barnard from Nashville, Tennessee. It has two lobes of gas like the wings of a butterfly or a bow tie. No central star has been detected. In spite of its shape, it is classified as a planetary nebula because of its size and spectrum.

David Evans and the late David Thackeray of the Radcliffe Observatory in South Africa have estimated that only 10% of planetary nebulae are circles with the classical "planetary" shape of SHAPLEY 1 or the RING NEBULA. On the other hand, only a quarter are totally irregular. Most, 70% have a "bipolar" symmetry, as shown by NGC 6302. It appears that the explosion which generated the planetary nebula was somehow funneled into two opposite directions, like a long balloon bursting at both ends. Perhaps the explosion took place in line with the poles of rotation of the star or along the field of its magnetic poles.

NGC 6302 is very red (Plate 13), partly due to the reddening effect of the passage of its light through the interstellar medium. The nebula lies in the galactic plane near the direction of the galactic center: this is a particularly dusty area of sky, and the interstellar dust does indeed cause the stars in the area to appear red. Additionally, however, NGC 6302 is very red in its own right, partly because of the hydrogen-alpha line which it emits, but mostly because of very strong nitrogen lines in its spectrum. Thus, even to astronomers so near to it that its light would have free passage to their telescopes, it would still appear about as deep a red as it does on our photograph.

Observations made recently by Manchester astronomer John Meaburn indicate that the explosion which generated NGC 6302 was particularly violent. Whereas in conventional planetary nebulae gas is outflowing at a few tens of km/second, in NGC 6302 velocities as high as 400 km/second occur. Other evidence of violent events in its nucleus is provided by the spectroscope which reveals highly ionized atoms near its center. For example, iron atoms have been stripped of six electrons. To do this requires energies three or four times higher than is available from the central stars of most planetary nebulae.

NGC 6302 AAT

Tycho's Star

Renaissance supernova

Although *nova* means "new," a supernova is in fact a dying star. The energy of its explosion is far in excess of that of ordinary novae, which is why the phenomenon is called a *super*nova.

Most stars fade relatively quietly into white dwarfs, perhaps puffing off a planetary nebula as a dying gesture. Massive stars do not go gentle. They rage against the fading of their light, exploding with an energy which for a few brief days can outshine the galaxy in which they occur.

Stars of a large enough mass to produce supernovae die in each galaxy the size of ours once every 30 years or so. A seventh of these become visible to the naked eye, judging by the fact that five have been noticed in the last millennium. But the exploding remnant gas clouds which the galactic supernovae leave behind can be identified in our Galaxy, and are known to astronomers as SNRs (supernova remnants). Most of our knowledge about supernovae themselves comes through observing their occurrence in the many galaxies outside ours, but also from historical records of supernovae in our Galaxy, such as Tycho's Star.

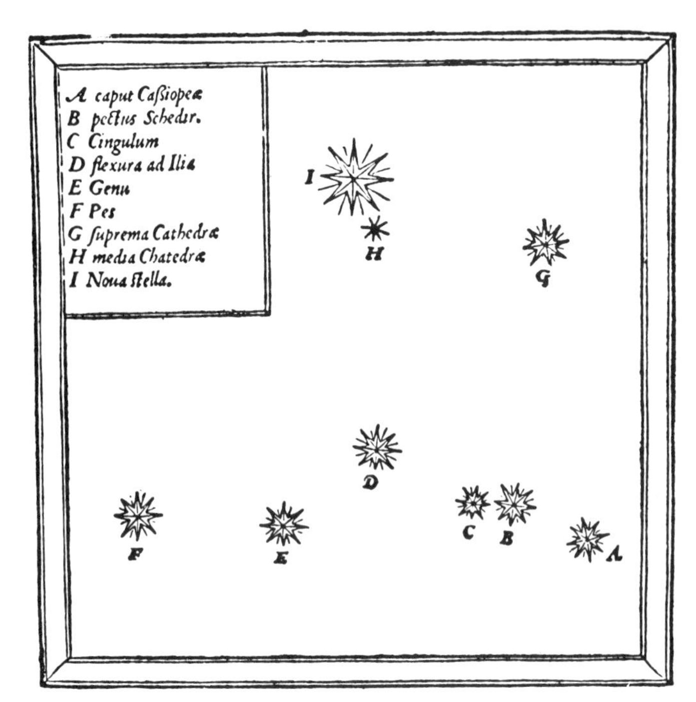

Tycho's Star

Although the first observation of the brilliant star which appeared in 1572 was made by a Sicilian mathematician named Francesco Maurolyco, the star was named after Danish astronomer Tycho Brahe, since his observations of it were more systematic. The star is one of only five supernovae seen in our Galaxy in the last 1000 years.

Tycho's supernova reached the apparent brightness of Venus in November 1572 when it was visible in daylight. It gradually faded throughout 1573 and became invisible to the pre-telescopic astronomers in March 1574. It was particularly important in the history of astronomy in that it remained fixed in the sky in relation to the other stars of the constellation Cassiopeia which surrounded it. Thus it had to be a true star, not a planet or a comet. Yet many people thought that the stars were eternal. The changes they saw in the supernova of 1572 and Kepler's supernova of 1604 hastened the revolution in astronomy which culminated in the general acceptance of the copernican sun-centered theory of the solar system.

Tycho Brahe made a series of detailed measurements of the position of the star with a new sextant-type instrument which he had just finished constructing. Other astronomers of the day, Digges and Mästlin, also made measurements of its position. Thanks to them, the position of the supernova is well known.

In the same area of the sky in 1952, radio astronomers R. Hanbury Brown and Cyril Hazard found an intense radio source. They were deliberately searching for the remnant of the supernova. Their discovery was thus different from the accidental discovery of the radio remnant of the CRAB NEBULA. The radio remnant of Tycho's supernova has the form of a hollow shell caused by the still expanding pressure waves from the explosion of 1572.

In 1959, Mt. Palomar astronomer Rudolph Minkowski found extremely faint optically visible filaments in the region of the shell, still moving outwards. In 1967, x-rays from the hot gas (at a million degrees C) were discovered by a rocket-borne x-ray sensitive telescope. Tycho's supernova is the clearest example of a supernova and its remnant, with a direct line through 400 years of astronomical observations.

Kepler's Supernova

Second supernova of the Renaissance

Only 32 years after Tycho saw his supernova, another appeared, peaking 1 November 1604. Thus in spite of the fact that astronomers have recognized no supernova in our Galaxy in the 300 and more years since then, the observers of the Renaissance could have seen two in a lifetime. The supernova happened to occur a few degrees from Mars, Jupiter and Saturn, a planetary conjunction being studied for astrological reasons. Thus the supernova was noticed before it reached maximum brightness, the first recorded observations being by I. Altobelli in Verona and a physician whose name is unknown.

Tycho's pupil Johannes Kepler, who became famous by discovering the laws of planetary motion from Tycho's careful observations, first saw the supernova on October 17, and made measurements of its brightness and position, and it has since been known as Kepler's supernova. The brightness measurements show that Kepler's supernova was of Type I, the most commonly observed kind. The positional meas-

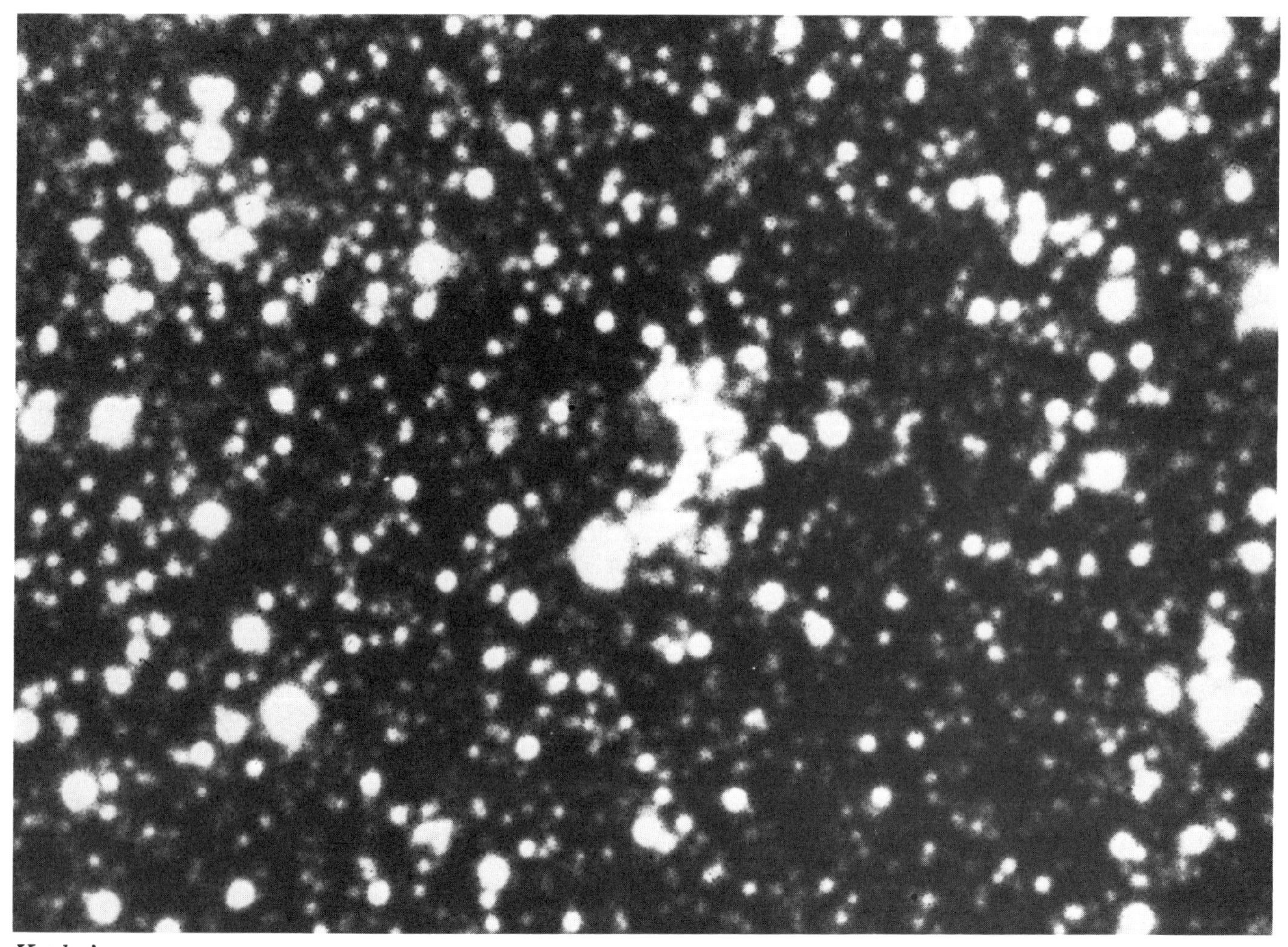

Kepler's supernova remnant AAT

urements show that at that place in the sky there now exists a shell-shaped nebula and radio source known as 3C 358, the remnant of Kepler's supernova.

Supernova 1972e

Supernova in a distant galaxy

In May 1972 a new star appeared in the sky close to the galaxy NGC 5253. The fifth discovered that year, it was called supernova 1972e. It was the fourth brightest supernova to be seen for 200 years; the second brightest also occurred in NGC 5253, in 1895. 1972e was found by Charles T. Kowal, the man who was then carrying out the now-defunct Palomar Supernova Search. Kowal repeatedly photographed 38 areas of the sky with the 48-inch Schmidt Telescope in order to discover supernovae in other galaxies.

Supernovae occur so rarely in our own Galaxy that most of what we know of supernovae themselves comes from studying ones in other galaxies. In a typical galaxy a supernova occurs every 20 to 30 years, though none has been seen in our own Galaxy since 1604.

Supernovae are of at least two sorts. 1972e was of the most commonly discovered variety, Type I. Type II's never occur in elliptical galaxies. They are thought to be caused by the explosions of massive stars. Type I's occur both in spiral galaxies like NGC 5253 and in elliptical galaxies, but their progenitors are unknown.

The spectacular brightening of a supernova can be ascribed mainly to the enormous increase in the size of a star in a rapid explosion which takes place deep within it. Expanding at 10-20,000 km/sec, the star's surface area increases a millionfold in a day. Unlike a nova explosion such as NOVA PERSEI (p. 100) a supernova explosion disrupts the whole star. The central region collapses to a black hole or neutron star.

January 4, 1959

May 6, 1972

April 24, 1973

Supernova 1972e in NGC 5253 *Kowal*

Vela Supernova Remnant

Brightest, but unseen supernova

The Vela supernova remnant (Plate 17) is a tangled web of filaments covering 6° of the southern constellation Vela. It appears rather undistinguished at first, but attention was drawn to the optical nebula after radio astronomers, particularly Doug Milne in 1966, had mapped three strong radio nebulae called X, Y and Z, in the area. All of them, and particularly the one called Vela X, showed the characteristics which radio astronomers had learnt to look for in SNRs. The chief characteristic is synchrotron radiation, produced by electrons gyrating in magnetic fields, and distinguishable from the thermal radio noise by its spectrum and polarization. Astronomers were therefore convinced that Vela X at least, was a SNR.

Maps of the radio nebulae were superimposed on photographs of the sky, made at Mount Stromlo Observatory, and filtered to show only the red color of

Vela supernova remnant UKSTU

glowing hydrogen. In that area of the sky the maps showed a half shell of lacy filaments reminiscent of the VEIL NEBULA (p. 156). Milne identified this as the optical SNR, and estimated its distance at around 1500 light years. Optical astronomers can see only half the expected shell, because absorption of light by intervening small grains of dust hides the other half. Such small particles have no effect on radio waves, however, and radio astronomers can readily see the whole shell.

The Vela SNR generated great excitement in 1968 when Australian radio astronomers discovered a pulsar—a pulsating radio source—within the outlines of the SNR, and quite near its center. A pulsar is distinguished from other radio sources by its very regular and rapid radio pulses, caused as a beam from the tiny, rapidly spinning star sweeps across our line of sight like a lighthouse beam. The pulse period of the VELA PULSAR is 89 milliseconds, giving 11 pulses per second. Vela was the first pulsar to be connected with a SNR, although Fritz Zwicky had been predicting since the 1930's that supernovae would produce compact stars which could spin very fast. The more famous discovery of the CRAB PULSAR in the CRAB NEBULA, a known SNR, followed quickly afterwards. Thus astronomers assume that the Vela SNR and the Vela pulsar were formed in the same explosion.

Astronomers have tried in vain to measure the rate of expansion of the Vela SNR; the nebula is growing too slowly. Even so, this represents an upper limit on the expansion, and from this astronomers estimate the age of the Vela SNR at greater than 10,000 years.

There is some evidence in the spectra of stars which lie by chance within the SNR that the SNR is expanding. The SNR shell is plowing into the surrounding cold interstellar material, and is building up an invisible shell of cold gas surrounding the hotter bubble of filaments. The cold shell contains such elements as calcium, which characteristically absorb the light of stars within the bubble. Traces of the calcium in the cold shell have been found in the light from two such stars.

The supernova which produced the SNR and pulsar probably occurred some 12,000 years ago, judging by the rate at which the pulsar's period is changing. It would probably have been the brightest supernova ever seen, since the SNR is the closest to us. No clear records of it are known. American researcher George Michanowsky has suggested that certain Sumerian tablets which refer to a great star in the south represent an oral tradition of the supernova, but his thesis is not widely accepted. He has also suggested that the cosmic particles from this supernova bathed the Earth with mutation-inducing radiation. These particles, he says, sparked off the intelligence of the human race. It is an interesting speculation, but is just that—a speculation.

Puppis A

Lit-up cosmic junk

Puppis A is a strong radio source in the southern sky. It has a shell-like structure and in the area is a motley collection of red filaments of nebulosity. It is the remnant of a supernova which went unrecorded in the distant past. The red filaments are unusual in that, although photographed on the same exposures that record H-alpha light, their H-alpha light is very weak. The light which exposes the photographs is from nitrogen, which emits on closely similar wavelengths to H-alpha. What we see is not the normal interstellar gas pervading the space between the stars, nor even individual bits of the exploding star. The filaments probably represent the gas lost by the massive star before it exploded. Massive stars go through a red supergiant stage like Betelgeuse or Antares before becoming supernovae.

The low surface gravity enables the atmospheres of such stars to leak off into space, aided by winds and storms on the stars' surfaces. The material leaked from a red supergiant star is expected to be nitrogen-rich, since nitrogen is one of the elements manufactured in its evolution. The subsequent supernova explosion sweeps over the mass lost from the star and causes it to shine.

Crab Nebula

Messier's first nebula

The Crab Nebula, approximately 6000 light years distant, is the remains of the supernova of AD 1054. It was discovered by John Bevis early in the 18th century

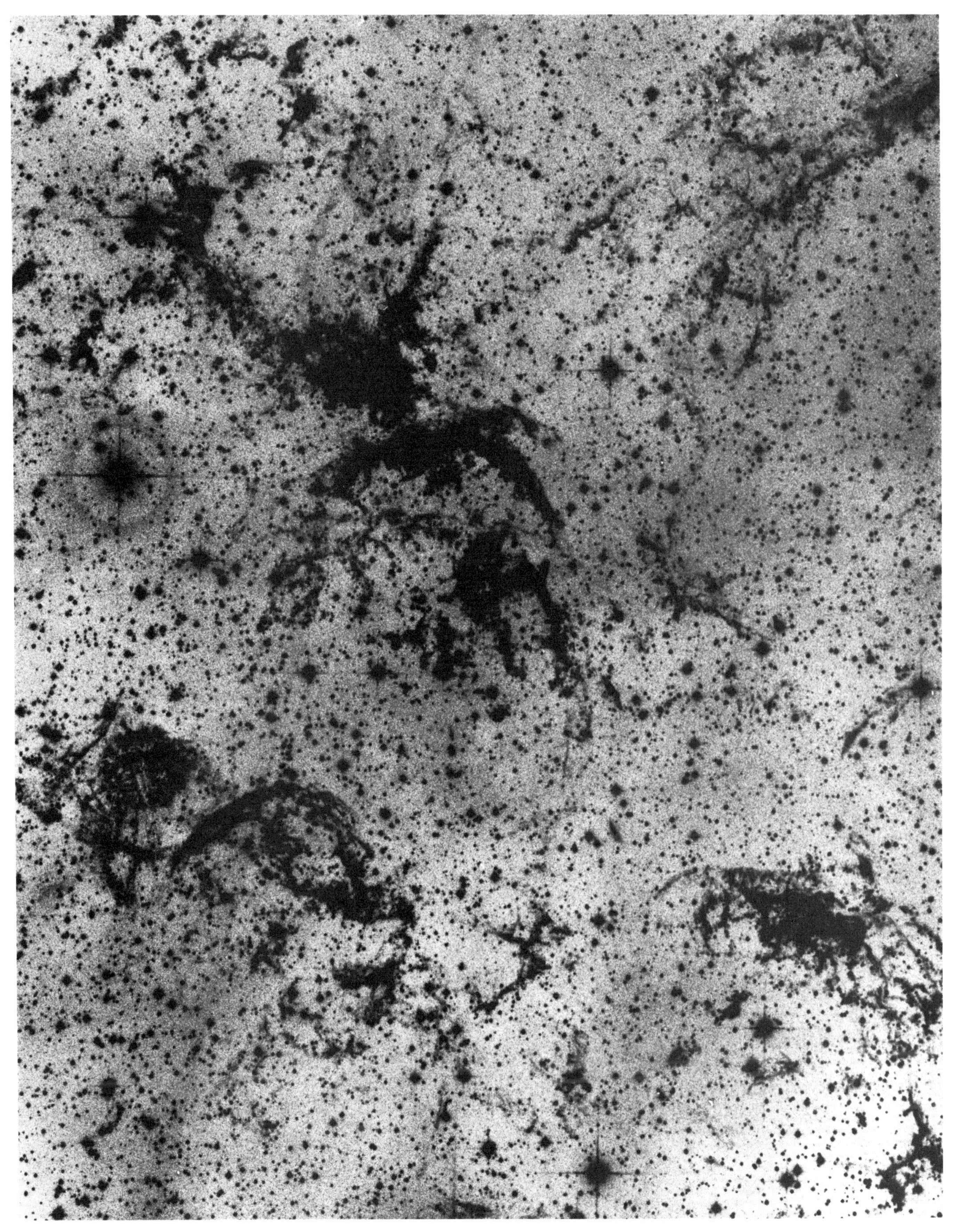

Puppis A AAT

Crab Nebula (Messier 1) *Hale*

and was rediscovered by Charles Messier in 1758 while he was searching for the expected reappearance of Halley's comet. It is first in Messier's catalogue of nebulae.

The Earl of Rosse observed the nebula through his three-foot telescope in 1844. On the basis of an imaginative sketch, he named it the Crab Nebula. In color pictures (Plate 15), the nebula appears as a network of red filaments surrounding an elliptical football of pale, white light. The white light is synchrotron radiation and is evidence of a mixture of very hot gas and magnetic fields permeating the central region of the nebula. The network of filaments is heated by the synchrotron radiation.

The Crab Nebula is one of the strongest radio sources known to astronomers. Radio astronomers call it Taurus A: it was the first radio source to be identified with an optically known object, the identification being made by John Bolton in 1948. To pinpoint Taurus A he designed a special radio telescope on a clifftop facing the sea so that it received radio waves from the Crab both directly and by reflection from the sea. Such a radio telescope can determine the position of a radio source with the same precision as a single radio telescope whose diameter is equal to twice the height of the cliff. Modern large instruments make this technique unnecessary, but by increasing the pointing ability in such a way, Bolton was able to pinpoint Taurus A as the Crab Nebula.

The filaments in the Crab Nebula are the thrown-off remains of the exploded star of 1054. They are speeding from the center of the nebula at 1500 km/second, so fast that even in a single generation of astronomers the angular size of the nebula slightly, but perceptibly, grows larger. The expansion rate closely ties in with the known date of the supernova, 1054.

The filaments are fleeing directly from the center of the nebula, a region of intense activity where the nebula irregularly brightens and fades. Several stars lie in the center of the Crab (see Plate 15) and the waves of brightness which surge through the nebula seem to originate from a particular one. This 16th-magnitude star was found in 1968 to pulsate and it is undoubtedly the stellar cinder formed by the supernova explosion. This star is the CRAB PULSAR and provides the energetic electrons which cause the Crab Nebula's synchrotron radiation. It is the powerhouse of the nebula.

The powerhouse is also a pump. Careful analysis of the expansion rate of the nebula actually puts the date of the explosion within ten years of AD 1140, significantly different from the known date of 1054. The discrepancy is curious because it is the reverse of what one would expect. Normally, the rate of expansion of a supernova remnant (indeed, of most explosions) slows down as the explosion crashes into surrounding material. If, long after such a slowing explosion, you estimated when it had occurred, you would estimate a time earlier than the actual time. If, as in the case of the Crab, you estimate a time for the explosion which is later than the actual time, this implies that something is speeding the expansion. In the case of the Crab, the pulsar is pumping energy into the nebula, speeding the expansion rate.

In 1963 a region near to the Crab Nebula was found to be emitting x-rays, a form of radiated energy thousands of times more energetic than light. The rocket-borne x-ray telescope which discovered this could not give images clear enough to tie down the source of x-rays precisely to the Crab. The Crab was proved to be the x-ray source only when an x-ray telescope was launched by rocket at the exact time that the Moon passed in front of Crab Nebula. The x-rays were eclipsed by the Moon at exactly the same time as the nebula, proving that the two were identical. Since the x-rays faded gradually instead of abruptly disappearing, astronomers could deduce that the x-rays came from the whole nebula rather than from the central star.

Thus the invisible Crab Nebula, in x-ray and radio, mimics the visible one; all these Crabs are caused by the same synchrotron process acting on the speeding electrons supplied by the central pulsar.

Crab Pulsar

Most rapid pulsar

At the center of the CRAB NEBULA lies one of the most

remarkable objects in the heavens. It is the Crab pulsar, first discovered as a source of very rapidly pulsing radio waves, but later found to be sending out light flashes in step with the radio pulses.

The Crab pulsar is distinguished for several reasons. It is the most rapid pulsar known with a period of 33 milliseconds, giving 30 flashes per second. This is so rapid that the eye cannot pick out the individual flashes from the optical object. It was the first pulsar to be identified at both radio and optical wavelengths, and it is the clearest example of a pulsar linked with a supernova remnant, the Crab Nebula itself.

As with all pulsars, the Crab's pulses are very regular indeed. One of the few ways in nature in which such regularity can be maintained is by the orbital motion or rotation of stars. Astronomers believe that pulsars are small rotating stars in which there is a lighthouse-like beam pointing into space. As the pulsar rotates, the beam sweeps across the Earth and the pulsar flashes.

To rotate so fast, the pulsar must be small. In fact, since its equator cannot rotate faster than the speed of light and its period is 33 milliseconds, the Crab pulsar can be shown to be less than about 1500 km (1000 miles) in radius. Contrast this with a sun-sized star which has a radius of nearly a million km. Astronomers believe that the Crab is a star just 10 km in radius, made up of neutrons combined from the original protons and electrons of the parent star, It is very dense; one matchboxful would weigh 10 million tons.

The neutron star was evidently formed by the collapse of an ordinary-sized star in the supernova explosion of 1054. In fact, this is why it rotates so rapidly. All stars rotate, typically with periods of days, weeks or months. If a star collapses, its rotation speeds up, just as a pirouetting ice skater can speed her spin by pulling her outstretched arms to her sides. The collapse of a star in a supernova explosion is so great that the speedup in period is many millionfold.

The energy radiated by the pulsar lighthouse is so great that it brakes the rotation of the star. In fact, the energy lost as the pulsar slows is the power source for the beam itself. Gradually the pulsar period lengthens. Over a million years or so, the Crab's period will increase to about one second, bringing it into line with the majority of the known pulsars.

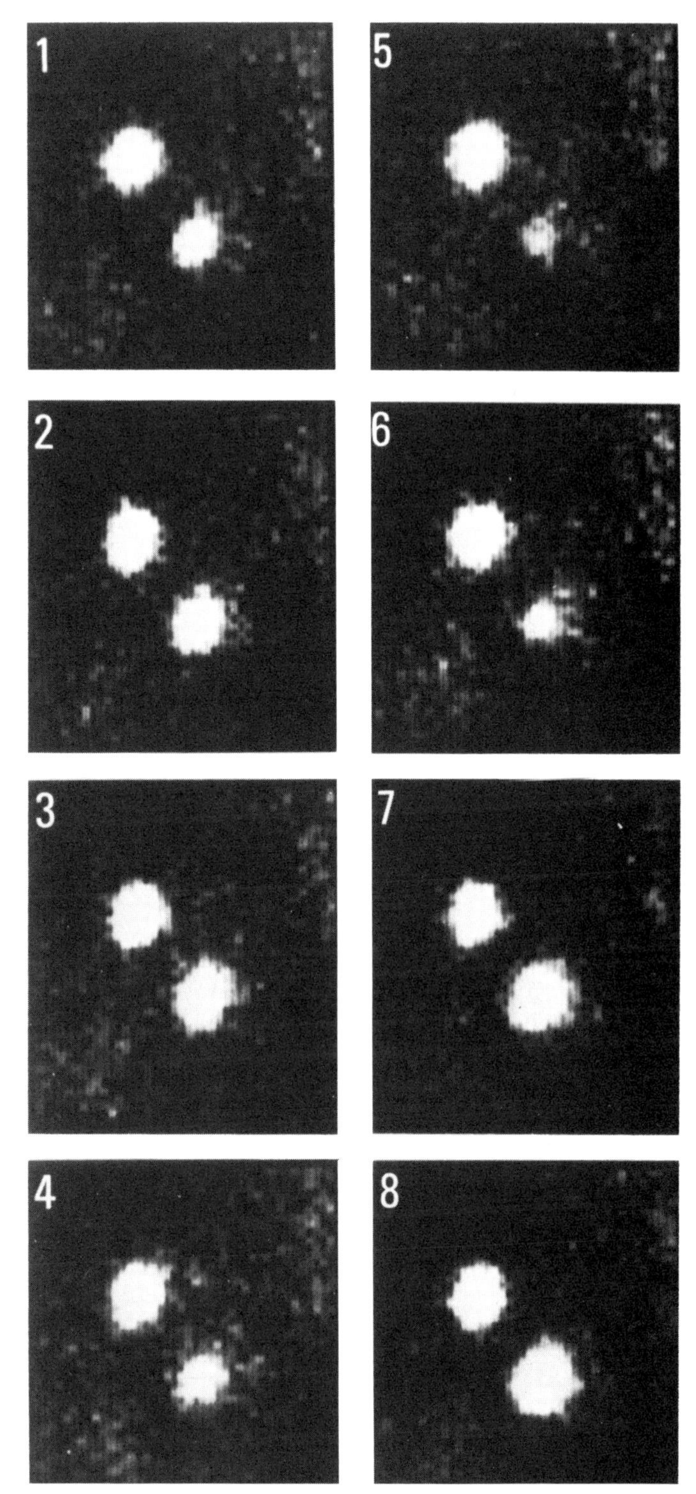

Crab pulsar AAT

Vela Pulsar

Faintest star measured

The Vela pulsar lies in the center of the Vela supernova remnant. Its rotation period of 89 milliseconds is the third shortest of all the more than 300 pulsars now known; it is slowing down, however, at a rate which indicates that its age is 11,800 years. The Vela pulsar was originally discovered by radio astronomers in 1968 but it was not until 1977 that it was found by optical astronomers; at 24th magnitude it is the faintest star whose brightness has been measured. It is visible as the flashing star on our pair of photographs. It was found by recognizing its light cycle at the same period as the radio pulses, though the light curve has a curious double peak in contrast to

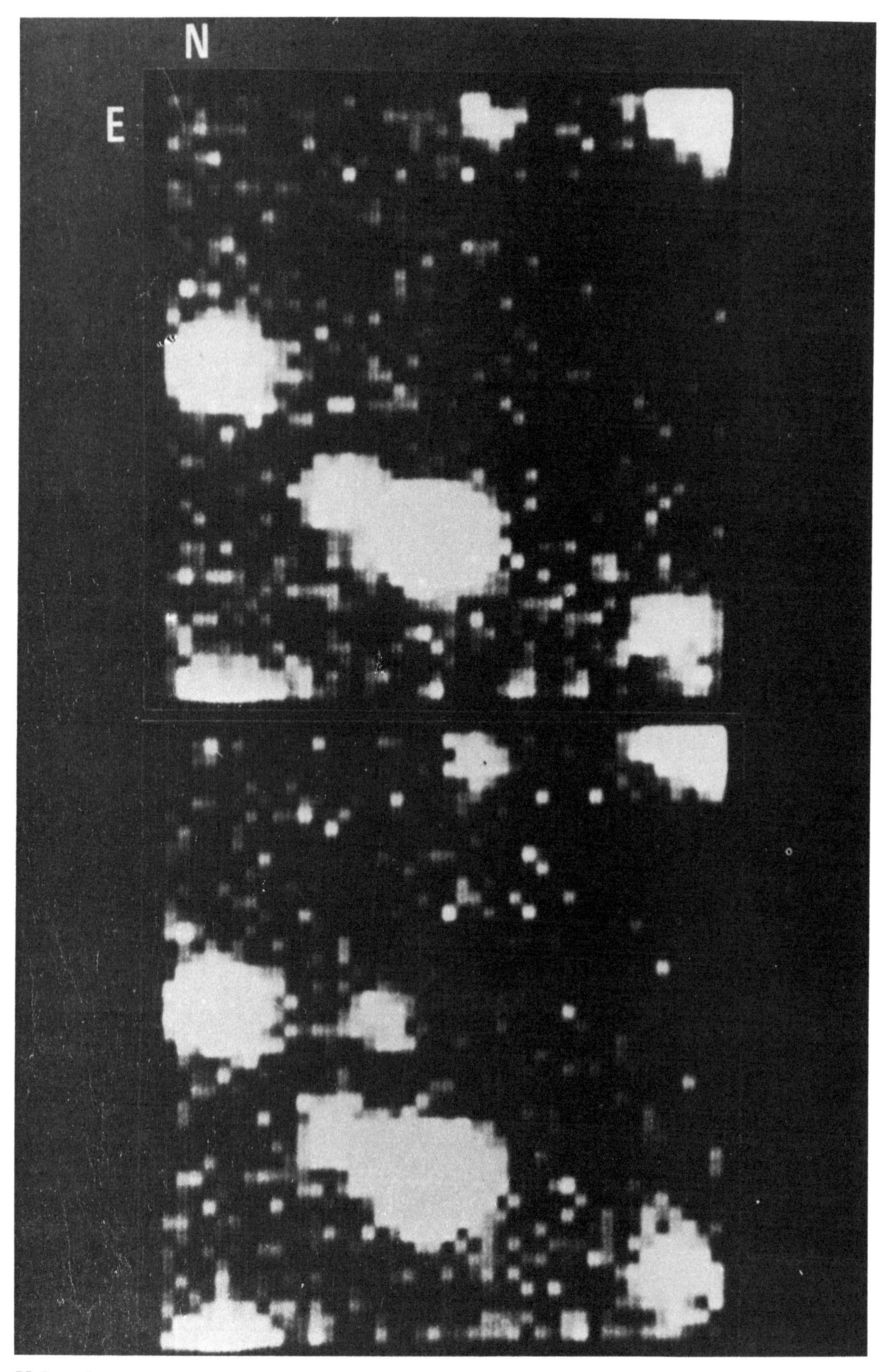

Vela pulsar AAT

the single radio peak.

So far, only the Vela pulsar and the CRAB PULSAR have been observed at both radio and visible wavelengths. It can be no coincidence that they are both of very short period, and are both connected with comparatively young supernova remnants. Many other pulsars have been examined for traces of light pulses, but without success. It would appear that as a pulsar slows, the optical object fades and only the radio pulses remain.

G 292.0 + 1.8

Crab Nebula of the southern skies?

This nebula is the faint remains of a recent but never-recorded supernova. We reproduce the picture of it as a negative so that the stars appear black. This improves the visibility of the faint gray smudge which, under most circumstances, would be completely overlooked. Finding this nebula is an example of how significant the slightest detail can be in astronomy once someone's attention is drawn to it. The supernova remnant was discovered as a strong radio source (G 292.0 + 1.8 is its radio catalogue number). W. Zealey of the U.K. Schmidt Telescope Unit found the hot optical nebula by taking a series of deep pictures at the radio location. It turns out to be about 15,000 light years away, in the Milky Way constellation of Centaurus.

The nebula is like the dog which did not bark in the Sherlock Holmes story: the most significant fact about it is what is not there. In the spectrum of nearly all nebulae the characteristic colors or spectral lines emitted by hydrogen are present, normally as the strongest lines in the spectrum. This is simply because hydrogen is the most abundant element in the Universe. The spectrum of G 292.0 + 1.8 is remarkable in that it reveals no detectable hydrogen or helium, but only oxygen and neon.

Astronomers believe that supernova remnants are the remains of exploding stars, and that the stars which explode are often very massive. Such massive stars shine initially by converting hydrogen into helium. Later they convert the helium into oxygen and neon,

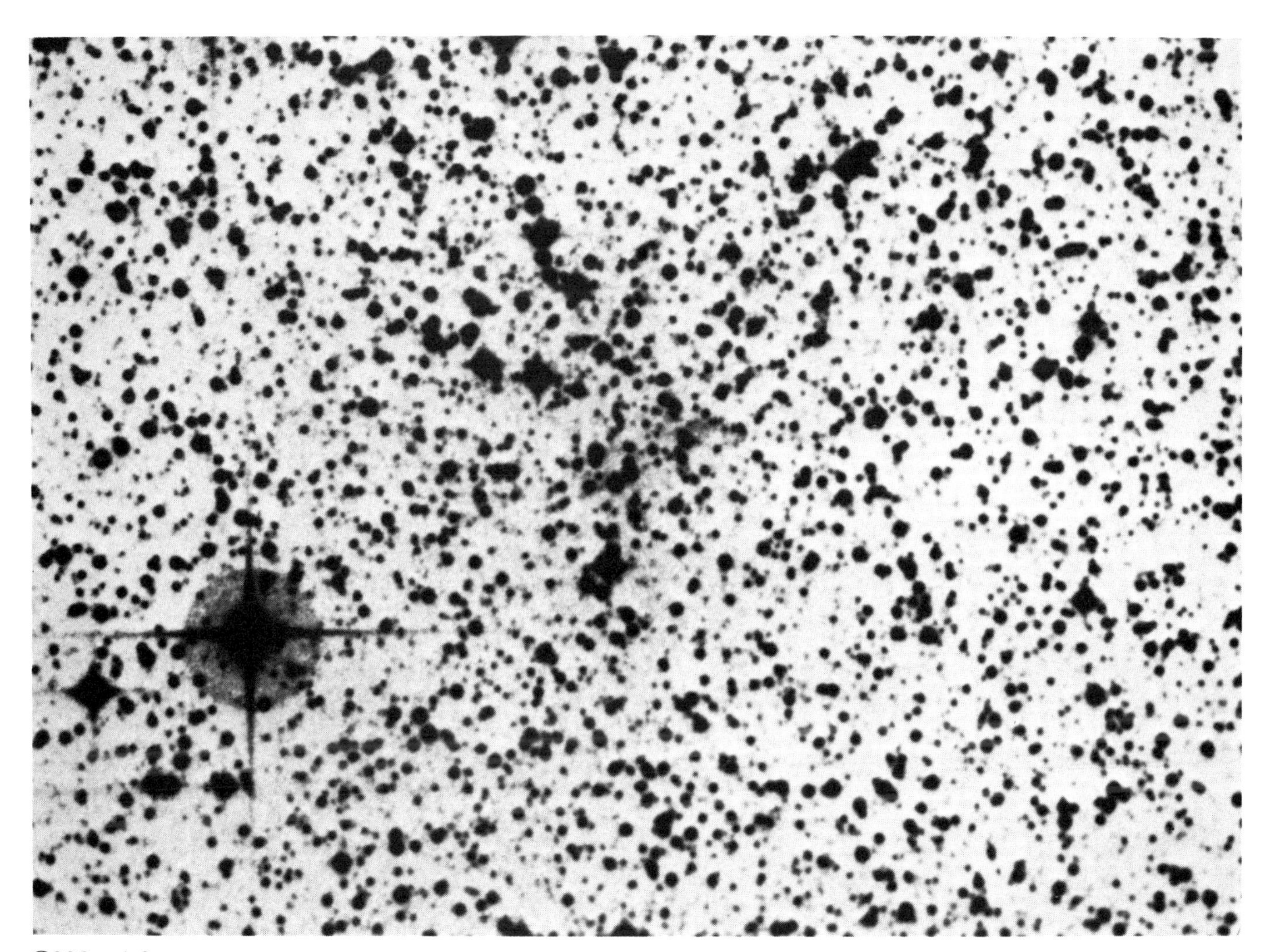

G292 + 1.8 supernova remnant UKSTU

and subsequently into calcium, argon, sulfur and so on up the table of elements as far as iron. Possibly, the nebula G 292.0 +1.8 is the remains of a star so massive that it had processed all its hydrogen to oxygen and neon but not yet to the heavier elements. At that stage the star exploded, sending its fragments dissipating into space. The explosion must have occurred relatively recently, since the remains of the star have not had time to become mixed with the vast quantities of hydrogen in interstellar space, as is the case with most supernova remnants. The map of the supernova remnant made by radio astronomers looks like that of the Crab Nebula.

The latest measurements show that G 292.0 +1.8 is a rapidly expanding supernova remnant, speeding by up to 1000 km/sec from its birthplace. Astronomers calculate its age at no more than 1500 years. It is likely to be more than 200 to 300 years old, since no bright star like TYCHO'S SUPERNOVA has been noticed in the sky like during the European colonization of the three southern continents. Perhaps an archaeological inscription or pictograph will one day be discovered, referring to a brief vision of a new star, a source of wonder to a forgotten people, distracted momentarily from their problems of subsistence.

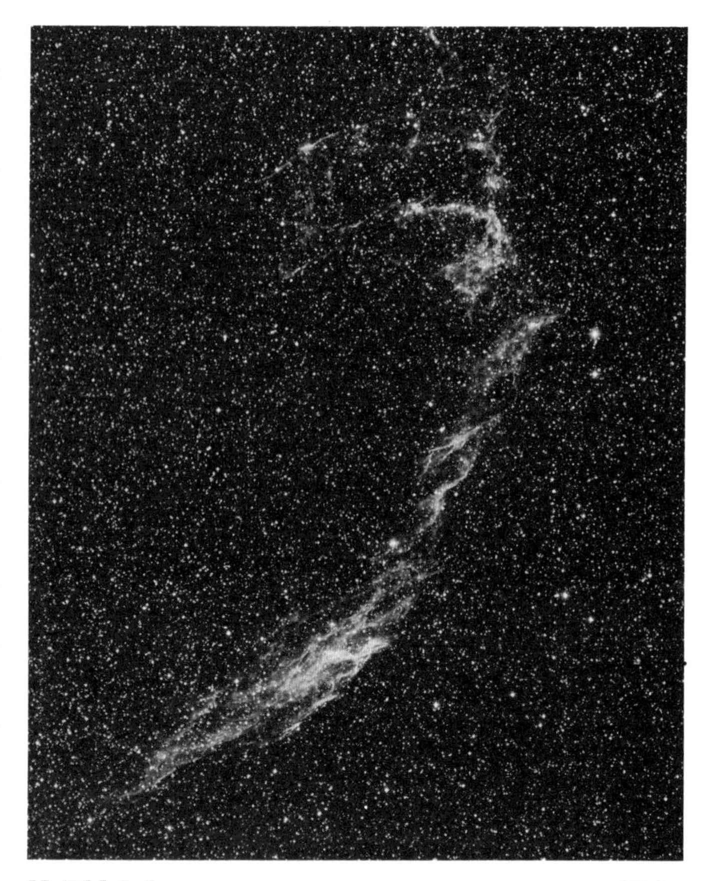

Veil Nebula *Hale*

Veil Nebula

Old supernova remnant

The Veil Nebula (Plate 16) was discovered by Sir William Herschel in 1784. It is also known as the Cirrus Nebula because of remarks made by John Herschel about its resemblance to cirrostratus clouds. It comprises NGC 6979, 6960, and 6992 to 6995. The whole is called the Cygnus Loop because it is a giant shell of diameter 3°. The similarity between the component parts was noticed by Lord Rosse and the way each linked into a shell was confirmed by beautiful photographs taken by Rudolph Minkowski with the 48-inch Palomar Observatory Schmidt Telescope.

Radio maps of the area overlaid on the Cygnus Loop showed clearly that there was a correspondence between the shell seen by radio astronomers and that seen by optical astronomers. Because the spectrum of the radio shell had the properties recognized to belong to supernova remnants, the identification of the Cygnus Loop as an SNR was certain.

The Cygnus Loop is expanding at the rate of six arc seconds every 100 years. At its distance of 2500 light years this is a speed of 116 km/second which would mean that the Cygnus Loop is 160,000 years old if it has expanded at an even rate since it was born. This is naive, however, because a supernova remnant of this age must have swept up an enormous mass of the interstellar material. The effect is like a snowplow moving into a snow bank. It will suffer an enormous braking force, and it is now moving slower than at first.

Evidence of the snowplow effect can be seen at NGC 6960 which runs north-south near the star 52 Cygni. On the east, inside the shell, can be seen many faint, background stars. On the west outside the shell are far fewer because the interstellar dust has been compressed here, making it denser so that it obscures the light of the stars beyond.

Bearing in mind the present slowdown of the Cygnus Loop, a more realistic estimate of its age is 30,000 years. It is one of the oldest known supernova remnants. It will not be long (on an astronomical time scale!) before its slowdown is complete and it has merged with interstellar gas.

The gas, enriched with elements like those formed in G292.0 + 1.8, will journey a few times around the Galaxy before becoming part of some new stars and some yet-to-be-formed solar systems. In the same way, supernovae which occurred in the first few billion years of our Galaxy made the elements which became the material from which our Sun, its retinue of planets and ourselves formed. The outcome of these long-past supernovae is the subject of the final section of this Catalogue.

3. The Solar System

Sun

We know more about the Sun and its immediate environment than we know about any other star. This is not surprising as the Sun dominates our view of the Universe—indeed we live inside its atmosphere. This star, our star, travels around the Galaxy together with a multitude of lesser companions bound to it by gravitational forces. The largest of the orbiting bodies we call planets, most of which are in turn orbited by smaller moons or satellites. These objects, together with many lesser masses, are known collectively as the solar system. Planets are not unique to the Sun. Large planets can be detected around some of the very nearest stars like BARNARD'S STAR, and it is highly probable that planetary systems accompany a large proportion of stars in this and other galaxies. In some young stars like MWC 349 we find evidence of large amounts of gas and dust lying in flattened disks, and these we believe to be the earliest stage in the formation of planetary systems.

It so happens that conditions on one of the Sun's planets were favorable to the establishment of a special kind of chemistry with large and complex molecules capable of replicating themselves and adapting to different environments. This phenomenon, which may be rather rare in the Universe, we call life. It is not possible to assign a meaningful probability to its occurrence on any other planet around any other star. Since some of the chemicals which can form the primitive building blocks of life have been found by radio astronomers in clouds of molecules in space, it seems unlikely that life is unique to our solar system. However, we have no means of detecting its presence elsewhere in the Universe unless that life has developed sufficiently to generate, intentionally or otherwise, radiation which can be distinguished from cosmic sources. Thus we have a highly biased view of the distribution of life.

The solar system is almost confined to a single plane, the ecliptic, and evidently formed from a disk-like nebula. The mechanism by which it formed is fairly well understood. A cloud of gas collapsed under its own gravity to form a central condensation. Prior to its collapse the gas must have had a certain amount of swirling rotation. We can measure the swirling by a physical quantity called angular momentum which depends both on the rotation speed and size of an object. According to physical laws, angular momentum can never be destroyed. As a gas cloud becomes smaller, it must spin faster to conserve its angular momentum. In the same way, a pirouetting ice skater speeds up when she pulls her arms in to her side.

Disk of the Sun KPNO

The rapid rotation prevented the solar nebula from total collapse into one object. A core did form, rotating fairly slowly, which eventually became the Sun. The remaining gas lay much farther out, retaining most of the angular momentum of the original cloud. Although this prevented the gas from falling in towards its axis of rotation, nothing stopped it shrinking in a direction along that axis. Under its own gravity, therefore, it collapsed to form a disk. The collapse modeled, in miniature, the formation of a galaxy.

A certain amount of chemical sorting appears to have taken place in the solar nebula before it began to break up into smaller clouds. Individual clouds themselves subsequently collapsed to form the various members of the solar system. The details of the formation of the solar system, and especially of the chemistry which went on in the collapse, remain a controversial topic among cosmogonists—those who study the formation of the solar system. While it is relatively easy to understand why the planets all lie close to a plane, it is by no means firmly established why those which formed near the Sun are small and dense objects with solid surfaces while the more distant planets are large and mostly gaseous.

From our platform in the midst of the solar system, the major planets appear brighter than the stars, and have therefore assumed great importance in folklore and mythology. But the planets have no internal light sources, and are made visible only by reflecting sunlight. Thus to appreciate the insignificance of the planets, one must compare them not with the stars, but with the Sun. As seen from Earth, the brightest planet is Venus: however, we receive more than a billion times as much light from the Sun as from Venus. The Moon enjoys an even more inflated prominence in our view because of its extreme proximity; in fact not only is it smaller than any planet, but it has a darker surface.

In short, planetary systems are insignificant in the scheme of the Universe although highly significant to the scheme of Man. They account for a negligible fraction of the Universe's mass, and contribute practically no radiation to it. They are merely a by-product of gaseous nebulae whose role was to absorb excess angular momentum. Thus they did at least allow the formation of the stars to which they are subordinate.

The Sun is just a star, one of a hundred billion inhabiting our Galaxy alone. Nor does the Sun's physique endow it with any special significance within that throng. It is a star of no great stature, low in the spread of luminosities and in the range of stellar dimensions, and dwarfed by almost one-half of our Galaxy's members.

Because it lies more than 100,000 times closer to Earth than any other star, the Sun assumes importance not only in our everyday lives, but to astronomers too. We know much more of the Sun than of any other star. And, we are aware of a multitude of the Sun's characteristics which defy explanation or exceed the bounds of our current comprehension. In studying the Sun, astronomers are made aware of the deficiencies in their understanding of the more distant members of the Universe.

Being a star, the Sun is an example of the fundamental building blocks of our Universe. It formed, 4.5 billion years ago, as the nucleus of a cloud of gas which was collapsing under its own gravitational attraction. Once the collapse began, the characteristics of the resulting star were preordained. The mass of gas which collapsed into the star was two million trillion trillion kg. At the center of the collapse, the infalling gas became heated by the gravitational energy it bore. Steadily the temperature rose. After some 14 million years, the central regions became sufficiently hot to trigger nuclear processes, turning hydrogen into helium and releasing energy in doing so. Once begun, this nuclear process was unquenchable. It continues today, and will do so for perhaps a further four billion years, halting only when the supply of hydrogen is almost exhausted. The heat and light we receive from the Sun is a by-product of this nuclear holocaust, and to supply the vast outpouring, five million tonnes of hydrogen are consumed every second.

The outward percolation of energy acts to counter the inward pull of gravity. An equilibrium has resulted. If the Sun were to shrink, its interior would become hotter, the nuclear reactions would be hastened, and the increased energy output would reinflate the Sun to its present size. Conversely, if the Sun tried to expand, gravity would dominate the dwindling nuclear resources and enforce a return to equilibrium. The conditions of the equilibrium—the 700,000-km radius of the Sun, its surface temperature of 5500°C, its central temperature of some 15,000,000°C, its density of 1.4 times that of water—were determined predominantly by the initial mass of the collapsing gas cloud whence it formed.

The young Sun was a star, involved in a flattened nebula like MWC 349. That nebula formed the planets, and from one of these we now view the Sun in middle age. Later, as its hydrogen becomes depleted, it will evolve into a red giant star, swelling to engulf the Earth and the inner planets. The outer layers will be shed in a series of hiccups, perhaps to form a planetary nebula (p. 140 onwards), then senility will take hold: the remnant Sun will fade gradually into oblivion, passing through the white dwarf stage on its way. That is a saga of the far future; the story which will unfold over the next few pages is that of an average star in that stable phase which characterizes the middle of its life.

The Sun is entirely gaseous. Its average density, in excess of that of water, might lead one to expect that liquid or solid portions lie beneath its surface. Instead, however, the gases of its interior are highly compressed but yet are too hot to have been liquefied. The central density is 160 times that of water, and the pressure is 250 billion times that on the Earth's surface. Yet the material is still technically a gas.

At formation, hydrogen was the most abundant gas, as elsewhere in the Universe, and accounted for three-quarters of the Sun's original material. This proportion obviously changes as hydrogen is burned, and within the very central regions virtually all the hydrogen has been converted to helium. The outer portions have not yet participated in hydrogen burning. Moreover, there has been only a little mixing of the innermost and surface materials. When we view the Sun's surface, and determine its chemical composition, we are sampling material little changed from the primordial gas. Astronomers have measured the chemical composition of the Sun, and can thus estimate that of the initial solar nebula from which Sun and planets formed. In addition to 78% by weight of hydrogen, they find 20% to be helium, while only 2% remains for all the heavier elements such as oxygen, carbon, nitrogen and iron. The gas from which the Earth and all its inhabitants were formed included even less of the heavy elements, and thus differed greatly from the present chemistry of, for instance, the readers of this book.

From our standpoint almost 150 million km away, we see the Sun as an almost circular disk about half a degree across (Plate 17). It is the only star that we can directly distinguish from a point of light. The details of its surface structure shown in the next few photographs are only inferred to exist on other stars.

Sunspots

Sun as a semi-regular variable star

Anyone who has never seen the Sun through a telescope might imagine that it shows a featureless disk. But as the previous photograph showed, its face is often littered with dusky spots, appropriately named sunspots. (Here we must include the customary warning never to look directly at the Sun using even the smallest telescope or binoculars.)

It is rare to find the Sun completely devoid of spots. Sometimes a great many groups and individuals appear, so that about 1% of the surface is dark. Large spots are big enough to be recorded by the unaided eye, though it need hardly be said that the Sun is too bright to be studied without damage to the human eye. Naked-eye sunspots are best seen on misty days just at sunrise and sunset, when the Sun's light is dimmed by passage through a long path in the Earth's atmosphere. Detailed records of naked-eye sunspots were kept in China and Korea over a period of several thousand years. The invention of the telescope nearly four centuries ago led to an improvement in the recording of sunspots, and for the last two centuries the records are complete. Study of these records shows that the degree of spottiness undergoes great variation.

Most striking is the 11-year cycle. The number of spots visible on the Sun reaches a maximum roughly every 11 years. Between these maxima are low points during which few spots are seen. The maxima them-

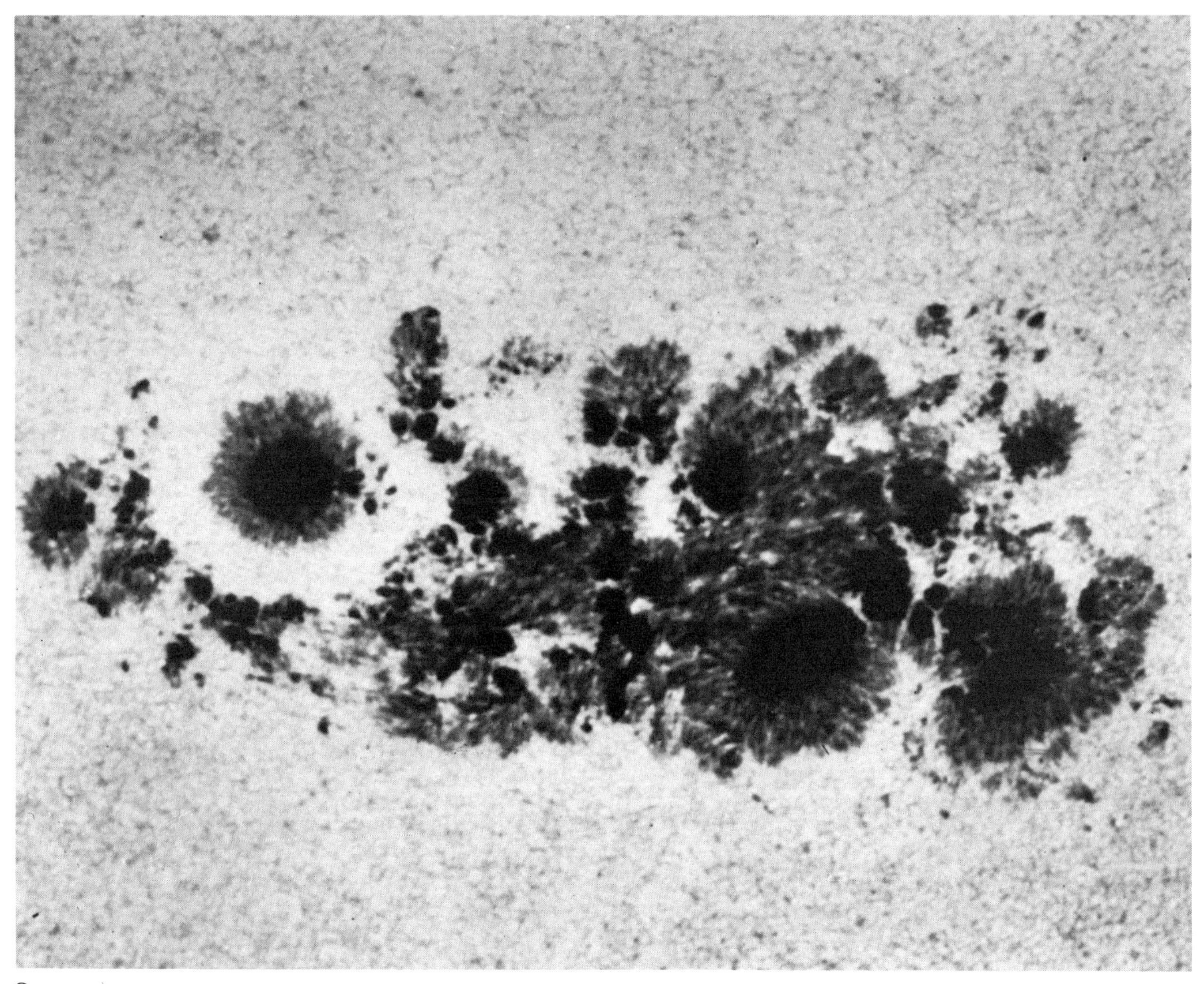

Sunspots CSIRO

selves are not evenly spaced, nor are they of the same intensity. At one maximum the Sun may display twice as many spots as at neighboring maxima. In short, the Sun is a semi-regular variable star of period about 11 years and amplitude less that 1% (0.01 magnitude). Such variability could not have been detected on any other star.

In addition to the 11-year period, Oriental records strongly suggest that there have been occasional periods of a few hundred years when no naked eye sunspots occurred. We cannot be entirely sure these are not simply eras when observation of the Sun was not fashionable, and astronomical historians are seeking corroboration of these results.

Sunspots comprise two regions, an umbra, which is the dark central portion, and a penumbra which is of intermediate brightness and usually appears streaky. Small spots can be composed entirely of umbra or of penumbra. The umbra appears to be depressed below the level of the Sun's surface. Thus when spots are seen towards the edge of the Sun, the umbra appears displaced within the penumbra in the same way that the floor of a crater is not central within its rim when viewed at a shallow angle.

Sunspots appear black on photographs such as this, but in fact are some of the brightest objects in our sky. The center of the darkest spot is many thousands times as bright as the brightest portion of the full Moon. However, the surrounding surface of the Sun is so much brighter that spots appear dark by contrast.

In order to understand the origin of sunspots one must know something of the Sun's magnetic field. We are familiar with the effect of a magnetic field on Earth, and mariners have long made use of its properties with the magnetic compass. The Sun has a magnetic field only slightly stronger than the Earth's. However, the Sun is entirely gaseous. Its gases seethe and bubble, as the next photograph will show. Whereas the Earth's field is stable, that of the Sun is continuously being distorted and convolved by the motions of the gas. If we could sprinkle iron filings around the Earth, they would align into a pattern very similar to that around a bar magnet. A similar experiment on the Sun would show a more complex structure. In places the field would have large loops, rising outwards in curving arches with spans of 100,000 km or more.

Where the loops of magnetic field cross the surface, sunspots form. Within the umbra of a sunspot the magnetic field is intense, almost 10,000 times as strong as that of the Earth's surface. This is so strong that the magnetic field is not easily pushed around by the seething gases. Sunspots last from a few days to a few months, therefore. Within the umbra of each spot the gas cannot readily transport the Sun's inner heat to the surface. The temperature is thus lower by as much as 2000°C. Being cooler, the umbra radiates less light, and that is why sunspots are darker than their environs. The magnetic field also thins the gas within the umbra, and thus it is clearer. In this way an illusion of depth is created, so that the spots resemble craters.

If the loops of magnetic field were simple structures, we should see two spots in each group—one where the loop comes out and the other where it reenters. The two spots would be opposite magnetic poles, one north, one south. Indeed, simple spot pairs are often seen with these properties. But the magnetic fields are very complex, and in large groups such as the one shown here a multitude of spots forms around individual concentrations of the magnetic field. Nonetheless, a complex group still divides into two regions of opposite polarity. The magnetic fields which intertwine between these two regions account for the swirling and streaky nature of both the penumbrae and the surrounding solar surface.

During the course of the 11-year cycle the latitudes of the spots change. At first, after minimum, the spots occur at high latitudes. Steadily the regions of spot formation migrate towards the equator. This behavior has been explained in complex analyses of the magnetic field. When the next cycle begins, the spots appear with opposite polarity. Thus if north pole spots lay to the right in one cycle, south pole spots would lie to the right in the next cycle. The full solar magnetic cycle is therefore 22 years long.

Sunspots take part in the rotation of the Sun. Day by day they move across the Sun's disk. By measuring their position on successive days we can determine the rotation period of the Sun. We find that at the solar equator the period is about 26 days. Near the poles, 37 days is required for one rotation. The fact that the equatorial regions spin faster accounts for the unusually complex twisting of the magnetic fields. In all probability, the inner core of the Sun rotates much faster than the surface we see, and this drags the equator round to overtake the poles.

Granulation

The seething Sun

This photograph shows details of the surface of the Sun near a sunspot group. It is included to show that the Sun's surface, the so-called photosphere, is by no means smooth. Rather, it has the appearance of a coarse rice pudding. This cell structure is known as granulation.

Essentially all the energy given out by the Sun is generated deep within its interior. The nuclear combustion gives rise to radiation of extremely high energy: gamma rays. These propagate outwards at the

Vela Supernova Remnant

Sun's disk

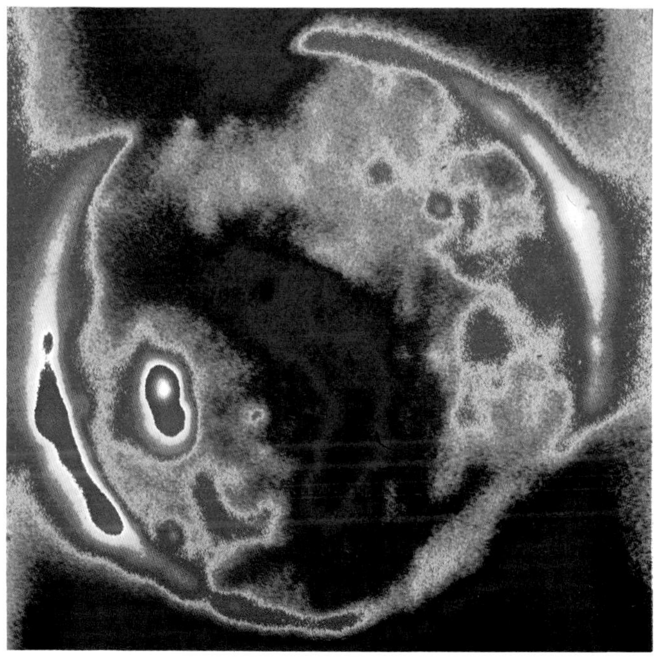

X-ray Sun

PLATE 17

Development of a radio flare on the Sun

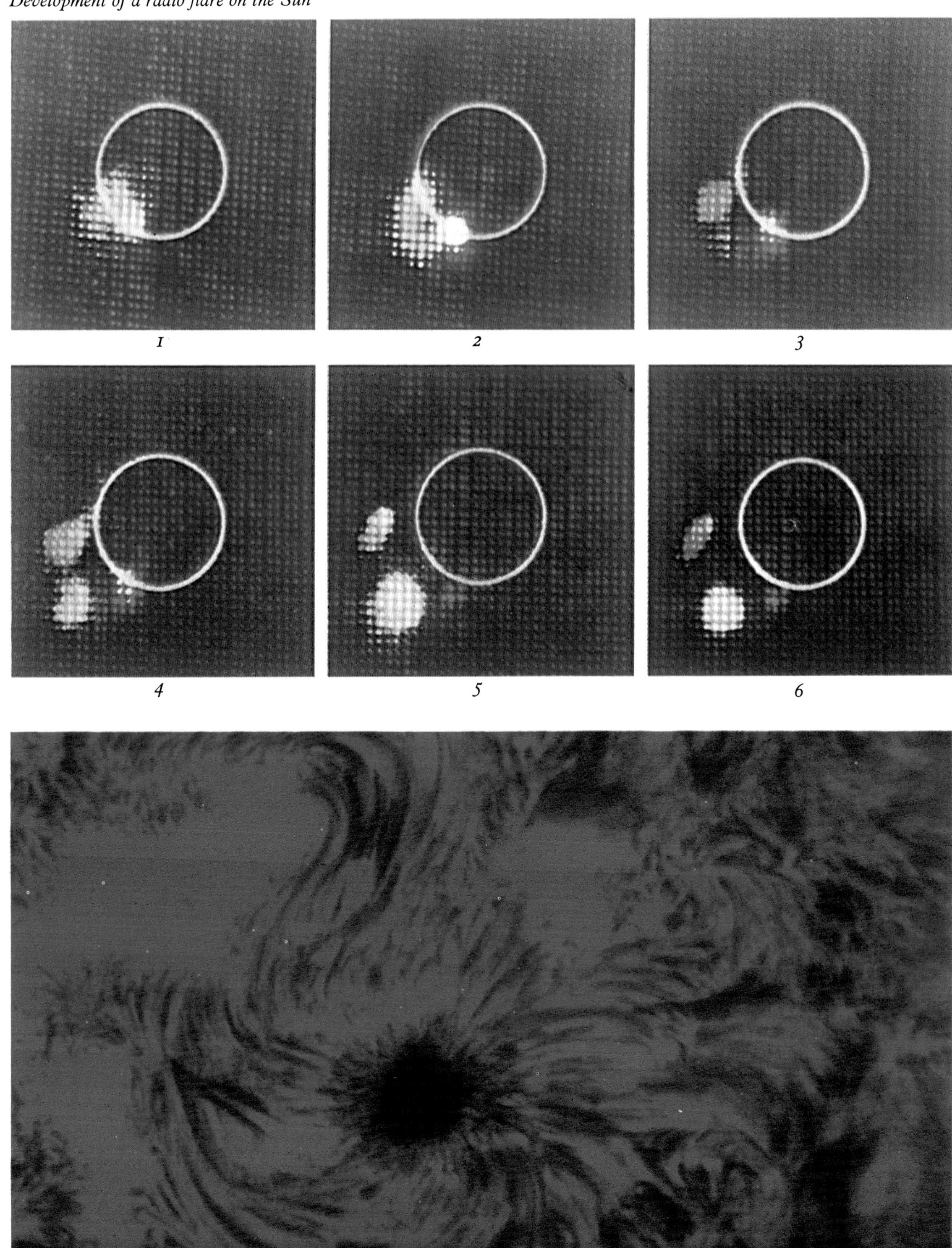

Sunspot in H-alpha light

PLATE 18

Jupiter's Great Red Spot

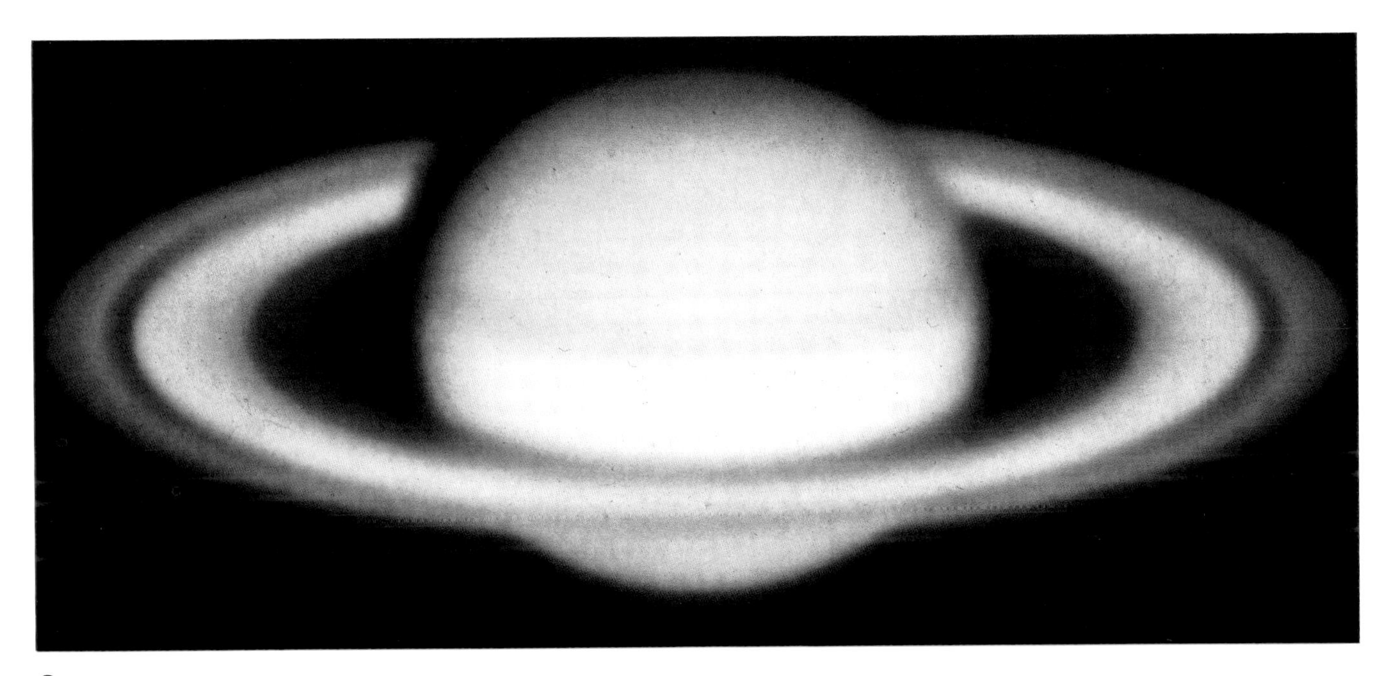

Saturn

Jupiter with Red Spot

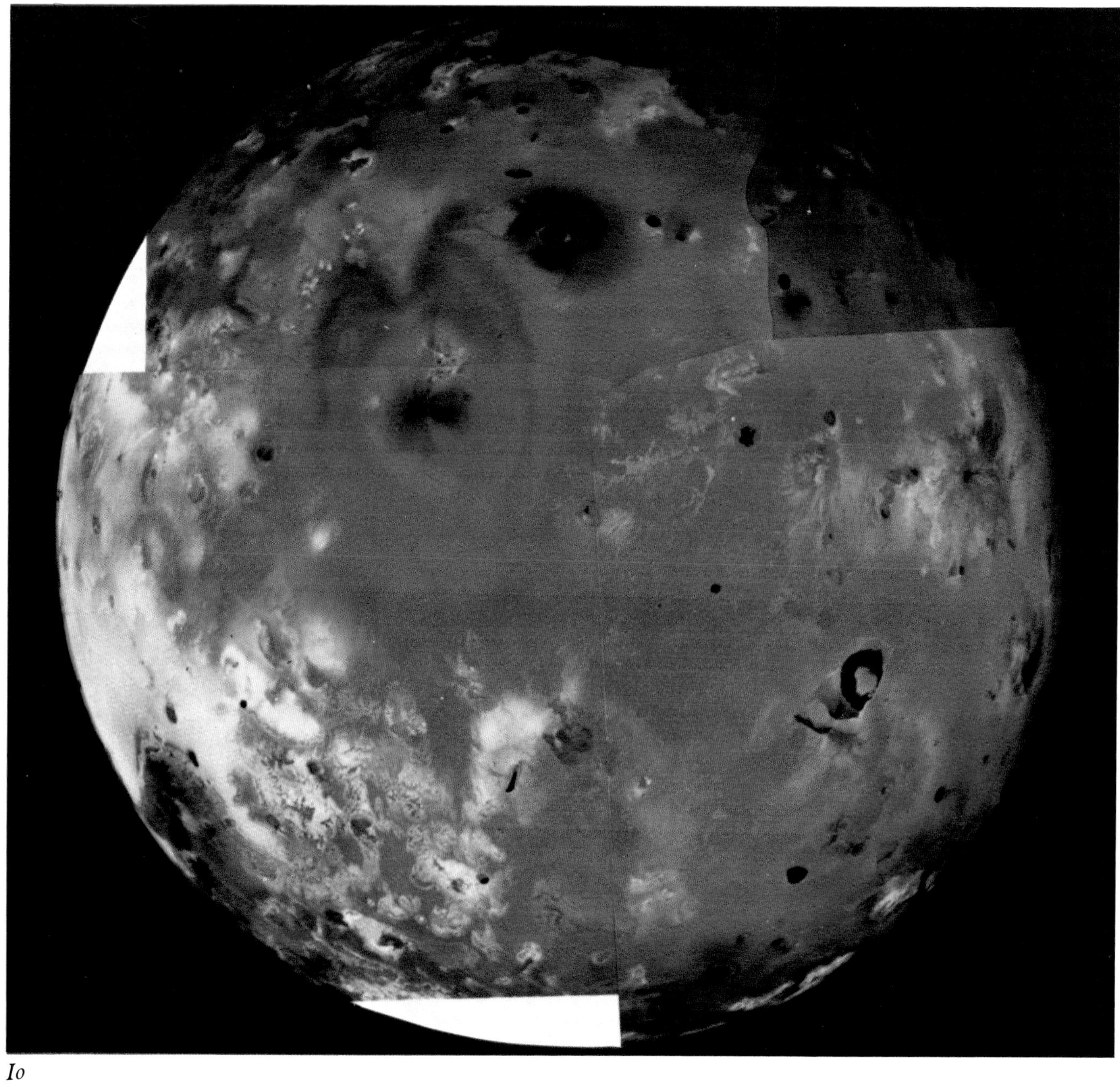

Io

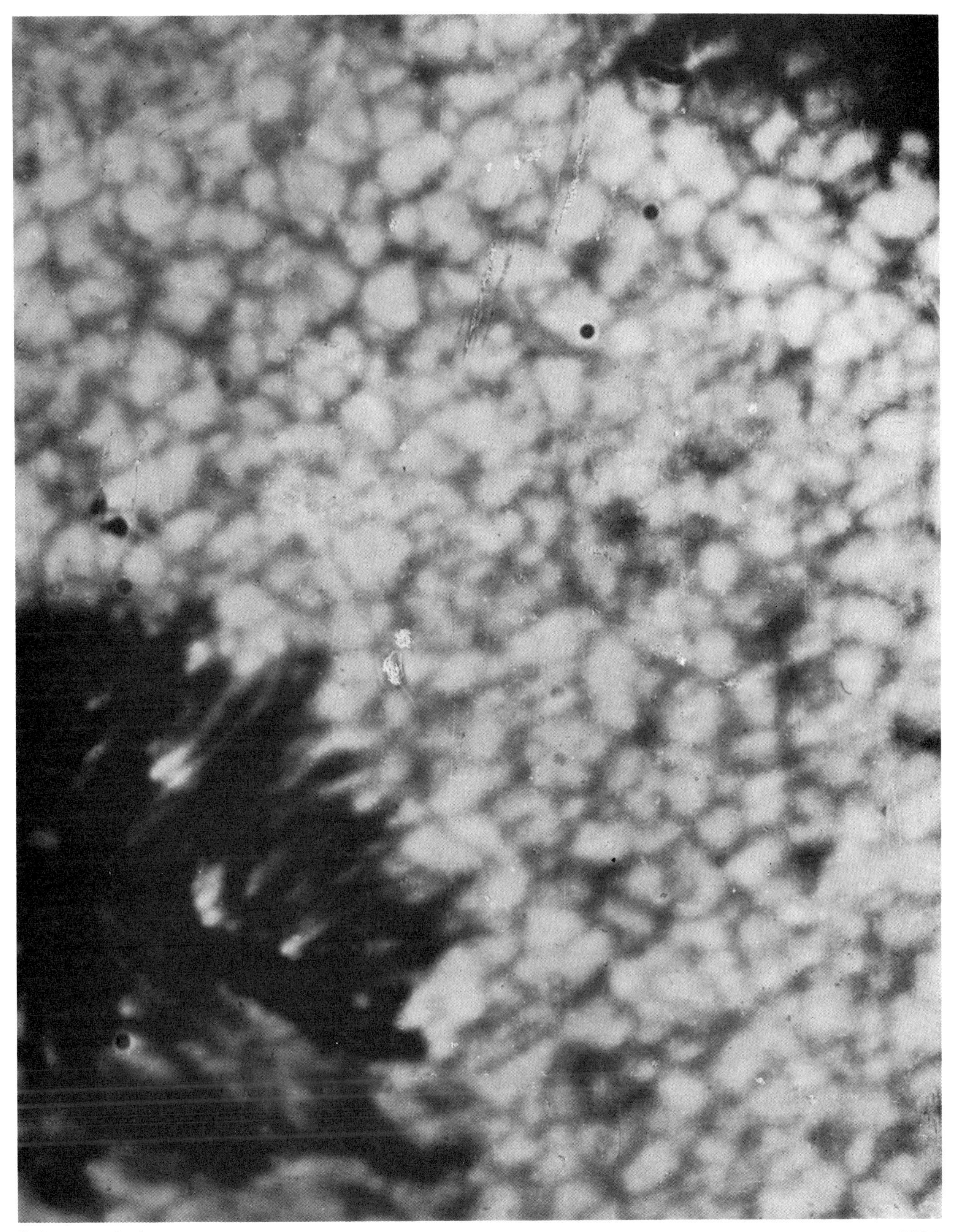

Solar granulation CSIRO

speed of light, but before traveling even one centimeter they are absorbed by atoms of the gas. This is the first of a succession of absorptions that each parcel of energy must endure on its tortuous journey to the surface. The absorptions occur so readily that the outward flow is slowed to an effective speed much less than a millionth that of light. With each absorption the radiation loses energy, so that gamma rays become degraded to x-rays, then to ultraviolet radiation and, finally, to visible light.

But that is not the whole story. At a point about 100,000 km below the surface the energy finds a better means of escape: convection. Locally hotter blobs of gas, being less dense, rise through the Sun's outer layers. Having risen they cool, releasing their excess energy, then sink again to make room for the next blobs. Convection within the Sun occurs much as in a heated pan of water. However, the distances involved are so large that the convection pattern does not extend right to the surface. Instead there is a series of turbulent, convective layers, and with each layer outwards the convection cells become smaller. The outermost layer has the smallest convection cells, which we see as granulation. Having reached the surface, radiation once more becomes the easiest way for energy to escape, and light and heat is emitted. Had the radiation been able to travel at this speed for all its journey from the center of the Sun, it would have taken a mere five seconds; in practice, its journey over one solar radius takes a great many years—according to calculations, an incredible ten million years or more!

The granulation cells can be measured as nearly 2000 km across, and each lasts typically 10 minutes. An underlying, larger pattern of convection is also seen, with cells more like 30,000 km across lasting typically 20 hours. This is called supergranulation. A terrestrial analog of granulation is the pattern of small, fluffy fair-weather cumulus clouds often seen on sunny afternoons. These are produced in exactly the same way, but by circulating air patterns. If we wish to take the analogy further, we could liken supergranulation to major thunderstorms.

Needless to say, granulation cannot be seen on any other star. The existence of convection on other stars is, however, inferred from spectroscopy by the mixing to the surface of elements produced deep within the interiors and by other, less direct observations.

The Corona

The Sun's atmosphere

One of the greatest boons to our understanding of the Sun is the remarkable coincidence that the Sun and the Moon appear roughly the same size in the sky.

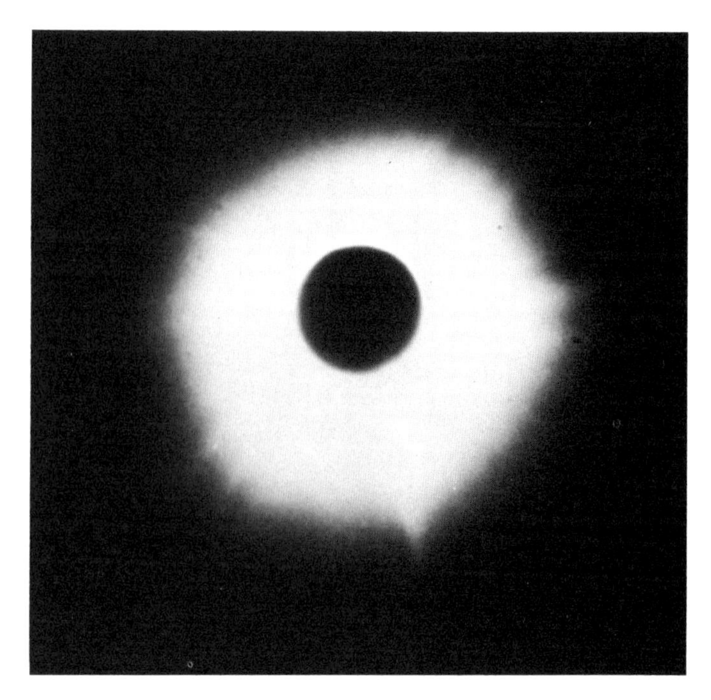

Eclipse of the Sun, 1976 DAA

Moreover, the Moon in its complex motion around the Earth periodically passes exactly in front of the Sun. It casts a shadow which falls on the planet as a circular spot usually about 200 km in diameter, and this spot sweeps across the Earth's surface as the Moon moves in its orbit. An observer standing within the path of the shadow will witness, weather permitting, one of the most spectacular of all natural phenomena: a total eclipse of the Sun.

If seen when the Sun is high in the sky, a total eclipse is an emotional experience. The slow darkening of the sky as the Sun is progressively covered stirs thoughts of evening, but at entirely the wrong time of day. Instead of lengthening, shadows remain short and gradually, eerily, melt into a uniform grayness. The horizon colors as at sunset, but dons its chromatic robes at all points of the compass. At the onset of totality the lighting fades with alarming rapidity, as at the movies before the main feature. Birds and animals become disturbed and confused; unwary humans too. With the Sun totally obscured, the sky darkens sufficiently for bright stars to be seen, and the atmosphere of the Sun appears.

The corona, which is the name given to the Sun's atmosphere, is illustrated in two photographs, both taken during total eclipses. One, in color (Plate 19), reproduces fairly accurately what the human eye sees. The other (p.163), in black and white, was taken through a filter so constructed as to record detail over an extensive area.

The color photograph demonstrates that the corona is a very pure white. At its brightest the corona is about as intense as the full Moon, or one-millionth as

Solar corona *High Altitude Obs*

bright as the Sun itself. It is therefore normally completely swamped by daylight, and cannot be studied except at a total eclipse or from above the Earth's atmosphere. Since most total eclipses last less than five minutes, the amount of information which Earthbound astronomers can glean on the corona is not great.

We do know that it changes shape. At the maximum of the eleven-year sunspot cycle the corona appears almost circular. At sunspot minimum it is markedly elongated, with its long axis parallel to the Sun's equator. At times between the extremes of the eleven-year cycle the corona tends to show streaks and plumes, as in these photographs. The streaks are often positioned above major sunspot groups or other surface disturbances. The streakiness itself, and the preferred location above sunspots, indicates that the corona is controlled by the Sun's magnetic field.

The temperature of the Sun falls steadily from the central furnace of 15,000,000°C to a surface value below 6000°C. It might therefore be assumed that the corona is cooler still, but surprisingly this is not the case. The solar temperature structure reaches a minimum at the visible surface. The corona is very hot, as was finally demonstrated by Edlén in 1942. Within 3000 km of the Sun's surface the temperature has risen to 2,000,000°C, and this is maintained throughout much of the corona. The Earth and inner planets are saved from being scorched by the heat of the corona only because it is extremely tenuous. Indeed, the Earth is actually imbedded in the corona. At our distance from the Sun, 150 million km, the coronal temperature is still 100,000°, but the gas is so rarefied that there are only a few atoms per cubic centimeter, a figure which should be compared to the 30 million trillion molecules per cubic centimeter of the Earth's sea-level atmosphere.

Unlike the Earth's atmosphere, that of the Sun is constantly escaping into space and being replenished by material ejected from the solar surface. The rate of loss of material from the Sun is nearly one million tonnes per second. At the Earth's distance this streams past at about 400 km/sec, and is known as the solar wind. The particles of the solar wind, which are mostly individual protons and electrons, take five days to travel to us from the Sun.

Although few in number, these particles interact with the Earth's atmosphere and magnetic field in many complex ways. For example, they make the upper atmosphere glow faintly, which in part accounts for why the sky is never completely dark at night. Aurorae are even more dramatic manifestations of the solar wind.

We cannot directly observe coronae around other stars, but in some cases we infer the existence of denser (and hence brighter) coronae than the Sun's.

The Chromosphere

The Sun in hydrogen light

Between the bubbling surface and the superheated atmosphere lies a region of the Sun known as the chromosphere. The name derives from its rich crimson color (*khroma* in Greek) which can be seen briefly as a total eclipse starts and ends. The chromosphere is a mere 3000 km thick; it therefore clings very close to the edge of the solar disk and is completely covered by the Moon around mid-eclipse.

There is a 2,000,000°C temperature change across the chromosphere. Moreover, from top to bottom the gas density decreases by a factor of 10,000. Add to these factors the constant seething of the surface beneath, and it will be evident that the chromosphere is a violent and turbulent part of the Sun. It has been likened to the froth atop boiling jam.

Because of the activity, the chromosphere contains ionized gas, rather like the ORION NEBULA (p. 102) and other H II regions. Its crimson coloration (Plate 18) is none other than that of hydrogen, so it is possible to study the chromosphere by observations restricted to the red hydrogen line. The two photographs accompanying this section were all taken in the light of ionized hydrogen.

Seen in the hydrogen line, the Sun appears much more active. Not only do we see sunspots, but many more streaks and blobs appear, some of them resembling flames. These are jets of glowing gas thrown out by the Sun and known as spicules. Large spicules can

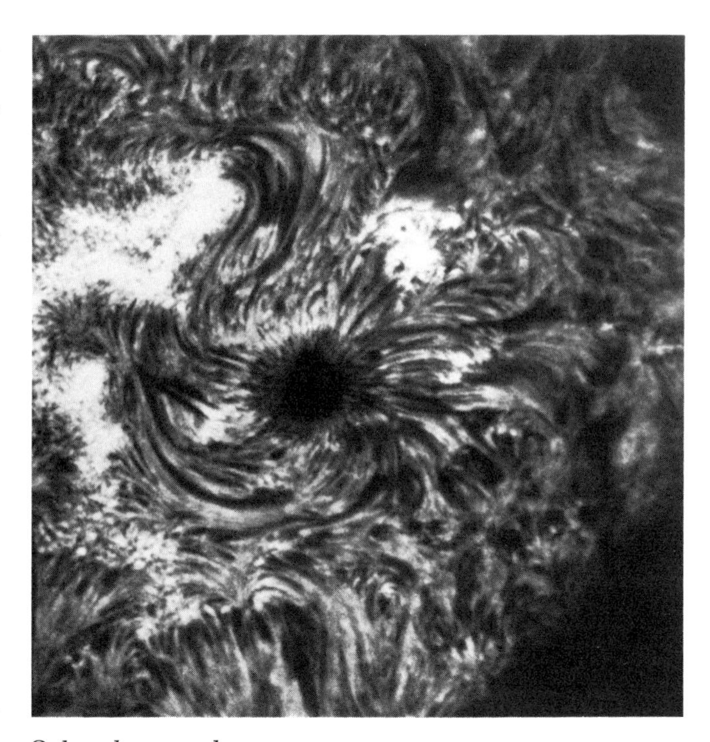

Solar chromosphere CSIRO

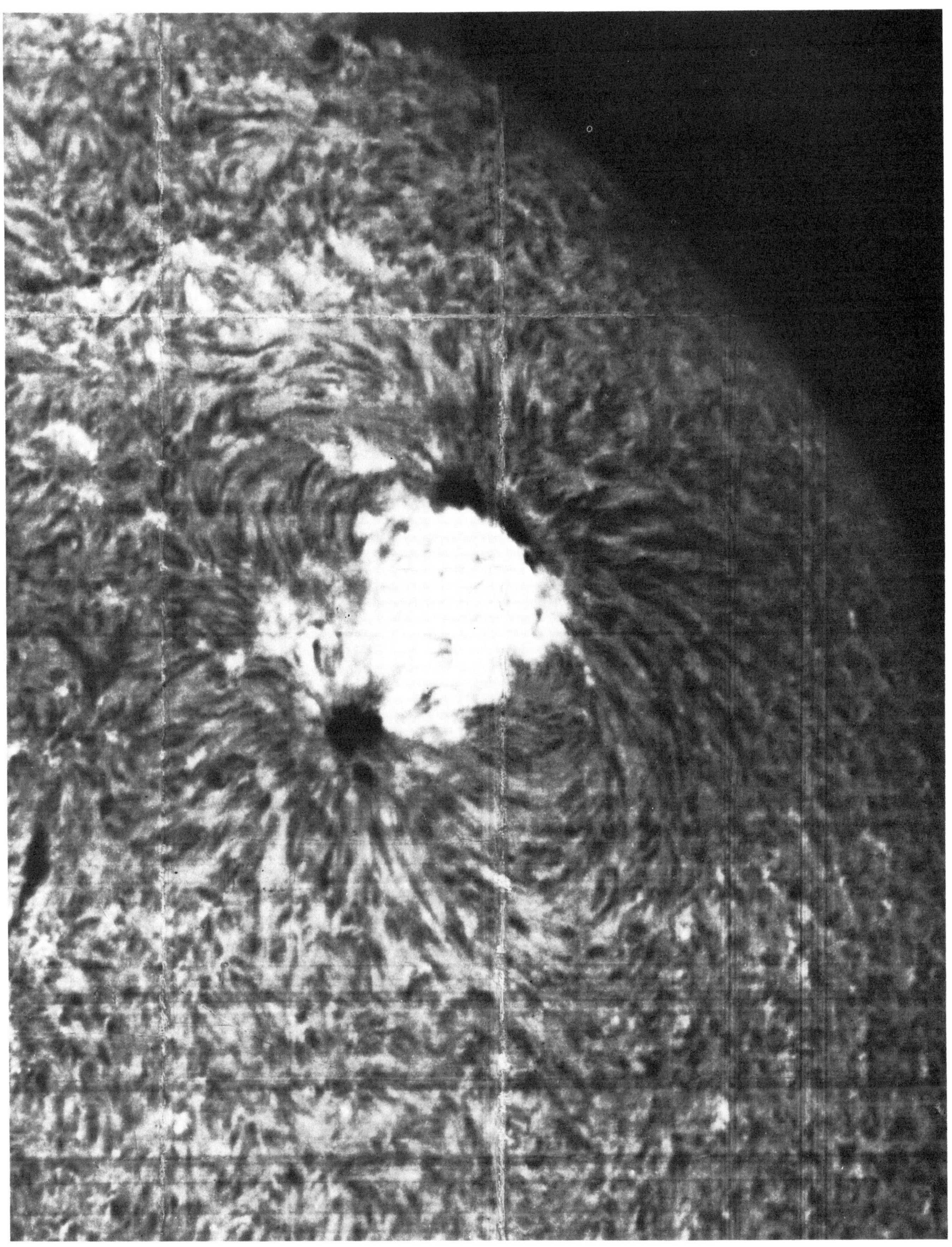

Solar chromosphere CSIRO

stretch 10,000 km from the surface, and persist for about three minutes. They occur where the magnetic field is slightly enhanced between the cells of supergranulation, and may be the major way the Sun loses material into the corona and solar wind.

Sunspots are particularly prominent in hydrogen light. Even before a sunspot group appears, a bright area known as a plage develops. The dark spots grow within the plage, which therefore forms a bright border to them. The plage remains as a scar for some days after the spots disappear.

Many other stars are known to have chromospheres, and in some cases the chromosphere is so well developed that the bright hydrogen lines can be recorded with a spectroscope even when superposed on the light of the star's entire photosphere.

Prominences

Solar projections

Although not usually visible to the unaided eye, prominences are revealed at a total solar eclipse just as is the corona. On these occasions a pair of binoculars shows them well as crimson flame-like projections, though few will be seen at sunspot minimum. Their color immediately suggests that ionized hydrogen accounts for much of the radiation they emit, and that photographs taken by its light might show them well. Indeed this is the case, and with the additional aid of an opaque disk to block off the Sun's direct light it is possible to photograph solar prominences in daylight. The photographs shown here were taken in this way. Moreover, photographs of the whole Sun taken in hydrogen light show prominences on the disk, where they appear dark because of the brightness of the Sun's radiation against which they are seen.

Two types of prominences occur. Long streamers or curtains projecting from the Sun are the commonest. These are long-lived, and can survive for as long as six months. They tend to change little with time, being controlled by stable portions of the magnetic field, and are therefore called quiescent prominences. Loop prominences are variants which comprise one or more curved arches, touching the Sun at both ends of their span. Loop prominences also follow magnetic field lines, but are more intimately connected with sunspots, where loops of the solar magnetic field break through the surface. The second type of prominence is called eruptive because it is ejected from the Sun in a violent burst. The largest eruptive prominence ever observed, in 1946, was traced out to almost one solar

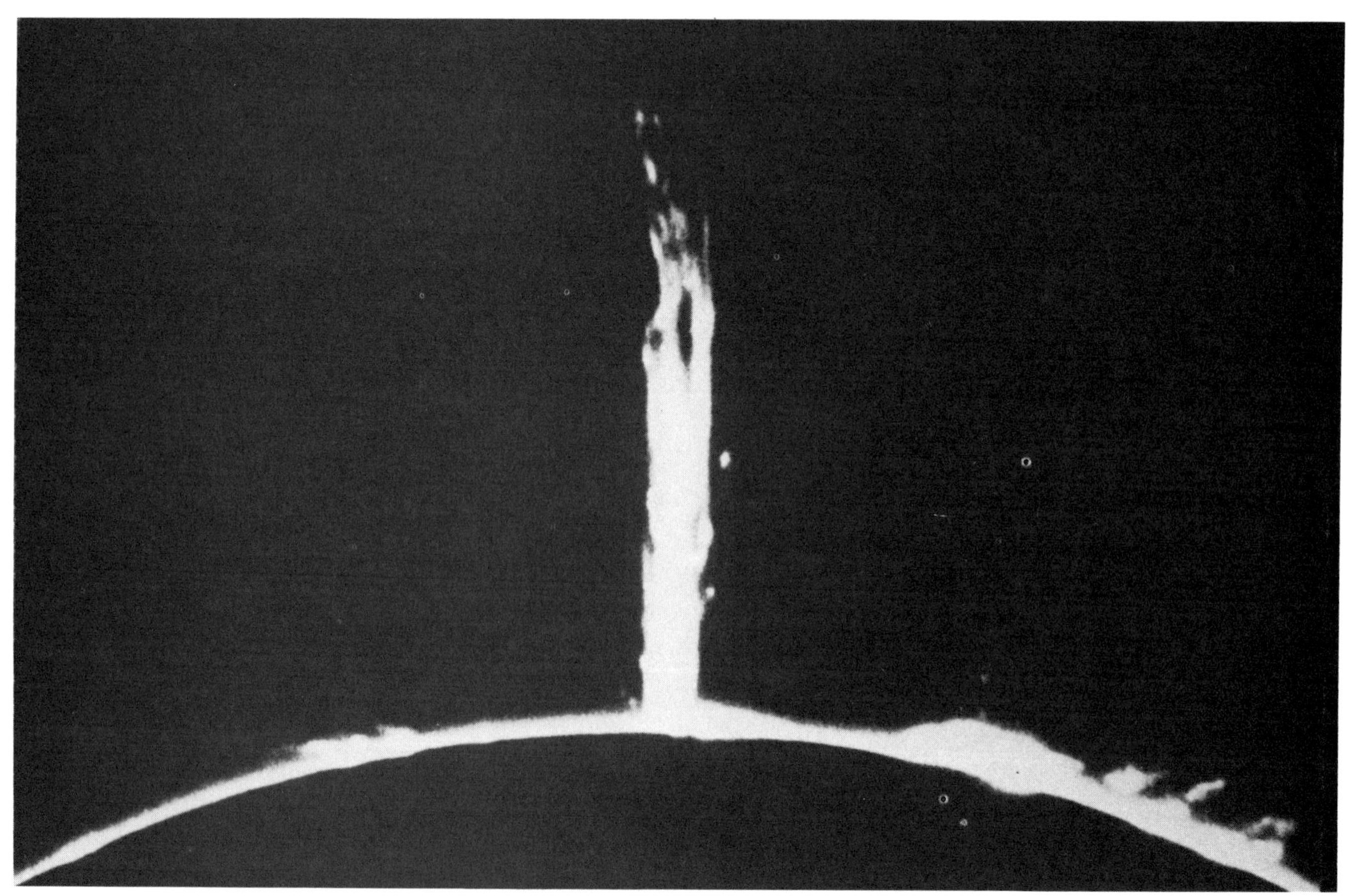

Jet prominence CSIRO

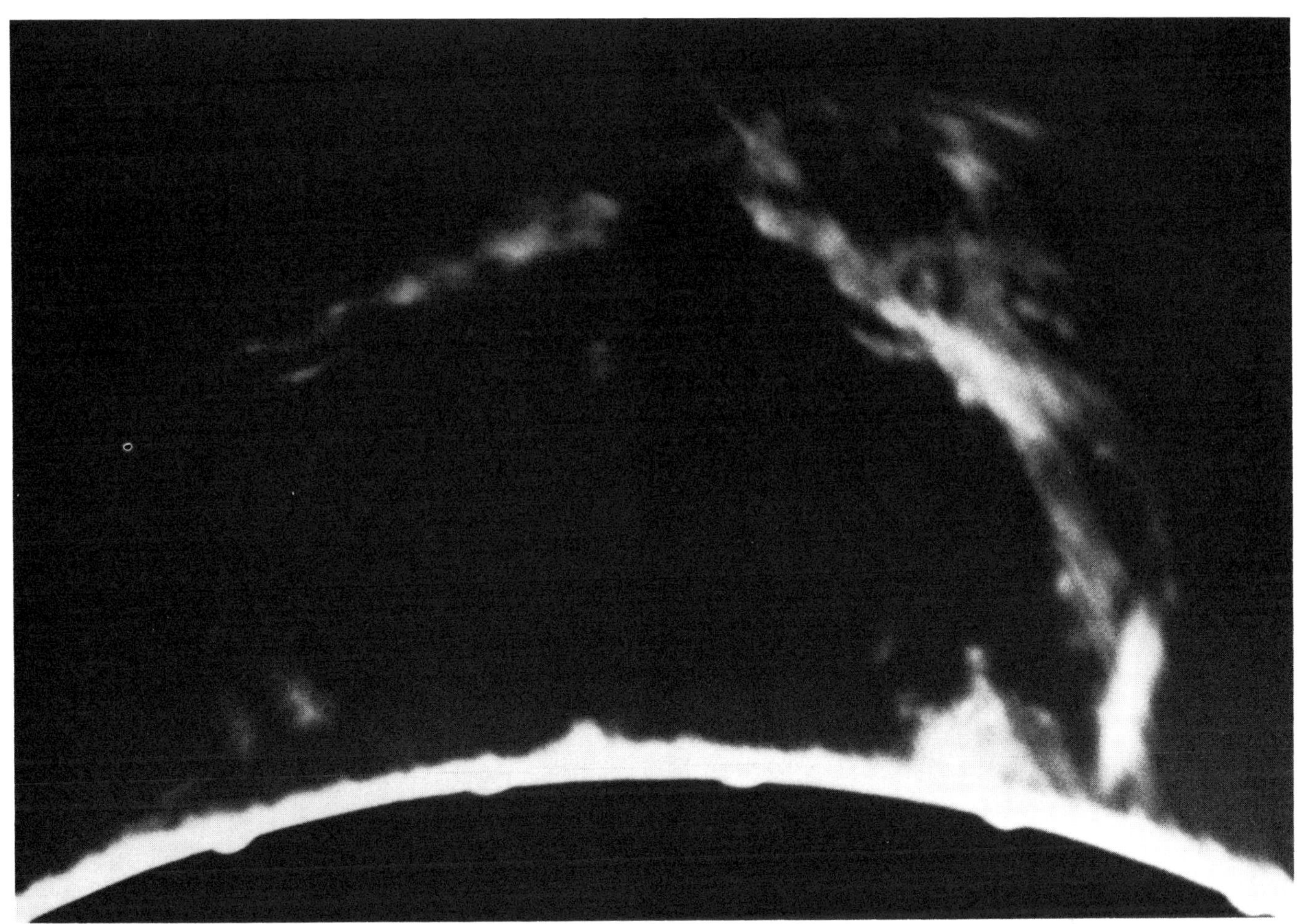

Arch prominence *Hale*

radius from the Sun's surface before dissipating. The whole eruption took only two hours, though the prominence had been observed in a quiescent phase for several months before.

It might be thought that prominences are merely larger versions of the spicules just described, and that they therefore represent regions where material is being injected into the Sun's corona. Despite the evidence of eruptive prominences, however, astronomers believe that quiescent prominences are infalling. They represent the paths by which cool material which the corona can no longer support falls back to the Sun's surface down the funnels of magnetic field lines.

Prominences are not known to exist on other stars, but astronomers know of no reason to suppose they are unique to the Sun.

Flares

The active Sun

In a way not entirely understood, the Sun stores up extra energy which it releases in violent, almost explosive events known as flares. The energy is released through the chromosphere and into the corona, whence it escapes as a ripple in the solar wind. Most flares develop in a matter of minutes and subside over an hour or two. Since magnetic energy is involved, it will come as no surprise to learn that flares are usually associated with sunspot groups where the strongest magnetic fields occur.

The accompanying sequence of four photographs shows the successive development of two flares from the same sunspot group. All photographs were taken in the light of ionized hydrogen, which accounts for the strongly textured image of the Sun. The photographs were made about half an hour apart.

In the first (upper left), an intensely bright patch at right of the major sunspot group is a region emitting the red hydrogen line particularly strongly. This indicates a strong disturbance of the chromosphere, and is a weak flare in progress. This bright region has entirely faded by the time of the second photograph (upper right): only the dark spots and brightish, surrounding plages are seen. However, the second photograph shows instead an eruptive prominence being ejected from much the same region. The

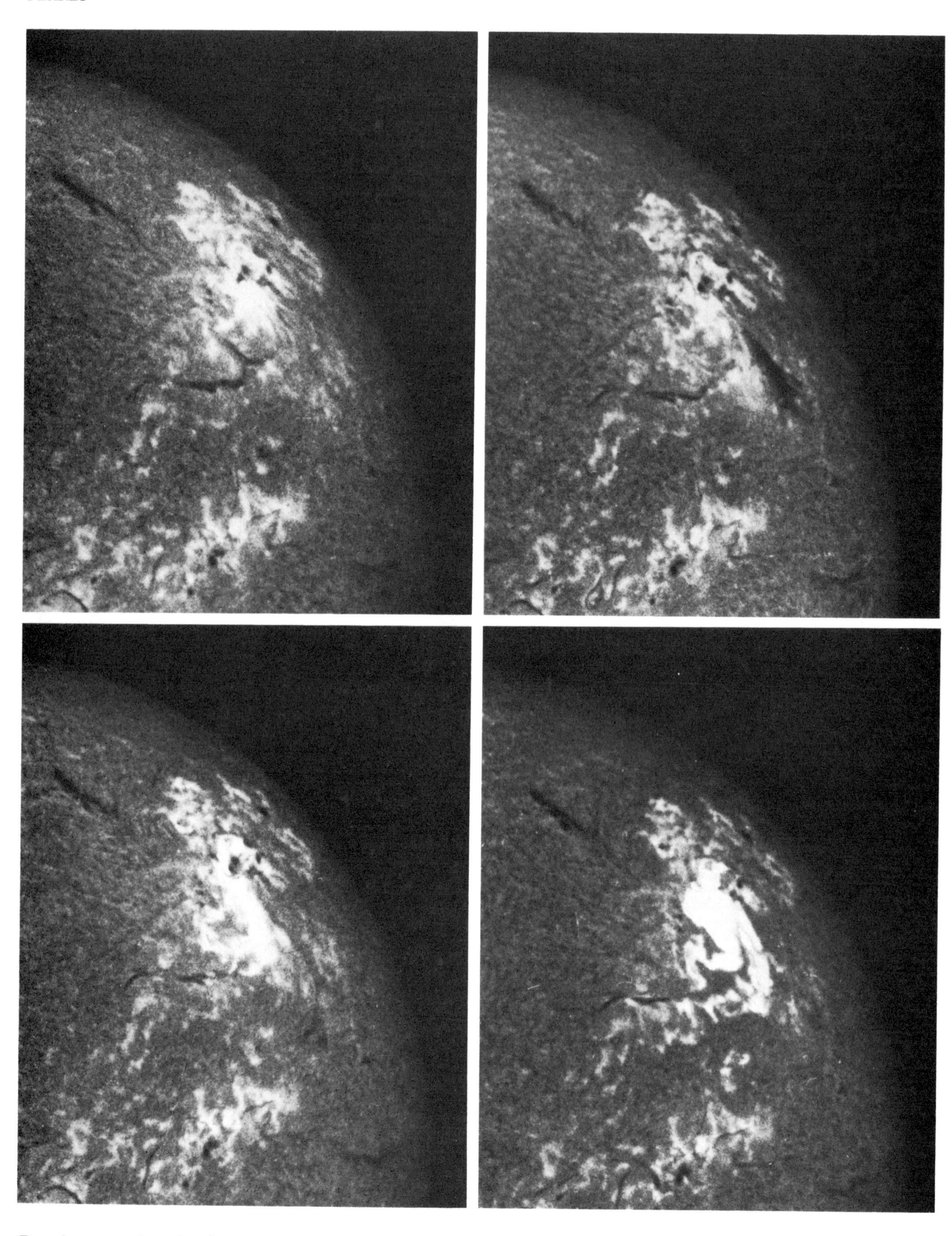

Development of a solar flare *Lockheed Solar Obs*

prominence is seen as a dark, rather blurred streak stretching to the right. This prominence appears dark, like others on the photograph, only because it is silhouetted against the much brighter surface of the Sun rather than against the sky. The gas in the prominence is moving at over 200 km/sec across the Sun. Since the motion is not entirely outwards, the prominence is not quite of the normal eruptive type; such variants are sometimes called surges.

The third and fourth photographs show the development of a major flare. By the time of the fourth photograph there can be no denying that something energetic is occurring. This flare accounts for a large fraction of the energy radiated by the whole Sun in the red ionized hydrogen line. Flares of that magnitude could be detected on other stars by very careful measurements, but few such observations have been attempted, and astronomers are therefore unsure how common flares are on other stars. However, a class of objects known as flare stars exists (such as PROXIMA CENTAURI, p. 84) in which activity resembling solar flares clearly occurs. Flare stars are smaller and cooler than the Sun, and the flares are enormously more energetic, causing the entire visible output of the star to increase by a large factor. If our Sun underwent such dramatic flaring activity, life on Earth would probably be destroyed.

A Radio Flare

The violently variable Sun

At the time of sunspot maximum in 1935, excessive numbers of hisses and crackles were heard overlying short-wave radio broadcasts. Before long it was discovered that these hisses emanated from the Sun. Thus was born solar radio astronomy. This sequence of photographs shows the progression of a solar flare at radio wavelengths. Such flares still interfere with terrestrial radio communications.

Today we know a great deal about the Sun's radio properties. Astronomers distinguish three basic forms of solar radio radiation. The first of these is the quiet Sun component, corresponding to the basic radiation of the disk of the Sun. At radio wavelengths the photosphere, which is the surface we see, is not detected because the outer atmosphere is opaque. At the shorter radio wavelengths the corona is opaque and the Sun appears as much as 2° across. Thus the temperature of the Sun measured by radio astronomers ranges from about 10,000°C to over 1,000,000°C, according to the wavelength in use.

As well as the radiation of the quiet Sun there is a component produced by solar activity. At first this component was attributed to the visible sunspots. Later, however, the regions of enhanced radio emission were found to be longer-lived than individual sunspot groups, and to correspond to plages. Optical photographs in hydrogen light show plages to be bright portions of the chromosphere, and much the same effect is seen in radio maps.

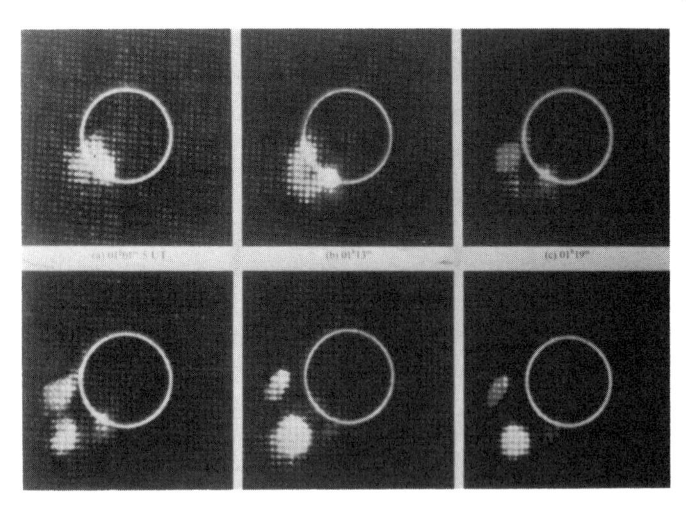

Radiograph of a solar flare CSIRO

The third component of the Sun's radio output was the first to be recognized and is the more impressive by far: flares, which interfere with radio reception. The appearance of flares at optical wavelengths was shown in the previous set of photographs. These are a mild manifestation of the phenomena radio astronomers encounter, for a single radio flare can increase the Sun's radiation a millionfold in a few seconds. In the visible spectrum a flare appears as a sheet of light spreading through the chromosphere, radio studies indicate that a disturbance tears through the corona, like an erupting prominence. When eruptive prominences are ejected they attain speeds of 200-300 km/sec. The radio-emitting flares travel through the corona at ten times that speed! The temperature of the radio flares can be as high as 1000 trillion degrees—almost a meaningless figure.

The six maps in Plate 18 show the progression of a radio flare seen in opposite radio polarizations—the radio equivalent of using polarizing spectacles turned at different angles. The two angles are shown as red and blue, and represent different portions of the magnetic field. The white circle represents the visible size of the Sun, and by comparison with this it can be realized that the radio flare has progressed to a great distance in the 30 minutes between the first and last map.

The direct radiation from a radio flare reaches the Earth in a little over eight minutes, at the same time as the visible light, and this radiation produces immediate radio interference. In addition, the flare ejects a stream of fast-moving atomic particles which reach the Earth about one day later. These particles interact with the Earth's atmosphere and magnetic field in several ways. Most obvious is a breakdown of the layer high in the atmosphere which reflects radio waves; at such times long-distance radio communication becomes impossible for a few hours. Another manifestation is the appearance of bright aurorae in the Earth's polar regions. Similar, but much less pronounced effects can be detected as a result of smaller ripples in the solar wind caused by the passage of large sunspot groups across the center of the Sun's disk.

As might be expected, flare stars can also be detected at radio wavelengths. Only in recent years have radio telescopes and receivers become adequately sensitive to detect other stars. All are weak sources, and emit radio waves by mechanisms not active on the Sun. Radio emission like the Sun's could not be detected in other stars, except during the most intense flares.

The Sun in X-Rays

Holes in the corona

Plate 17 shows a false color picture of the Sun taken by an x-ray detector flying above the Earth's atmosphere. It is typical of the many such images produced over the last decade of x-ray astronomy.

At the very short wavelengths of x-rays, just as at the longest wavelengths used by radio astronomers, the solar corona is opaque. In this picture all the bright areas are portions of the corona. Towards the edges (limbs) of the Sun we look along a great length of the corona, and thus see a greater brilliance. Because the corona rapidly becomes more tenuous with altitude, the brightness drops quite sharply outwards from the Sun, falling to only 1% of its maximum x-ray intensity at about twice the solar radius. We see mostly the details of the inner corona, just above the chromosphere. The poles of the Sun (top and bottom in this picture) appear dark because the corona there is colder and less dense.

In most respects the x-ray Sun and radio Sun are very alike. There is stronger x-ray emission above sunspots and plages, where the magnetic field is most intense and the corona therefore hotter. Several such areas are apparent on this picture. Also there are x-ray flares, as in the radio, during which time the Sun may brighten a hundredfold in as short a time as a few seconds. One might therefore ask why astronomers study the Sun in x-rays, an expensive process because of the need to escape the Earth's atmosphere, when existing radio telescopes monitor the Sun's vagaries daily. The simple answer to this is that x-ray pictures are far more detailed than radio maps can be. The improved detail on x-ray images has led to several discoveries, for example that the corona contains small

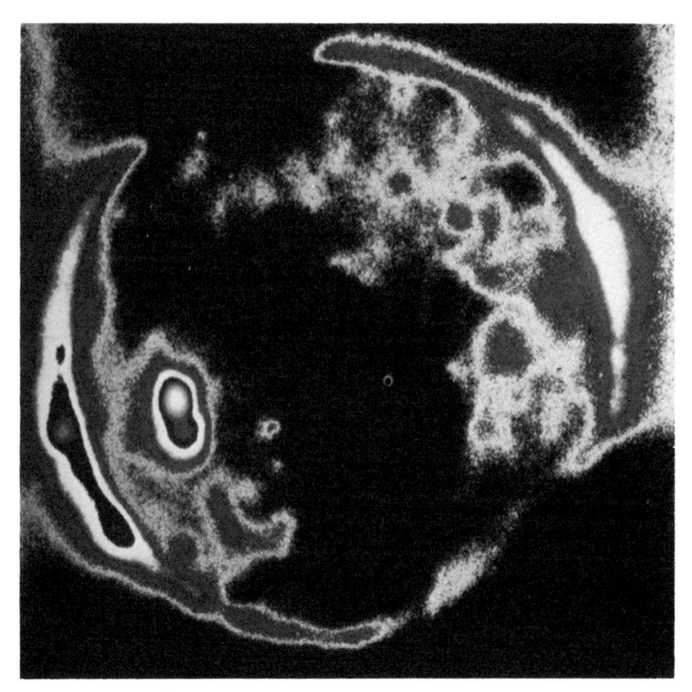

The X-ray Sun AS&E

hotspots only a few tens of kilometers across, but with temperatures in excess of 20,000,000°C. These spots are, in fact, the hottest parts of the entire Sun.

The most interesting discovery from x-ray observations has been of the phenomenon called coronal holes. In 1972 two groups of U.S. astronomers, led by G. L. Withbroe of Harvard and Martin D. Altschuler of Boulder, independently recognized that the darker portions of x-ray photographs had especial significance. The irregular dark area in the center of this picture, joining the darker poles, is at least in part a large coronal hole. Here the magnetic field is weak, making the corona transparent. But the transparency of the corona allows the solar wind to escape more rapidly, carrying more energy. Thus is explained a phenomenon long known, namely that radio interference and aurorae can sometimes be noted on Earth at times when no sunspot groups, plages or flares have been recorded.

X-ray emission has not been detected from any other single star. All the x-ray sources described earlier in this book appear to be binary stars, where the x-rays are generated by mechanisms not active on the Sun.

The Giant Planets

Jupiter

Largest planet, failed star

Of all the solar system bodies, Jupiter is the largest. Its

Jupiter *Pioneer 10*

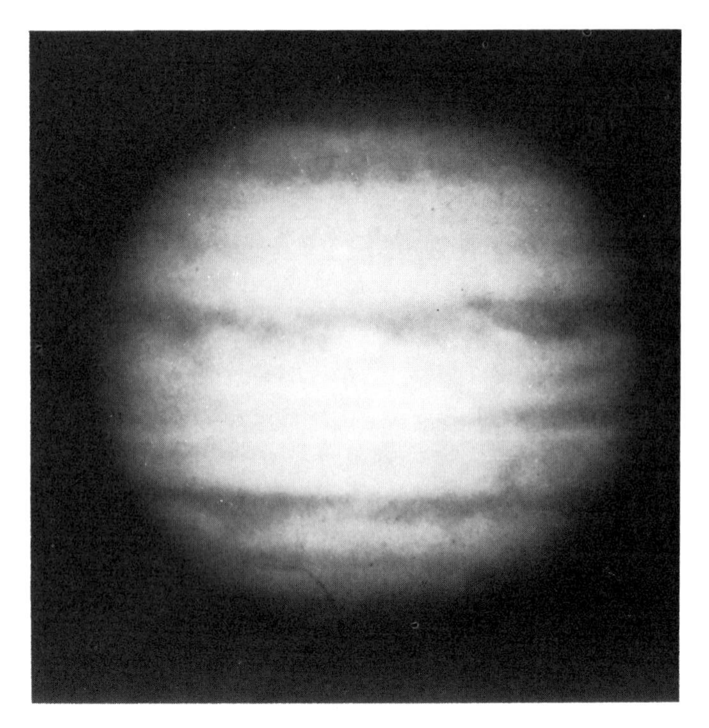

Jupiter's disk *Hale*

diameter, 142,200 km is about one-tenth that of the Sun. Since its density is similar to the solar value, Jupiter's mass is about 1000 times less. In fact the planet has about a tenth the mass of the least massive stars known, placing it part of the way to being a star. The differences between Jupiter and a star arise simply because Jupiter was too small for nuclear processes to occur in its interior. The energy released when Jupiter condensed from the solar nebula was insufficient to raise the central temperature to that critical value at which the nuclear reactions which power stars could begin. Jupiter therefore offers a hint of how protostars might appear before they start hydrogen burning.

As the photo shows, the planet is streaked by alternate dark and light belts. They lie along lines of latitude. The streakiness is generated by Jupiter's rapid rotation. Jupiter spins on its axis in 9 hours 50 minutes, with a speed of over 90,000 km/hour at the equator. At higher latitudes the planet rotates more slowly, requiring 9 hours 55 minutes for one rotation, and thus shearing itself along the belts.

The change of rotation period with latitude, indicates that Jupiter is not a solid body. The surface we see is only a cloud covering, and the dark and light belts have slightly different compositions. They present a pageant of changing patterns. On a time scale of days or even hours, blobs and streaks merge and move. White spots in the darker belts have slightly different rotation periods and therefore drift slowly with respect to other features. On a time scale of years these spots form and vanish, shift in latitude, split into different components or even disappear altogether. In Jupiter's southern hemisphere, a persistent, ruddy oval, the Great Red Spot, has lasted for at least three centuries.

Jupiter's bulk is probably in the liquid state. While astronomers cannot entirely discount the existence of a small solid core, at most a little larger than the Earth, there is no doubt that the majority of the planet is composed of liquid hydrogen and helium, contaminated by traces of other compounds. The central temperature is about 30,000°C, but the hydrogen and helium are prevented from boiling by a pressure 30 million times higher than that of the Earth's flimsy atmosphere. About 1% of the mass of the planet resides in its gaseous atmosphere, in which the opaque clouds float over a sea of liquid hydrogen.

Hydrogen is the most abundant gas in Jupiter's atmosphere. Helium occurs in roughly the ratio of one helium atom to six hydrogen molecules which is about the same ratio as in the Sun. Carbon and nitrogen are also present combined with hydrogen as ammonia and methane. Both gases have less than one-thousandth the abundance of hydrogen. Other gases present in smaller amounts are water vapor, carbon monoxide, ethane, acetylene and phosphine. Hydrogen sulfide and ammonium hydrosulfide are expected to occur below the cloud tops.

The collapse of the portion of the solar nebula which produced Jupiter has not ended. The planet is shrinking at a rate of about one millimeter per year. This minute shrinkage is enough to release more energy than Jupiter receives from the Sun. This energy generation produces the unexpectedly high internal temperature of 30,000°C.

Deep within Jupiter the swirling of the liquids due to the planet's rotation generates a strong magnetic field. Even at the altitude of the visible clouds this is twenty times stronger than the Earth's field, but it has opposite polarity—a compass would point south on Jupiter.

Jupiter is a radio source. Of particular interest are the violent bursts of radio emission at very long wavelengths, which occur when the satellite Io is in one of two positions with respect to the Sun and the planet. The origin of those radio bursts is not understood, but they may be produced by massive Jovian thunderstorms.

Jupiter's Atmosphere

Cloud patterns

The photograph on the next page (top) illustrates the intricate pattern of Jupiter's cloud layers. What the photo cannot reveal is the ephemeral nature of these. features. Like the clouds we see above us in the Earth's atmosphere, those of the giant planet are ever-changing.

Jupiter Portrait *Voyager 1*

The features of Jupiter's clouds can be understood in terms of terrestrial meteorology. There are two visible layers. The upper layer comprises bright clouds and forms what astronomers call the zones, between the darker belts. These are a cirrus-like cloud composed of tiny crystals of frozen ammonia. They should be pure white, since ammonia ice is colorless. The yellow and orange hues they display are caused by impurities which have not been identified for certain. These clouds cover the whole planet but are partly transparent. Their temperature is about –150°C.

The belts are formed of a lower, opaque cloud layer which also covers the whole planet. The composition of these darker clouds is mostly methane ice, again with some impurities and these clouds, being lower, are much warmer, about –40°C.

Between the cloud layers is a region of Jupiter's atmosphere in which giant cumulus clouds rise, their anvil tops resembling the thunderheads on Earth. The cloud tops produce the bright spots which are seen to circulate within and at the edges of the bright zones. Some of the spots are short-lived, others last for many years.

Most of the clouds are massive cyclones, spinning around like their terrestrial counterparts. Because Jupiter rotates so rapidly, however, these cyclones are stretched out into long, thin ovals. The more robust cyclones remain oval spots, but most have been stretched all the way round the planet until their ends join up to form the belts and zones we see.

Every cloud and spot has its own individual rotation pattern about the planet, just as cloud patterns drift around on Earth. The patterns therefore drift past and interact with one another, introducing great complexity into the scene. Most rotation periods lie within the range 9 hours 50 minutes to 9 hours 56 minutes, with the faster rotating clouds nearer the equator. In 1970 a bright spot appeared for a few months with the shortest rotation period ever recorded: 9 hours 47 minutes 3 seconds.

The Great Red Spot

Hailstorm

The most famous meteorological feature of Jupiter is the Great Red Spot. It can be seen in Plates 19 and 20, and again in more detail here. It was first seen by Robert Hooke in 1664, and has been recorded ever since except for a few occasions when it briefly disappeared. Its color varies from very pale pink to a prominent brick red, and as in the case of other features of Jupiter's atmosphere, the reasons for this coloration are not certain.

The Red Spot changes in both hue and size. At its largest it measured almost 39,000 by 14,000 km, much greater than the diameter of the Earth. The rotation of the Red Spot about the planet is another variable feature. Measurements have now accumulated for sufficiently long to show that it rocks backwards and forwards relative to the surrounding features with a number of different periods. The most prominent of these is 90 days long. The combination of these different peregrinations have caused the Great Red

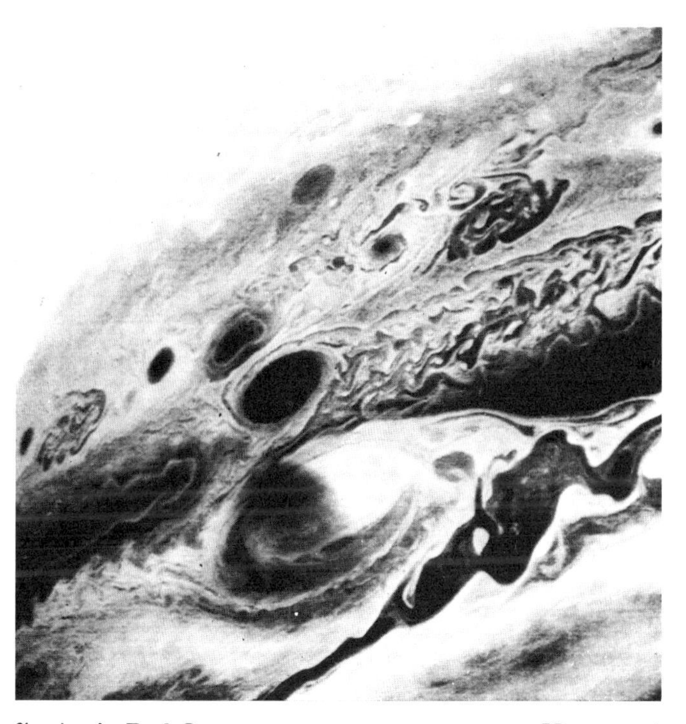

Jupiter's Red Spot *Voyager 1*

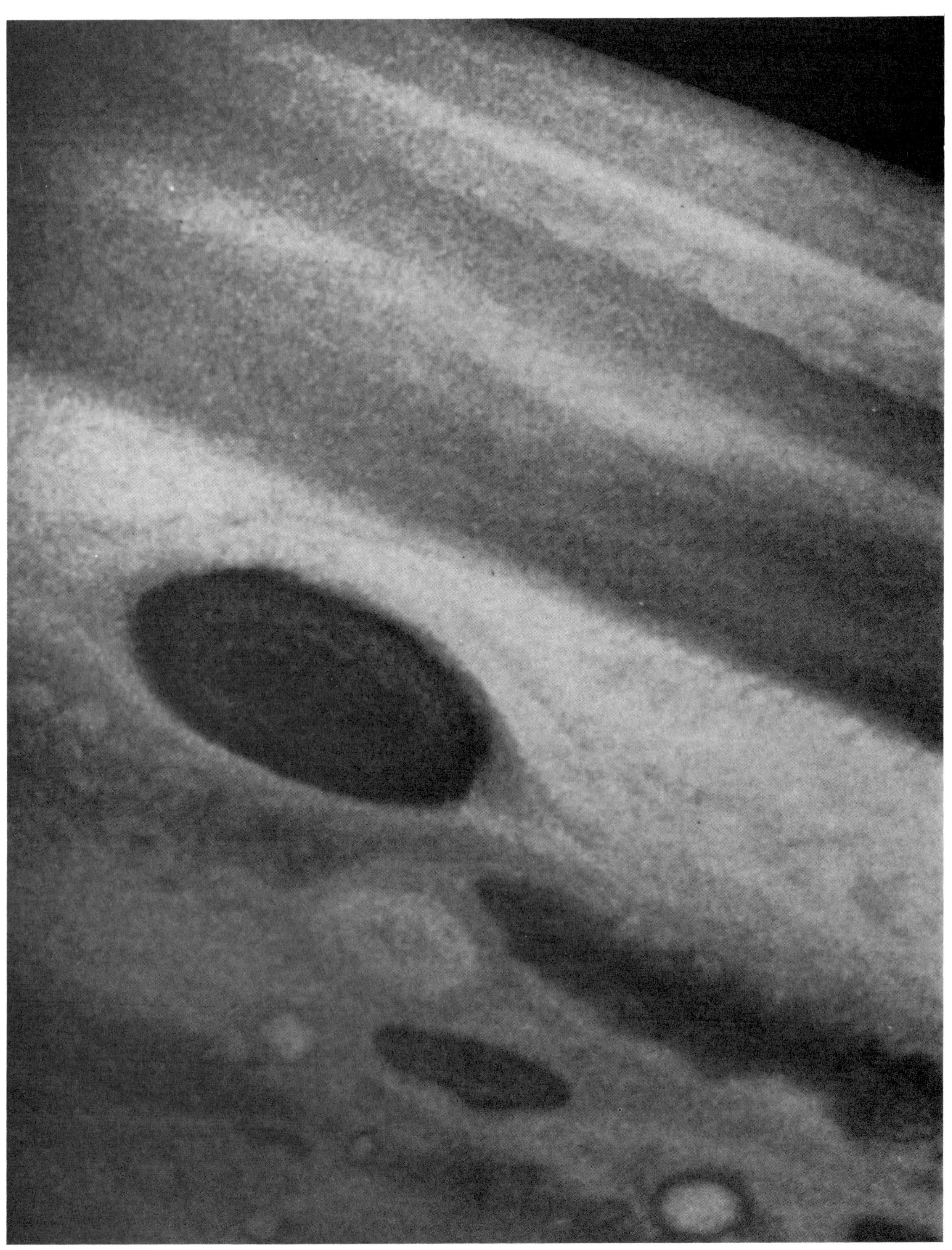

Jupiter's Red Spot Pioneer 11

Spot to travel several times around the planet, relative to its encircling material, since its discovery.

The Great Red Spot also rotates about its own center. At the periphery the period is 12 days, while a little in from the edge a period of 9 days has been measured. At the same time, material appears to rise in the center and slowly spiral outwards to the edge. At its center, the spot rises above the surrounding cloud tops by about 8 km.

It would be fair to state that astronomers are still perplexed by the Great Red Spot. The most credible explanation is that it is an enormous cumulus thunderhead rising through the cloud layers to form a cyclone of gargantuan proportions. The longevity of the Great Red Spot seems surprising, but it has been estimated that once established, such a vortex could continue for a very long time. So long, in fact that it might be 100,000 years before the theory could be verified by observations of the final dispersal of the feature. If, indeed, the Great Red Spot is a giant thunderhead, it may be shedding a continuous rain of Jovian hailstones: blobs of frozen ammonia whose destiny is to plummet into an ocean of liquid hydrogen.

The Satellites of Jupiter

Miniature solar system

Just as the planet Jupiter is almost a small star, so its collection of satellites is almost a miniature solar system. These bodies orbit the planet, bound by its gravitational attraction in the same way that Jupiter is bound to the Sun. Currently 14 satellites are known, and another is discovered every few years. However, all but the innermost five are tiny bodies traveling in irregular orbits at great distances from the planet. These are almost certainly objects which originated closer to the Sun and became bound to the planet relatively late in their history—that is, they were captured, rather than formed together with the planet. Since Jupiter is next to the asteroid belt it is frequently fed potential satellites.

Of greater interest are the inner five, in order from the planet: Amalthea, Io, Europa, Ganymede and Callisto. All five lie in the plane of Jupiter's equator and rotate so as to keep the same face turned towards the planet. Amalthea is faint and lies very close to the planet. It was discovered by Barnard in 1892. The other four are bright, and readily seen in small telescopes, which is why they were discovered soon after the invention of the instrument in 1609. The German observer Simon Marius was the first to see these satellites, and a few days later they were independently found by Galileo. They are usually known as Galilean satellites, and two of them are clearly seen on this photograph.

Little is known of Amalthea. It is a small body, estimated at 160 km diameter, and has an extremely red surface, probably redder than any other known object in the solar system.

Io (Plate 20, and below) has a diameter of 3640 km, slightly larger than the Moon. Its density is 3.5 times that of water, just slightly greater than the Moon's density. The 1979 Voyager 1 space probe showed, however, that Io is quite unlike the Moon. Io still carries at least eight active volcanoes ejecting sulfurous fumes which constantly re-form its yellow-brown surface and hide any meteoric craters. Broad salty plains are criss-crossed by cliffs and fault lines. The cones of large volcanoes are visible in Plate 20.

As Io orbits Jupiter, it carries round with it a cloud of gas which could be thought of as an extended atmosphere. A ring of hydrogen extends most of the way round Io's orbit, but more surprising is the presence of a cloud of sodium closer to the satellite. This emits yellow light, like some streetlights. It is thought that the sodium is released from a deposit of sodium chloride or sodium nitrate somewhere on Io's surface.

Europa is slightly less dense and considerably smaller than Io, measuring 3060 km diameter. It is not known to possess an atmosphere. Its surface shows no evidence of minerals but appears instead to be almost completely covered by water ice.

The largest of Jupiter's satellites is Ganymede, a body of density 1.9 times that of water. This is so low that astronomers are convinced that Ganymede is not solid rock but has an extensive mantle of frost. The frost is probably a mixture of methane, ammonia and

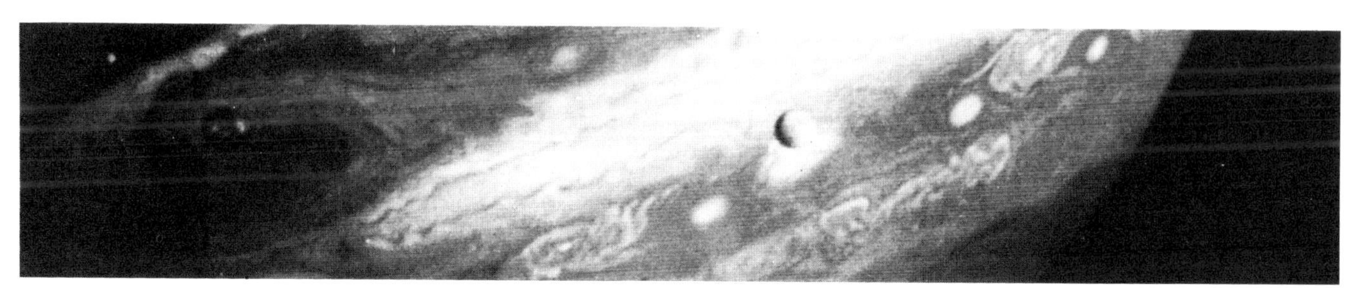

Jupiter's satellites, Io and Europa *Voyager 1*

water ices, and it is known that about half of the surface is covered by water ice. Radio astronomers, measuring radar echoes bounced off Ganymede, have deduced that the icy surface is pitted by numerous craters, just as are the rocky surface of the Moon and some of the inner planets. Ganymede has a bright north pole, which may be a large ice cap. The southern pole, by contrast, is dark, and there is another dark patch on the equator. Ganymede's diameter is 5220 km, sufficient to retain a weak atmosphere.

Callisto is even less dense than Ganymede, so it too must have a large mantle of ice. However, its surface seems mostly to be covered with soil, and there is little exposed ice. Water is present instead, chemically combined within the minerals. Callisto is the darkest of the Galilean satellites and has a bright polar cap which may be an ice of carbon dioxide, methane or ammonia. The diameter of Callisto is about 4900 km.

Saturn

Ringed planet

Saturn is distinctive because of its rings but the planet itself has as received comparatively little attention, mostly because of its similarity to its larger and more readily studied neighbor Jupiter. The spotlight falls instead on Saturn's rings.

Saturn is a slightly smaller body than Jupiter and it rotates more slowly. It too is a liquid planet sporting a gaseous, cloudy atmosphere. As on Jupiter, the rotation has resulted in the shearing of the weather patterns into belts (Plate 19), but these are neither so distinct nor so turbulent as Jupiter's belts.

The atmosphere is also similar to that of Jupiter, comprising mainly hydrogen and helium. Methane, phosphine and ammonia are also present, though the latter is scarcely a tenth as abundant as on Jupiter.

Jupiter's faintest satellite, J13 *Kowal*

Earth

PLATE 21

Thunderstorms over the Amazon

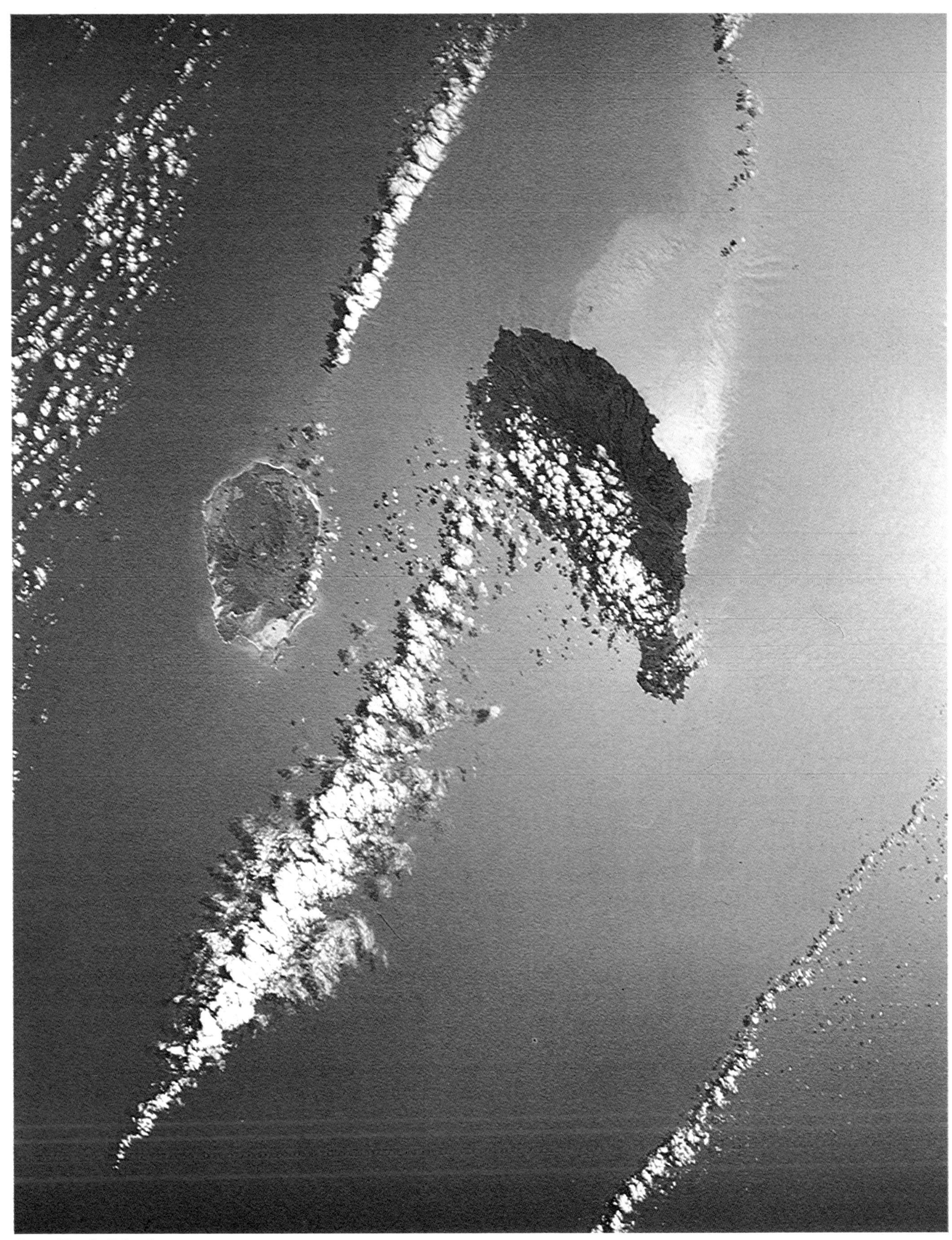

Cape Verde Islands

Red River

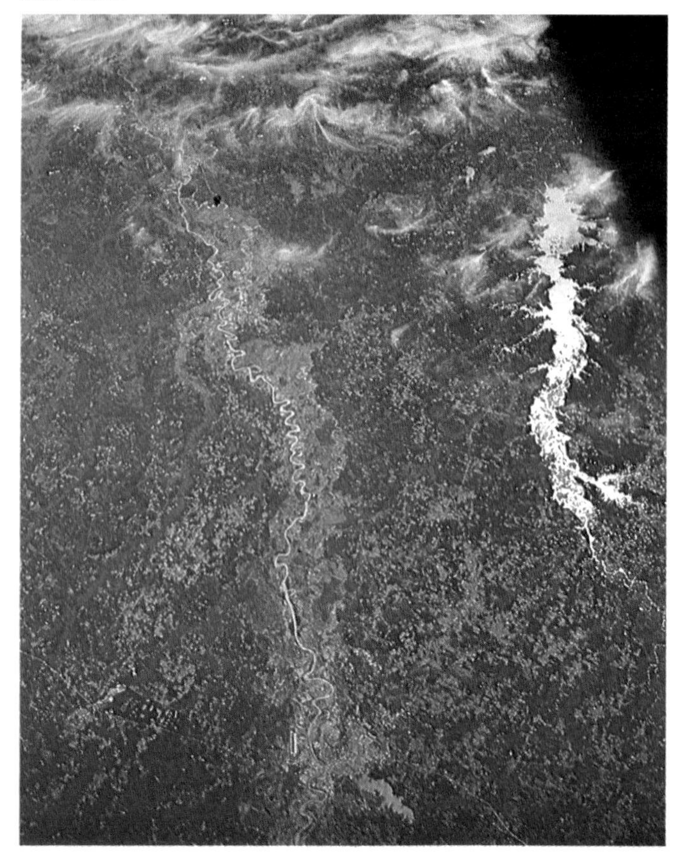

Himalaya Mountains

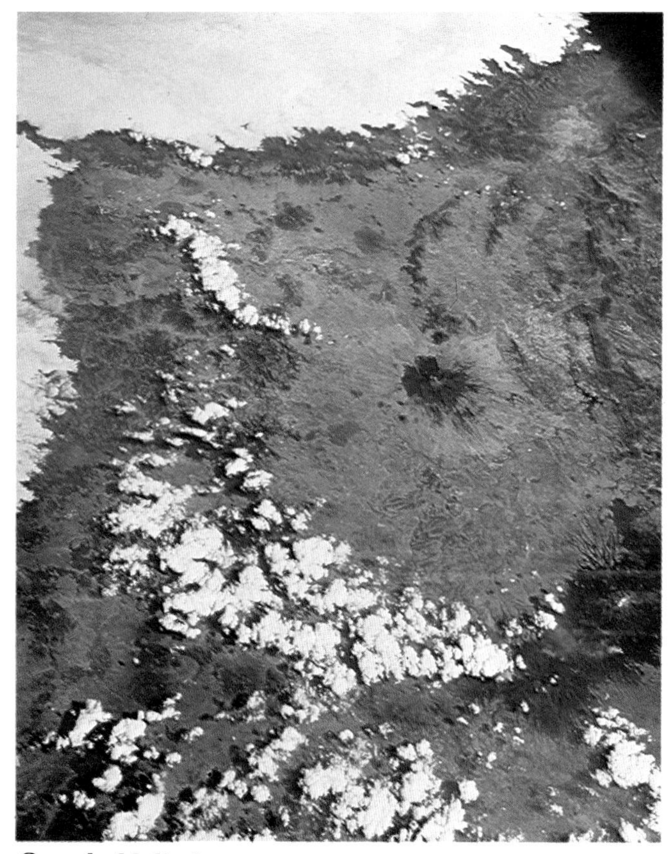

Cerro la Malinche

Richat formation

PLATE 24

Saturn *Hale*

Beneath the atmosphere is a ball of liquid hydrogen and helium. Because of Saturn's smaller mass, the liquid is not compressed to the same high density as within Jupiter. Moreover, though Saturn too is shrinking to generate internal energy, its central temperature is not so high. The density of Saturn is extremely low, lower than that of any other body in the solar system including the Sun. Indeed, its density is less than that of water, so in theory Saturn could float on water. Nonetheless, many astronomers believe that there is a dense, rocky core at the center of the planet, and that this core is the largest and most massive lump of rock in the solar system.

The Rings of Saturn

Local eddy

At first glance, Saturn resembles a spiral galaxy. The planet itself represents the nuclear bulge, and the role of spiral arms is played by the flattened disk which surrounds it. Our series of pictures shows the changing appearance of the rings of Saturn as we view them from above and below towards its changing position in the Solar System. The analogy is appropriate in more than just appearance.

Spiral galaxies form by the collapse of a gas cloud. One portion of the cloud becomes a spherical or ellipsoidal core while the remaining material is constrained by its angular momentum to remain more distant, and can collapse only to form a disk. Precisely the same pattern was followed by Saturn when it formed as a local eddy in the solar nebula. The pattern that occurs in galaxies, some nebulae and planetary systems has therefore repeated on the scale of an individual planet.

Indeed, the same pattern arose when Jupiter formed, but in that example the material condensed to form the major satellites. Saturn, too, has a well-developed satellite system. Why, then, did rings form rather than a satellite?

A clue to this comes from the closeness of the rings to the planet. Theory predicts that no satellite can exist very close to a planet. If such a body were to exist, at any given moment that part of the body nearer to the planet would need to travel faster than the remainder in order to counteract the locally stronger gravitational pull. In the attempt, the body would be torn apart. Only at distances from Saturn greater than the outermost rings could a satellite survive.

The fact that the rings lie within the disruption zone has led to the suggestion that one of Saturn's satellites moved too close to the planet and broke up to form the ring system. However, most astronomers now consider that the rings represent material which never condensed to form a satellite.

Not a great deal is known about the rings. Their diameter ranges from 77,000 km at the inner edge to 136,000 km at the outer. The thickness is believed to be about 2 km, and the rings are partially transparent. They consist of large numbers of particles probably a few centimeters across, each orbiting the planet independently. The composition of the particles is not known, but astronomers have evidence that water ice is the commonest material, at least at their surfaces. The ring particles must resemble large hailstones.

If Saturn resembles a galaxy's nucleus and the rings resemble the surrounding disk, why don't the rings form spiral arms? Because the particles are so crowded together, they must jostle one another from time to time. This jostling prevents the development of a spiral pattern. There is, however, a pattern to the rings: there are several gaps in which very few particles lie (see the pictures showing the rings at their widest open). These gaps are attributable to the satellites of Saturn. Any particles lying in one of the gaps orbit Saturn with a period which is in an exact ratio to the period of one of the satellites. This means that every few orbits, such particles are given a boost by the gravitational attraction of a passing satellite. That boost serves to move them inwards or away from the planet. The particles of the rings cannot spend more than a few orbits in the harmonic positions, whereas they may exist almost indefinitely at other distances from the planet. Hence the gaps in the rings are formed and maintained.

Saturn's Satellites

Few captured

Only one or two of Saturn's ten known satellites appear to be captured bodies; this reflects the fact that

Rings of Saturn *Lowell*

Saturn lies farther than Jupiter from the asteroid belt, where many small objects orbit the Sun. The remaining eight satellites, together with the rings, form the most complete system of small bodies orbiting any planet in the solar system.

The apparently captured satellites are Phoebe and Iapetus. Phoebe lies 13 million km from Saturn on average, and moves backwards relative to the other nine. It requires nearly two years to complete each orbit. Iapetus, which might be a natural satellite of Saturn, is distinguished by having two contrasting sides. One hemisphere reflects under one-tenth of the sunlight which falls on it, whereas the other reflects half. Iapetus is 1600 km in diameter.

In order outwards from the planet, the other satellites are Janus, Mimas, Enceladus, Tethys, Dione, Rhea, Titan and Hyperion.

Titan is by far the largest of these, and indeed is the largest satellite in the solar system, with a diameter of 5800 km. Its density is 1.4 times that of water, indicating that the body is made up of frosty material such as ices of water, methane or ammonia. Titan has a well-developed atmosphere, composed mostly of methane, but probably with some hydrogen, ethane and other gases too. The clouds cover much of the satellite. Titan is similar to the giant planets of the solar system, differing mostly in being too cold to be liquid.

The other satellites are too small to possess atmospheres. After Titan, Rhea is the next largest. This has a frosty surface, possibly covered with water ice. Its diameter is at least 1320 km.

Janus is the smallest and faintest satellite, and orbits only 23,000 km beyond the edge of the rings. Because it is so faint, Janus cannot readily be seen next to the glare of the rings, and was therefore not discovered until 1966, a time when the rings were seen exactly edge-on so that their light was almost completely eliminated.

Uranus

Herschel's planet

Although it is just visible to the unaided eye, the discovery of Uranus had to await the invention of the

Saturn's satellites *Lowell*

Uranus and its satellites *Lick*

telescope. Sir William Herschel spotted it during his routine sky surveys, and he recognized it as unusual because it appeared to him to be a small disk rather than a star—though at first he thought it might be a comet.

It was soon realized that Uranus was a planet on account of its slow motion through the sky. Uranus requires 84 years to complete one circuit of the Sun, and it therefore travels only 40 arc seconds through the sky daily.

The disk which Herschel recognized is only 4 arc seconds in diameter, about a five-hundredth of the apparent size of the Moon. This is too tiny for terrestrial telescopes to reveal much surface detail, and no photographs exist which show any certain features. Visual observers have, however, reported the existence of faint belts which resemble those of Jupiter and Saturn. The rotation period of Uranus seems to be about 24 hours but measurements are in disagreement about its precise value.

Through a telescope the planet appears blue-green. The spectrograph reveals that much of the yellow and red sunlight it reflects is absorbed by methane, a major constituent of its atmosphere. Hydrogen is the most abundant gas in Uranus' atmosphere, but has only minor absorption bands in the visible spectrum. There is almost certainly helium on Uranus too, but this has not been detected from Earth. Ammonia and other gases are believed to exist deeper in the atmosphere.

Most of the planets of the solar system rotate about axes which point out of the plane of their orbits. In the case of the Earth, the obliquity—the angle between the rotation and orbital planes—is about 23°. This tilt in the Earth's axis, is reflected in the apparent path of the Sun from the Tropic of Cancer to the Tropic of Capricorn and back each year. The obliquity of Uranus is 98°—greater than a right angle. This means that during every orbit of Uranus, the Sun appears to move from one pole to the other. Currently the Sun is shining down on to the north pole, which means that most of one hemisphere of the planet is in darkness. Unlike the Antarctic winters on Earth, which can last at most six months, the Uranus winter amounts to about 40 years. Another consequence of the obliquity of 98° is that the planet rotates in the opposite direction to most other planets.

Because of its great distance, the Sun and Earth never appear more than 3° apart in the sky from Uranus, so we cannot see more than 3° into the night hemisphere of the planet. We therefore have no indication of conditions there, but astronomers believe that the temperature will be only marginally lower than that of the sunlit hemisphere because of the rapid rotation and hence efficient circulation of the atmosphere. The sunlit side has been measured by infrared techniques at –218°C, about as cold as solid air.

Uranus has five known satellites, three of which are visible on this photograph. They orbit in about the same plane as the planet rotates. It is very likely that, like our Moon, they keep the same face towards the planet. If so, each satellite will experience the same seasons as the planet, and since the satellites have little or no atmosphere to help spread the Sun's feeble heat around their surfaces, their night-time hemispheres are probably among the coldest places in the solar system.

In addition to its five satellites, Uranus has a system of at least eight rings of particles. These have never been seen, but were discovered in 1977 when Uranus moved in front of a bright star. As each ring passed, the star was momentarily extinguished. Whereas Saturn's rings are broad and separated by narrow gaps, those of Uranus are narrow and widely spaced. Exactly why this should be so is not yet fully understood. The narrowness of the rings makes them very faint, and their proximity to the planet renders them unlikely ever to be photographed from Earth in visible light.

Neptune

Sampling the solar nebula

The planet Neptune resembles Uranus in so many ways that the two are normally paired together. When seen from Earth, both exhibit small green disks, possibly crossed by weak belts. Both have atmospheres composed predominantly of hydrogen, helium and methane, and both may have small solid cores. Their optical spectra are dominated by the absorption bands of methane, a compound of carbon and hydrogen, and indicate that the ratio of carbon to hydrogen in these planets is ten times greater than in the Sun, Jupiter and Saturn. This is believed to be representative of the chemical composition of the portion of the solar nebula in which they formed. Ethane has also been detected on Neptune, but not on Uranus.

Neptune rotates on its axis in about 19½ hours, and is inclined to its orbit at 29°. Thus during the course of its 165-year orbit of the Sun, Neptune experiences seasons just as Earth does, except that they are 41 years long! The mean temperature at the top of the atmosphere is extremely low, however, around −215°C. Nonetheless, this is 10° hotter than expected, from which astronomers deduce that Neptune, like Jupiter and Saturn, has an internal heat source.

Two satellites orbit Neptune. The outer is called Nereid and is an insignificant body probably captured late in the planet's history. Triton, the inner satellite, is a far more significant member of the trio, and is easily seen on this photograph. With a best-determined diameter of 3700 km, Triton is one of the largest satellites in the solar system. Surprisingly, it orbits in the opposite direction to all other major satellites.

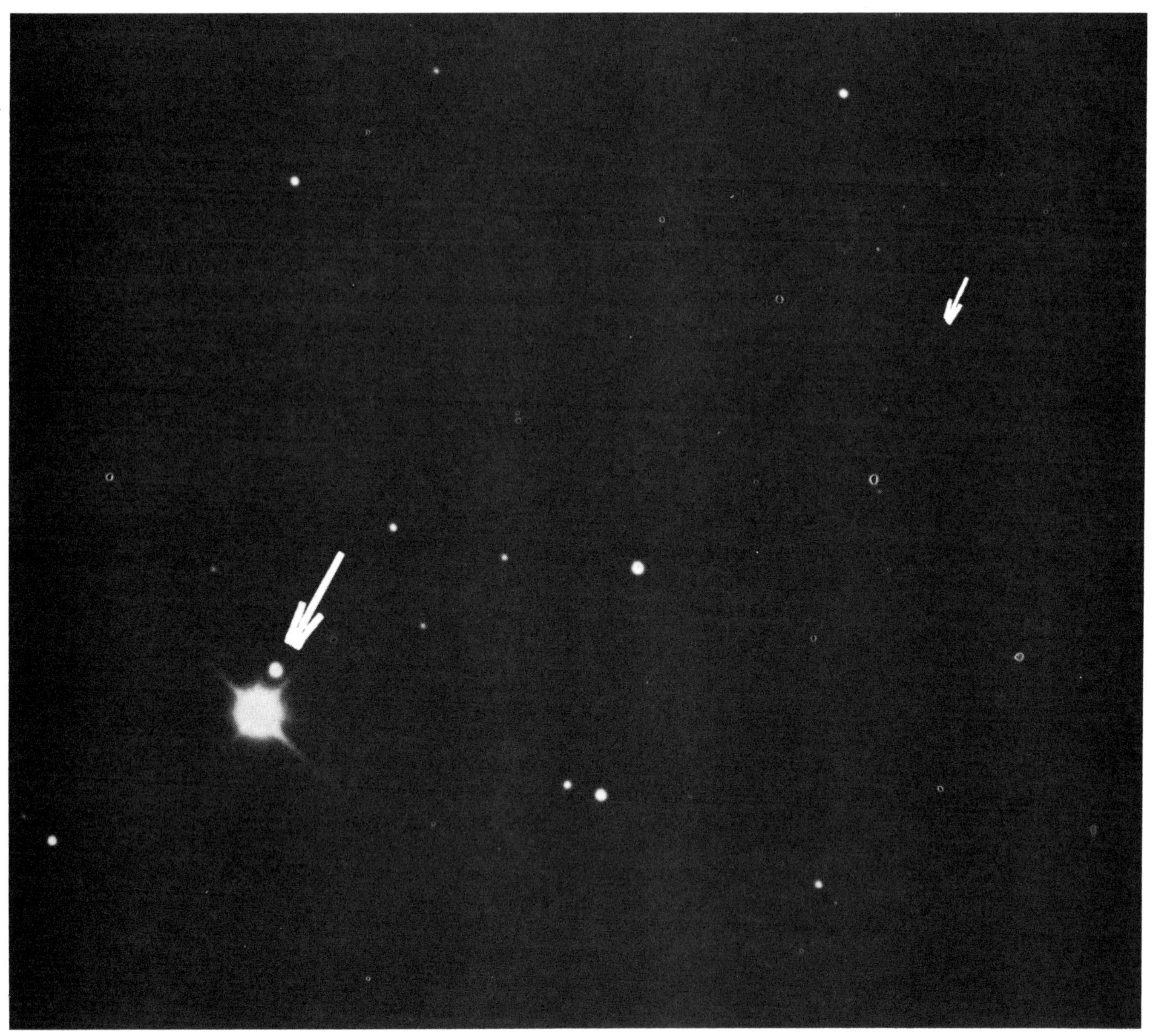

Neptune and its satellites *Lick*

Pluto

Outermost planet, a misfit

At the farthest reaches of the solar system lies the tiny planet Pluto. Insignificant though this object is, it is a glaring misfit in the orderly scheme of companions to the Sun. This is evident at first glance. Though on average little more distant than Neptune, Pluto appears 1000 times fainter. This is principally because of its diminutive diameter. No telescope has ever revealed Pluto as a disk, and it can be distinguished from a star only by its slow motion against the field stars. These two photographs, taken a day apart, clearly show Pluto's progress across a portion of the constellation Gemini. In its small diameter, Pluto resembles more closely the inner planets than the giant, gaseous balls of Jupiter, Saturn, Uranus and Neptune.

The orbit of Pluto also distinguishes it from all other planets. Whereas the planets follow almost circular orbits around the Sun, Pluto's is markedly elliptical. At its closest approach, Pluto is less than 30 astronomical units from the Sun and thus lies within the orbit of Neptune. (An astronomical unit is the Earth-Sun distance.) Closest approach next occurs in 1989. One half of an orbit later, in AD 2113, it will be almost 50 astronomical units from the Sun. Again, whereas the other planets orbit almost within a plane, the ecliptic, Pluto's orbit is greatly inclined. In 1973 Pluto lay 10 astronomical units above the ecliptic.

Discovery of Pluto *Lowell*

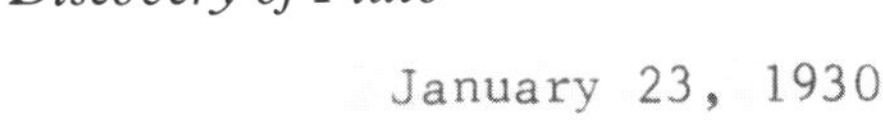

January 29, 1930

The nonconformist nature of Pluto has led some astronomers to suggest that it formed at the same place and time as Neptune. This theory states that Pluto was a satellite to that planet, and that during a close encounter with Neptune's other satellite, Triton, it was thrown off into its unusual orbit. For Pluto to have been a satellite of Neptune is quite plausible, since it is almost certainly a smaller body than Triton, but the energy needed to drive it into its present orbit is rather larger than might readily be available by this means. A more damaging blow to this hypothesis was the discovery, in 1978, of a satellite of Pluto. It would be almost impossible to contrive the ejection of a pair of bodies from Neptune's retinue. Thus it seems likely that Pluto formed more or less in its present orbit.

Pluto was discovered following a systematic search initiated by Percival Lowell. From irregular tides in the orbit of Uranus, calculations yielded the approximate celestial position of the hypothetical planet, and in 1930 Pluto was found near this position. Just how serendipitous was that discovery has only recently been realized: some of the early measured positions of Uranus, on which the calculations were based, were inaccurate and produced an entirely spurious prediction. We now know that Pluto is too small to influence the motion of Uranus to any appreciable degree. The pair of plates on which Clyde Tombaugh found Pluto by its motion between them is shown in our illustration.

No evidence has ever been found of an atmosphere on Pluto. This is not very surprising since, at its mean surface temperature of below –220°C, most materials would exist as liquids or solids. Moreover, most of those which remain gaseous at this temperature cannot be retained by so small a body, just as hydrogen rises and escapes from Earth. The sole exception is neon, a substance which could not be detected from Earth. The brightness of Pluto at infrared wavelengths, however, suggests that there is some form of ice on the surface, and frozen methane is considered to be the most likely ice.

The brightness of Pluto varies with a period of 6 days 9 hours, and this has been assumed to be the result of the rotation of the planet with this period. The size of this variation has been increasing since Pluto's discovery, while the planet itself has been fading slowly. This may be the result of a high obliquity, as in the case of Uranus. The situation was confused in 1978, however, by the discovery that photographs of Pluto dating back a number of years, taken with a variety of telescopes, show elongated images (p.184). This indicates that Pluto has a moon, orbiting it with the same period as its rotation, 6 days 9 hours. The orbital plane is inclined to our line of sight, so it does not seem possible to ascribe the brightness changes to the presence of this moon.

The U.S. Naval Observatory astronomer who discovered Pluto's moon, James Christy, has given it the provisional name of Charon. Judging by the relative brightness of the two bodies, the satellite may be as much as one-third the diameter of Pluto, making it the largest moon, in proportion to its main body, in the solar system.

Another surprise that comes from the discovery of Charon is that analysis of its motion makes possible a new estimate for the mass of Pluto. This comes out at only a four-hundredth that of the Earth. Combining this with a measured upper limit for the diameter of less than 3000 km, the density of Pluto may even be less than that of water.

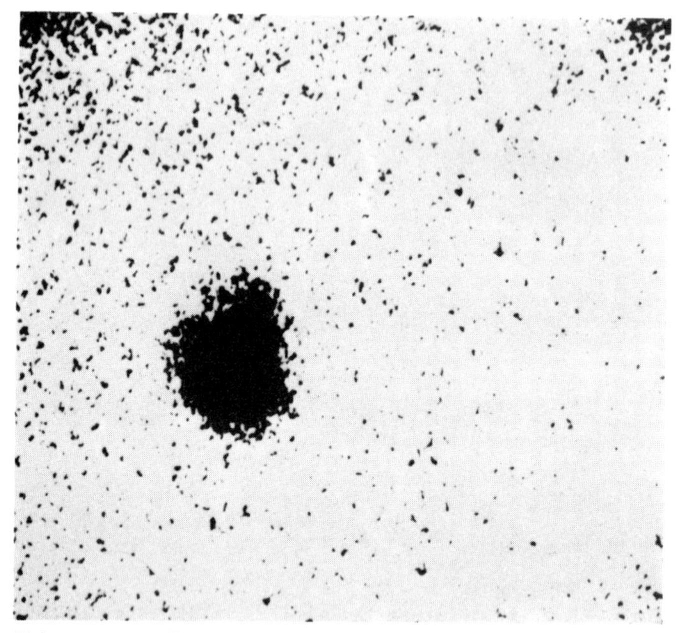

Discovery of Pluto's satellite, Charon USNO

The Terrestrial Planets

Earth

Third planet

The largest of the innermost four planets is called the Earth. Although it is the planet we know the most about, space vehicles can look back and see the planet as a whole (Plate 21). We can try to regard it as we see the other planets of the solar system.

Earth's diameter, about 12,500 km, is a scant tenth that of Jupiter. Unlike the giants of the solar system, however, it and the smaller planets have solid surfaces. Earth, Venus and Mars are sufficiently massive bodies to retain gaseous atmospheres, but these are only tenuous and represent a tiny portion of the masses of the planets. The lighter gases such as hydrogen and helium easily escape from Earth and are only slowly replenished from surface chemical and nuclear reactions. The major gases of the atmosphere of the Earth are nitrogen, oxygen, carbon dioxide and water vapor.

Earth *Apollo 10*

The mean surface temperature of the planet, 12°C, lies between the boiling point and freezing point of water, and there are therefore extensive basins filled with water, occupying two-thirds of the planet's surface—the oceans. The third which rises above the water level is rock, dust and soil, rich in silicate minerals. On this photograph the water appears blue and the rocky landmasses are brown. The white portions are dense clouds of minute water droplets, floating in the atmosphere a few kilometers above the surface. At any given time such clouds cover about one-half of the planet, but they move about, form and dissipate rather randomly, so that the whole surface can eventually be seen. At the poles the surface temperature is low enough for permanent sheets of solid water ice to have formed, and these too appear white on photographs.

The central third of the Earth's interior is rich in iron, and forms the core. Of this the inner 7% is solid and the remainder is molten. The central temperature is 4000°C. The heat required to maintain this temperature comes not from gravitational contraction, as in the major planets, but by the decay of radioactive materials in the rocks. The Earth is not contracting to any appreciable degree. Under the conditions of high temperature and pressure, the denser materials such as iron have migrated to the center of the planet leaving the lighter rocks at the surface as a slag: this process is known as fractionation. The slag accounts for most of the planet's volume and is called the mantle. The core of the Earth acts like a giant electrical dynamo as the molten iron rotates. This results in the production of a magnetic field.

The internal heat source does not confine its effects to the core but manifests itself in surface disturbances. Huge convective cells rise from within, circulate across the surface and disappear again, in some ways resembling solar granulation, but on a proportionately much larger scale. This process sends rafts of the crust material around the planet's surface, sometimes splitting landmasses, sometimes closing oceans. This crustal movement has recently become known as plate tectonics. Where plates collide, or where other minor deformations of the crust occur, crumpling of the surface can produce mountains a few kilometers high.

The Earth's heat also escapes at localized points, and here one might draw a loose parallel with solar prominences. Molten rock may rise close to the surface, producing hot ground and causing the ejection of heated water and steam, or it may break through the crust and outflow across small portions of the planet before cooling and solidifying. The ejection of molten rock is a process called volcanism. Volcanoes and volcanic ranges of mountains are formed by successive depositions of molten rock, and are often higher than fold mountains though having gentler slopes.

External agents also modify the Earth's surface. Tiny solar system bodies known as meteoroids, typically a few tens of meters across, are occasionally swept up by the Earth as it orbits the Sun. These crash to the surface, producing impact craters, some recent like the Barringer crater in Arizona, some more ancient like the southeast inlet of Hudson's Bay centered on Les Iles Belcher, the islands being the peaks of a central mountain.

The surface of the Earth is shaped by plate tectonics, crustal deformations, volcanism and meteoroid impacts, but it is also leveled by a process known as erosion. Water, both running as a liquid and alternately freezing and thawing, is the main erosion agent, but turbulence in the atmosphere—wind—also contributes. Erosion wears down exposed rock to dust and distributes this into low-lying areas. Thus even when sufficiently large meteoroids produce craters on the Earth's surface they are rapidly rubbed away.

Another result of the presence of liquid water on Earth is the development of a life system. Although life began to evolve in the oceans, it subsequently migrated onto the land, where it is more easily observed. The immobile forms of life, vegetation, progress through a cycle of color changes during the course of one orbit of the Sun; these color changes are readily detected from a considerable distance. In addition, the technologically most advanced of the mobile life forms, man, generates light (particularly at night), heat, radio signals, chemicals and nuclear reactions, most or all of which are readily distinguished from natural effects.

Thunderstorms

Convection in Earth's atmosphere

Water vapor is constantly evaporated off the oceans, and the Earth's atmosphere is nearly saturated with it. This means that at the slightest provocation, minute droplets of water will condense to form clouds. The provocation usually responsible is a cooling of the atmosphere, and this can be accomplished in a variety of ways. For example, warm moist air may move into a colder region; this occurs near Newfoundland, where a cold Arctic water flow meets a warm prevailing wind. Because the atmosphere becomes colder with increasing altitude, any upward motion of air also cools it and tends to produce cloud at the top of rising columns of air.

Beneath the canopy of cloud on Plate 22 is the Amazon jungle, a hot, humid portion of the Earth's equatorial regions. In the afternoons, air near the jungle top becomes heated so much that it rises in massive thermal eddies. In its upwards motion, the air cools; since its humidity is high, much water is condensed from it. Clouds cover the Amazon basin on most afternoons, and rain often falls from them. Where the thermals are strongest, the clouds rise particularly high; examples can be seen to the left in this picture.

In addition to carrying water aloft, convection moves electrical charges. High clouds such as these attain a strong charge, which they must eventually discharge to ground. The discharge is lightning, and the associated rain and clouds are thunderstorms.

On average there is one lightning strike per second somewhere on the Earth's surface. Studies from Earth satellites show that lightning rarely, if ever, strikes the oceans.

Thunderstorms over the Amazon *Apollo 9*

Cape Verde Islands

Peaking above the ocean

Plate 23 shows a group of islands in the Atlantic Ocean. The largest is São Tiago; to its left is Maio, and at right of the picture is part of Fogo. These are three of the volcanic archipelago known collectively as the Cape Verde Islands.

Although no longer active, volcanoes once erupted from this part of the ocean floor. By their steady outpourings of lava, they built up mountains high enough to protrude above the Atlantic's surface. Hence the archipelago was formed, and what we see today is only the very summits of a range of mountains which rise 6000 meters (20,000 feet) from the seabed. Fogo, the highest, almost attains an elevation of 3000 meters (10,000 feet) above the ocean.

Relentlessly, the ocean waves erode the shoreline, continually removing material and depositing it underwater. The sandy area to the north east of São Tiago is entirely submarine, and was generated by the eroding forces of the waves.

In this region the prevailing winds blow from the northeast (upper right). Having made a lengthy ocean crossing, the winds are moist. Hence as the air is forced to rise near the mountain peaks, cumulus clouds form. Long strings of cumulus clouds, together with their shadows, stretch southwestwards from the islands.

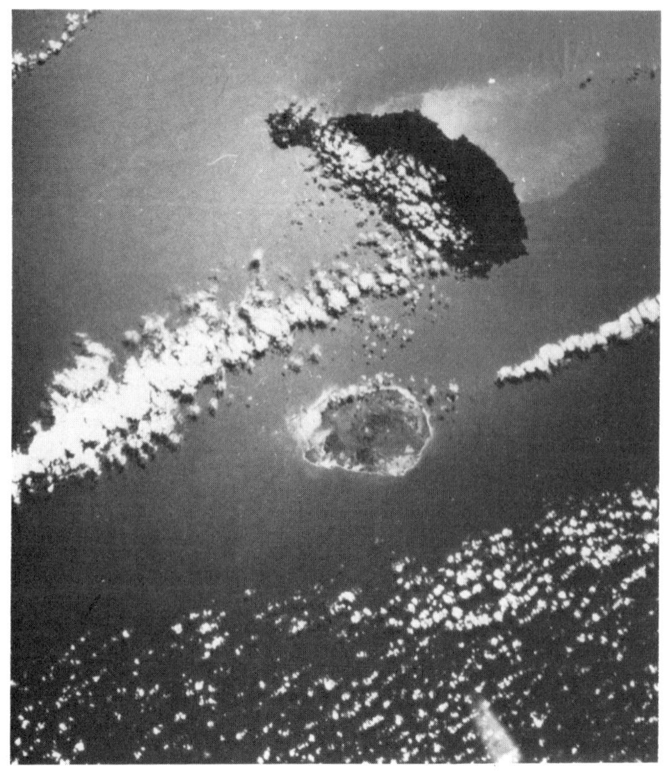

Cape Verde Islands *Apollo 9*

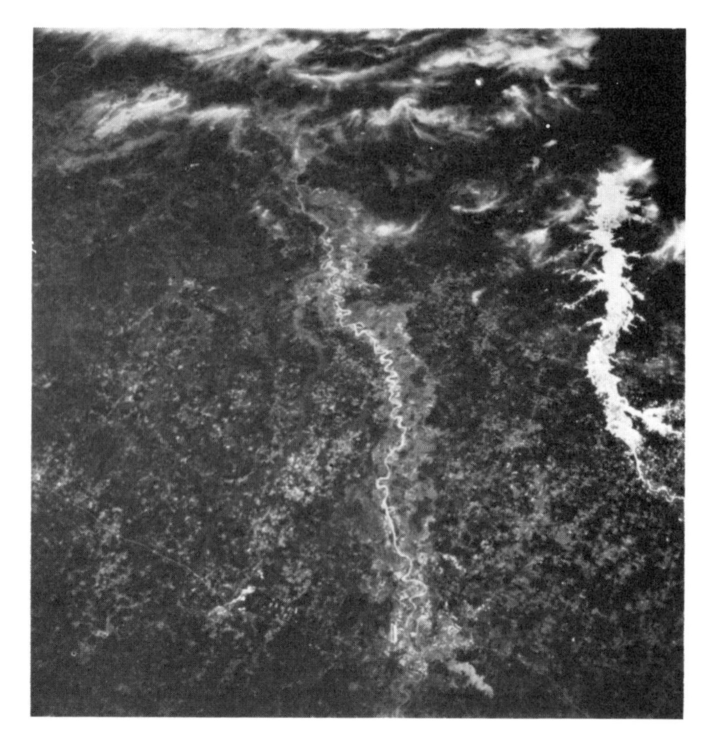

Red River *Apollo 9*

Red River

Evidence for flowing water

The most important difference between planet Earth and all other solar system bodies is the presence of surface water. Life depends critically on water's ability to convey nutrient minerals from place to place. Plate 24 illustrates another important facet of water, its ability to erode.

Dwelling on Earth, we are constantly reminded of water's powers. In mountainous regions, an entire hillside may be drastically altered by a major rainstorm. In cities, water may strip the surface off a roadway. A hand-held hosepipe easily removes dried mud from your car. We are impressed by steep-sided ravines and canyons cut out of solid rock by seemingly insignificant streams, with only time to assist. Yet many of us fail to appreciate how extensive is the eroding action of water.

Slow-flowing rivers cross gently sloping plains or course through broad, shallow valleys. It is easy to be misled into a belief that these cause no significant erosion. The truth is otherwise. In most cases hundreds, even thousands of meters of rock have been removed by the river from over a great area. Broad, shallow valleys were usually carved by the very rivers that meander along their floors, and represent an enormous amount of erosion. Plains are usually plains because rivers have coursed widely, sandpapering the relief. Rivers meander only because they can no longer erode downwards, having neared sea level, and instead erode sideways.

Plate 24 shows the meanders of a mature river which has helped to level the plains of Louisiana. It is the Red River, a tributary of the Mississippi. A similar photograph taken a thousand years hence would show very different meander patterns, for erosion is constantly changing the course of the river's banks, and the face of the Earth.

The Himalayas

Collision between landmasses

Because of the heat generated within it, the Earth is in constant upheaval. Its motion is that of boiling jam in a saucepan: warmer portions rise to the surface, are displaced by subsequent material, cool, and fall back within. If the saucepan is heated only at its center, there is a continuous flow outwards towards the edges. On a cold day, a skin of almost solid jam may form at the outer edges, and the convection will then cause the skin to pile up into wrinkles against the walls of the saucepan.

In exactly the same way, portions of the Earth's crust are rafted around the globe, and pile up into wrinkles against one another. Whereas the rise and fall time of boiling jam is a minute or less, that of the almost solid mantle of the earth is a few hundred million years. On that time scale the major landmasses have migrated great distances across the Earth's surface.

One of the fastest-moving portions recently has been the subcontinent of India. It broke away from a landmass which included Australia and Antarctica, and over the last few hundred million years has headed northwards, crossing the equator. Its passage has recently been resisted by a collision with the greater landmass of Asia. At the junction of the two landmasses the crust wrinkled, just like the skin on boiling jam. The wrinkle is minute: at its highest point it represents one-seventh of one percent of the radius of the planet. Yet Man regards this as a mighty range of mountains, and calls the wrinkle the Himalayas.

With several of the taller peaks more than 8000 meters high, the Himalayas stretch a significant distance into the atmosphere. At that height the air is much colder than near the ocean surface, typically below −20°C. Snow falls and remains permanently on the higher portions of the Himalayas, even though they lie only 30° of latitude from the equator.

On Plate 24, the pattern of ridges is outlined by white snow. The ridges were not themselves produced by the wrinkling of the Earth's crust; they have resulted from the continuous process of erosion. Glaciers, rivers, even wind and snow combine to wear the mountains down to level plains. We see mountains here only because the uplifting of the crust is proceeding faster than erosion. The valleys represent regions where erosion is almost keeping pace with uplift. Cloud fills some of the deeper valleys on this photograph; it appears slightly darker than the snow.

Himalaya mountains *Apollo 9*

Cerro La Malinche

Evidence for volcanism

Below the thin surface, the Earth is a restless, hot body trying to find ways to cool. In parts the temperatures are so high that rocks we think of as immutably solid melt to a syrup-like liquid. Such places often produce weaknesses in the 30-km thick crust, and at times molten rock will be squeezed by internal pressures to the surface.

The appearance of molten rock—lava—at the Earth's surface is described as volcanism. It can lead to a variety of events, some of which take place underwater and are not, therefore, easily observed. If the rock has an alkaline chemistry, it is free-flowing and can spread extensively before cooling. Some portions of the Earth's surface have been covered by such flows. Acidic rocks produce thick, sticky lava which flows poorly.

If volcanic activity continues in the same place for many eons, the volume of rock extruded is sufficient to build mighty mountains. Alkaline lava accumulates to form a shallow-sided, domelike mountain. Acidic lavas tend not to flow far before solidifying, and therefore build steeper volcanic cones.

Calderas or craters are usually found on volcanoes. The area of the caldera is that in which the rock melted at the last eruption. Outside that area, solid rock remained or was added to; within the caldera

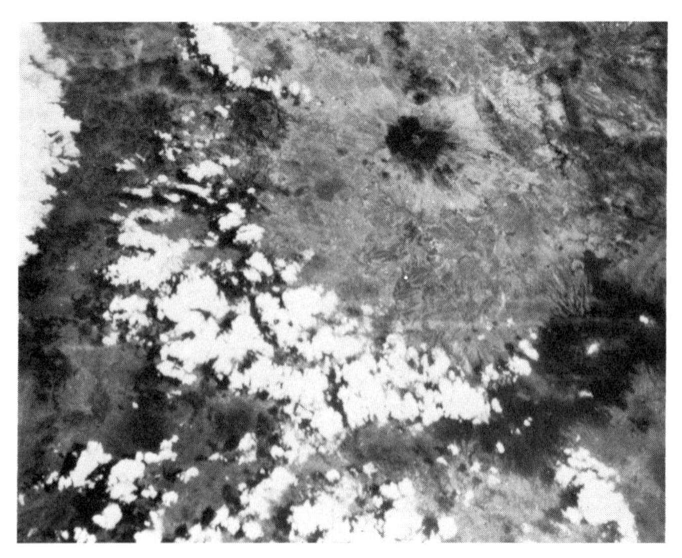

Cerro la Malinche *Apollo 9*

molten rock drained back a little way into the Earth at the end of the eruption.

Seen from above, volcanoes resemble chicken-pox scabs: shallow pimples on the vast abdomen of the Earth. When seen in profile, the same volcanoes are impressive mountains to man-sized beings. One need only recall Fujiyama, Egmont or Vesuvius. Plate 24 shows the Mexican volcano field. At center is Cerro la Malinche, a very extensive, shallow-sided volcano sporting a large, dark crater. To the lower right are the higher, steeper cones of Popocatepetl (5400 meters, 17,900 feet) and Ixtilcahuatl (5200 meters, 17,300 feet). Small snowfields lie on their summits, barely distinguishable from the cumulus clouds to the left. Smoke is rising from an eruption below Ixtilcahuatl.

The Richat Formation

Meteor crater

At the lower right of Plate 24 is a series of concentric rings of rock locally called Bou Tellis, but better known as the Richat formation. Beyond, in contrast, stretches the almost featureless expanse of the western Sahara Desert of Mauritania. Although described on some maps as a single mountain, the Richat formation is a circular ring of hills about 40 kilometers across and surrounding a crater.

Geologists have not explored the Richat formation in great detail, but they have unearthed extensive areas of a mineral known as coesite. This is a form of quartz which has been made in the laboratory only under extreme pressures—greater than 20,000 atmospheres. Not surprisingly, it is a denser mineral than normal quartz. Coesite has been found in only about 20 localities on Earth, and in every case it is associated with a crater such as the Richat formation. It is natural to associate craters with volcanoes, but in these examples there is no evidence of volcanic activity. Moreover, volcanoes cannot generate pressures high enough to manufacture coesite. Some other agent must be found to produce these craters; a meteoric origin is usually invoked.

Meteoroids are lumps of rock weighing upwards of a few kilograms, and periodically one may penetrate the Earth's atmosphere and crash to the ground. Most do little damage, but every hundred thousand years or so a large specimen arrives, capable of leaving a sizeable scar where it collides with the planet.

The Richat formation is such a scar. Within the last few million years a meteroid, probably more than a kilometer in diameter and a billion tons in weight, crashed into this part of Africa. In the course of the collision this meteoroid compressed sand sufficiently to create coesite, blasted out a great quantity of rock to form the crater, forced the bedrock to rise into an encircling ring of hills, and virtually annihilated itself.

A score of such craters is known on Earth, together with a greater number of possible candidates which include the SE inlet of Hudson's Bay in Canada. Nonetheless, astronomers believe that many millions of such craters have been formed by impacting meteorites. Most of these scars would have disfigured the first solid crust formed on the young planet Earth, some four billion years ago. At this time a period of meteoric bombardment beset all the planets. We can no longer locate the great majority of meteoroid craters on Earth. The slow irresistible forces of wind and rain, of ice and sunlight, of rivers and ocean tides, have totally obliterated them. Only the more recent survive particularly in dry and slowly eroding areas like deserts.

Water in fact covers two-thirds of the Earth's surface. Clouds of water droplets course across the

Richat formation *Apollo 9*

land and drop rain or snow almost everywhere. In most parts, saturated ground lies only a few feet below the surface. Nonetheless, there are portions of the Earth's surface which are extremely arid, and this, the Sahara Desert, is one.

Deserts mostly lie near the tropics. This is because of atmospheric circulation pattern in which warm, moist air rises at the equator, dries in the upper atmosphere, and descends to the surface at tropical latitudes. Rain here is rare. Clouds do not often obscure the Sun, which therefore produces considerable evaporation of surface water and desiccation of the soil.

When no water is present to give cohesion to soil, it breaks down into very fine particles of sand, a common feature of deserts. Generally, however, the amount of sand is not great. This is for two reasons: first, the absence of rain means that rock suffers little erosion to finer particles: second, wind carries away the smallest grains of sand. Therefore deserts are primarily regions of bare rock, and only secondarily areas of inland sand.

Sand's propensity for being moved by wind leads to the formation of dunes. These are large bodies of sand which migrate slowly across a desert as wind lifts sand from one side and deposits it at the other. Dunes usually have distinct crests which generate shadows when the sun is low. Aerial photographs of deserts, particularly when shadows are long, show regular patterns of dunes extending over large areas. These may form long, parallel ridges, close-packed collections of peaked dunes, fields of small crescent-shaped dunes, and many other forms.

The Surface of Planet Earth

Rock, water, air

The surface of the planet Earth comprises two distinct states of matter, solid and liquid, and it is for that reason unique. The boundary between the two is the only place known in the Universe where solid, liquid and gaseous forms of matter coexist stably; it is illustrated in Plate 25 (Hawaiian Islands).

As a consequence of the favorable conditions on Earth, living forms have developed, capable of motion, reproduction and evolution. It is difficult, if not impossible, to find any location on the planet at which evidence of life cannot be detected. In many places, as here, life is obtrusively evident. Photographs taken even a few seconds apart frequently illustrate the motion of life forms. Life is, however, restricted to the liquid oceans and to the surface and upper few meters of the solid land. It can be regarded as a purely surface phenomenon of rather little consequence.

The land is, inevitably, the more varied surface. Mostly its variety is an effect of erosion by wind,

Surface of the Earth (Hawaiian Islands) DAA

Surface of the Earth (Grand Canyon) DFM

flowing water and frost action. Without erosion the planet's surface would consist of bare rock plateaux. Erosion is extreme on the littoral strips adjacent to the oceans, and much of their length is occupied by sand, the ultimate product of erosion. Inland, the surface generally comprises larger units of rock, except in desert regions where wind erosion can also produce sand.

A characteristic of Earth rocks is stratification, and evidence of this abounds, as in Plate 26 (Grand Canyon). Stratification occurs wherever the rock has been produced by a succession of deposits. The individual strata can originate in a variety of ways: accumulation of sand on a beach; flows of lava from volcanoes; deposition of volcanic ash; accumulation of skeletons of marine life. Stratification is displayed only in the presence of erosion. Strata which were deposited horizontally can be raised or lowered, or tilted and bent by movements of the Earth's crust. The Himalayas, illustrated earlier, are a good example of strata having been lifted and twisted by crustal movements.

Venus

Brightest, invisible planet

At any given moment about half of the atmosphere of the Earth contains opaque clouds of water droplets; the surface of the planet is visible through the other half. Next in towards the Sun lies a planet of almost identical size, Venus. The atmosphere of Venus is permanently opaque, and the planet's surface is never seen from space. Until recently, therefore, all that was known about Venus referred to the top of its cloud layers.

Unlike those of giant planets, the clouds of Venus show no regular patterns. They have very high reflectivity, rendering Venus particularly bright. Only at ultraviolet wavelengths, as in this photograph, are shadowy ephemeral markings seen. At visible wavelengths few of these swirls are evident. Some of the markings persist long enough for their motion to be tracked: these show a rotation period of about 4½ Earth days. Only recently has it been demonstrated, by bouncing radar echoes off the surface, that the underlying planet rotates once in 243 days. The obliquity of Venus—the inclination of the equator to its orbit—is small, only about 3°. In contrast to the other planets with small obliquities, Venus rotates backwards.

The clouds contain droplets of liquid typically one micron (one-thousandth of a millimeter) in diameter. It is particularly difficult to determine the chemical composition of this liquid from a distance, and even the space probes which have penetrated the atmosphere to the surface of the planet have not provided all the relevant information. Therefore astronomers disagree about the clouds' chemistry. There is general agreement that water droplets are not the principal constituents of the clouds. Some of the chemicals which have been proposed include compounds of mercury, hydrated ferric chloride, simple hydrocarbons, and sulfuric acid. The last of these is most widely believed at this time. The acid might sometimes fall as rain on the planet, the most corrosive rain in the solar system.

The clouds float a considerable distance above the surface of the planet: the visible tops are some 60 km up. The atmospheric pressure there is only one-tenth that of the Earth, and the temperature is around −30°C, irrespective of whether it is locally day or night. The pressure and temperature increase steadily towards the planet. At Venus' surface conditions are extreme. The atmospheric pressure is nearly 100 times that at the Earth's surface and the temperature is close on 500°C.

The clouds float in a gaseous atmosphere that is more amenable to study. For terrestrial observers, however, there are difficulties in detecting on other planets those gases which are also present in the Earth's atmosphere. As reflected sunlight passes through Venus' atmosphere characteristic features due to, say, carbon dioxide appear in the spectrum of the sunlight. To reach an astronomer's spectroscope, the light from Venus must traverse Earth's atmosphere which also contains carbon dioxide. Thus the effects of Venus' carbon dioxide on the sunlight are confused with the effects of Earth's cabon dioxide. It was thus not known until 1932 that carbon dioxide is the major constituent of the Venus atmosphere, accounting for 97% of the gases near the cloud tops. Argon and nitrogen account for most of the remaining 3%; carbon monoxide, hydrogen chloride and hydrogen fluoride are also present in very small proportions: 50, 0.5 and 0.006 parts per million respectively.

The presence or absence of water vapor in Venus' atmosphere was the issue of a heated controversy until spacecraft first ventured near the planet. It is now known that little or no water exists near the visible clouds, but that small quantities exist at lower altitudes in the atmosphere.

Very little sunlight filters through the clouds. The light which does complete its journey heats up the planet's surface and is therefore reemitted as infrared radiation, heat. The atmosphere is totally opaque to infrared radiation, so the heat has extreme difficulty in escaping. This is the reason for the unexpectedly high surface temperature. The same principle operates in a greenhouse: glass lets light pass but is opaque to infrared radiation, so the heat is trapped. In astronomical parlance, the trapping of heat by a planet's atmosphere is called the greenhouse effect.

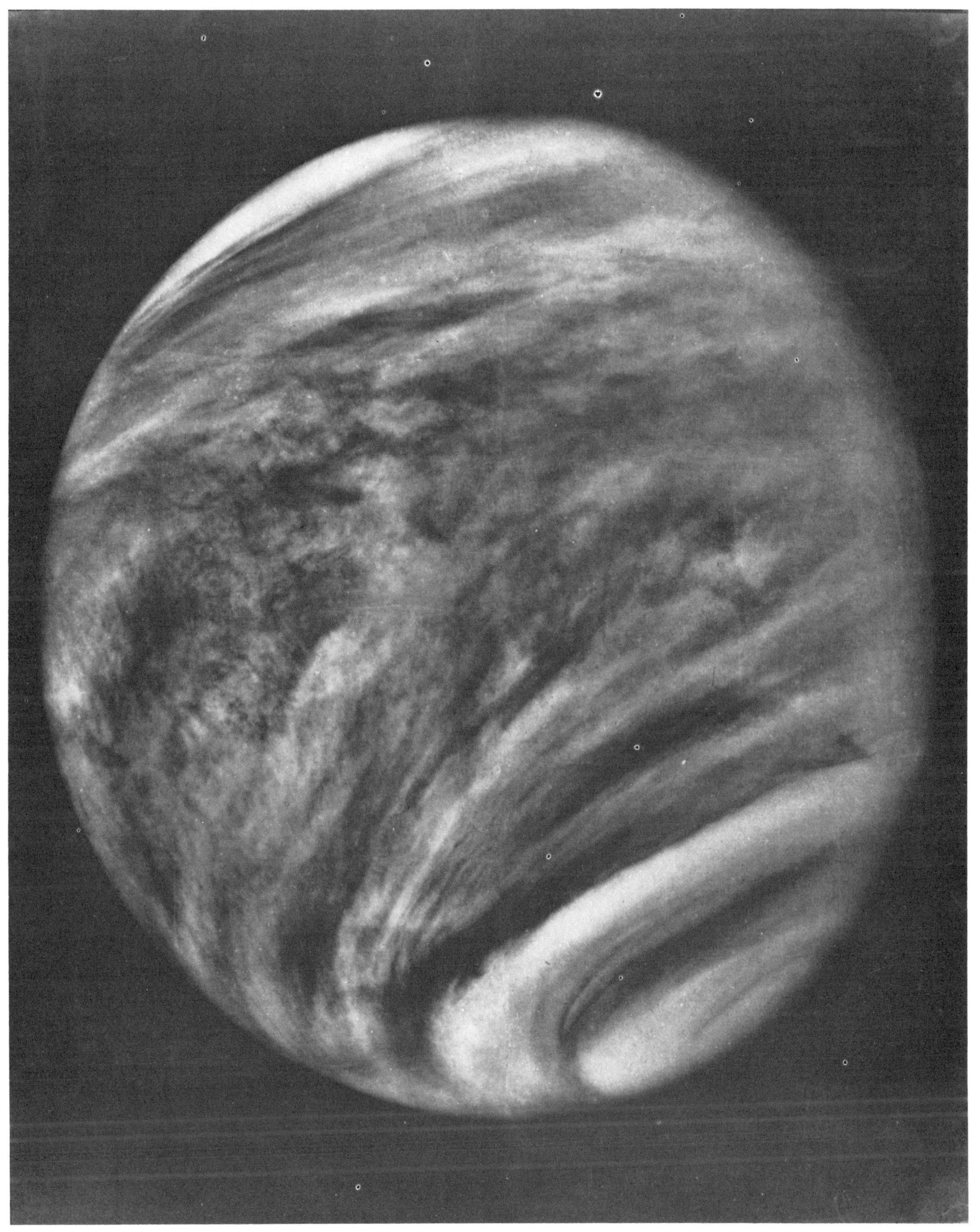

Clouds of Venus JPL

Curiously, the cloud is not equally opaque all over the planet. A more transparent hole, some 1000 km in diameter, exists at the north pole of Venus (by inference at the south pole too). It shows itself as an excess of infrared radiation escaping from the surface of Venus at the polar regions. Its cause is presumably a phenomenon connected with the way that the clouds of Venus are driven over its surface by convection and the planet's rotation.

According to the Pioneer Venus mission in 1978, the darker bands visible in Venus' cloud pattern seem to correspond to patches of sulfur dioxide. Sulfurous fires on the surface may be the cause of an increase of light observed by two spaceprobes as they approached the surface of the planet. It may be that the sulfur-clouds, the fires and the sulfuric acid "rain" constitute a "sulfur cycle" on Venus, akin to our water cycle.

Surface of Venus

Radar map

Because of its cloud covering, very little is known of the surface of Venus. We are not totally ignorant of its nature, however, as radio astronomers have been able to map the radar reflectivity of the surface. Distinct differences in reflectivity are found from place to place, and are attributed to a variety of surface coverings. Solid rock at or very near the surface reflects well; deep dust deposits do not.

These pictures show the face of Venus. The radar technique is not practicable near the planet's equator, hence the dark stripe on one picture. Areas reflecting the radar well are shown bright.

Astronomers have learnt from this radar map that there are no mountains higher than about 1000 meters (3000 ft), but that quite a number of craters exist. Since little erosion is expected on Venus, because liquid water cannot exist there, large meteoroid craters would remain visible much longer than on Earth.

Just north of the equator is a region of very high radar reflectivity. This may be volcanic in origin. Since Venus and the Earth are of similar size, similar amounts of internal heat should have been generated, and similar histories of volcanism might be expected. Indeed, it is likely that much of Venus' atmosphere was liberated by volcanoes, and there is no reason to suppose that volcanic activity has now ceased. One crater, 80 km in diameter, tops a shallow mountain 300 km across, and is believed by many astronomers to be a volcano. Whether it is active is not known.

Plate tectonics may also occur, for there is a hint that a linear trough runs across the unobservable equatorial belt and projects from each side. This might be a giant rift valley such as that in eastern Africa and the Red Sea or the ARIADAEUS RILLE on the Moon.

The 1978 Pioneer Venus mission largely confirmed the radio astronomers' picture of Venus, by close-up radar mapping of the surface. Mainly, Venus showed a rolling, Earth-like terrain. The hint in the radio-astronomy maps of a large rift valley south of the planet's equator was confirmed as Pioneer mapped a huge canyon, 4000 km (2500 miles) long, 240 km (150 miles) wide and 6500 meters (21,000 ft.) deep. The satellite, repeatedly orbiting Venus, also detected remarkably high lightning activity, confirming earlier results from the Soviet Verena 11 and 12 spacecraft. Small areas of Venus receive as many lightning strikes as the entirety of the Earth.

Mars

Hundred-million-year drought

Mars (inset Plate 27) is a smaller planet than the Earth and retains little atmosphere. Its surface is therefore easily studied and has been photographed in great detail by the space probes Mariner 9 and the Viking orbiters. The photographs, some of which are displayed on the following pages, have revealed a surprising variety of scenery.

Vast areas of Mars, particularly in the northern hemisphere, are desert plains strewn with ruddy rocks and soil. These are old lava flows, and give the planet its red hue. Other regions, mostly in the south, show the scars of bombardment by huge meteoric bodies which collided with the planet in the remote past: jumbled arrays of overlapping craters occupy the

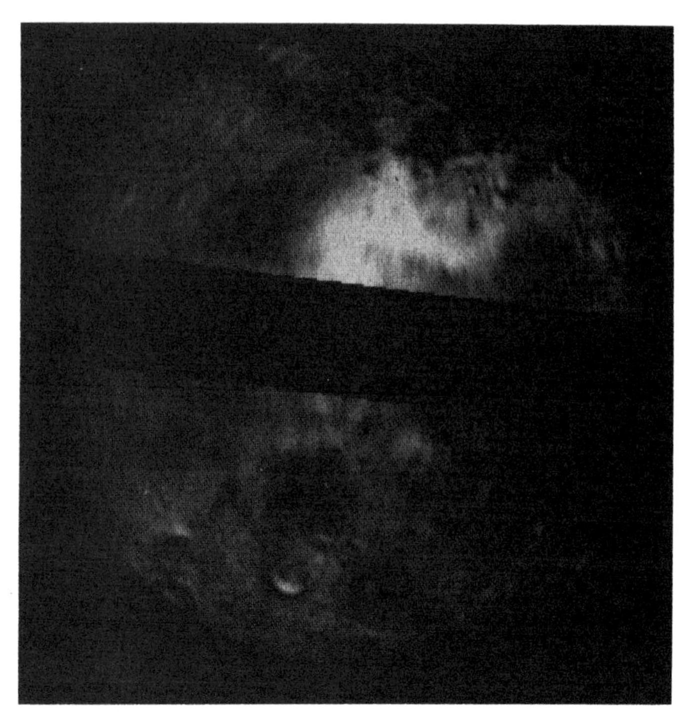

Radar maps of Venus JPL

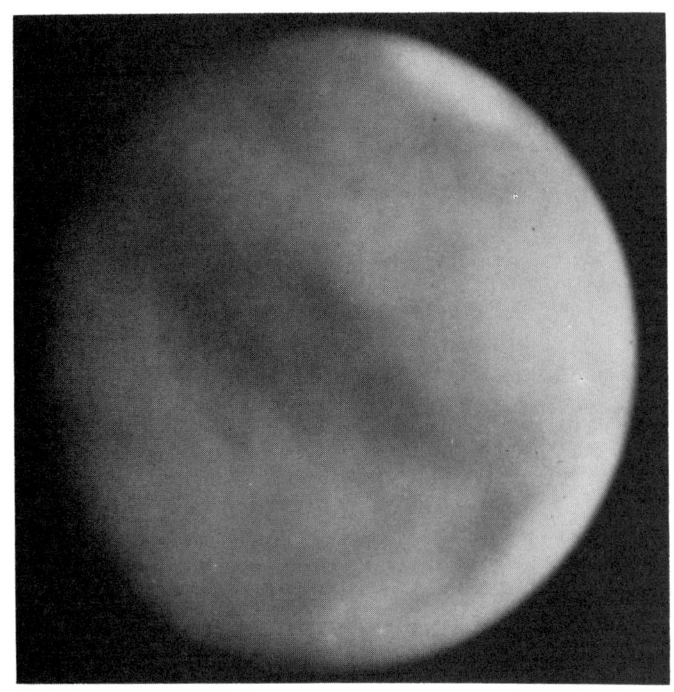

Mars *Hale*

highland plateaux which have lain almost unaltered for four billion years. Among the southern plateaux is Hellas, the largest impact crater in the solar system, 1800 km in diameter, and 3 km deep. Giant volcanoes rise from the plains; these have probably been extinct for about 100 million years. Fractures and faults in the planet's surface are easily seen. Most impressive of these are the VALLES MARINERIS, or Mariner Valleys, a massive canyon system 2500 km long. Chains of volcanic craters and extensive lava flows also testify to former violent activity.

No surface water exists on Mars. In the past, perhaps in a warmer era when volcanic activity threw vast quantities of gas into the atmosphere, water must have flowed, if only occasionally. There is ample evidence that water, or a similar fluid, eroded extensive portions of the surface.

Winds still occur in the tenuous atmosphere, and are sufficiently strong to lift sand from the deserts and generate global dust storms which temporarily obliterate the surface features. The dust is very fine: the particles are typically a micron in diameter. After major dust storms the redistribution of sand causes changes in the color and brightness of some portions of the planet. White clouds drift on the winds too, but tend to be shortlived.

The poles of Mars are covered with ice caps which expand and contract with the seasons just as do the Earth's. The martian polar caps contain rather little snow and water ice, however, but are largely formed from carbon dioxide, dry ice. Gaseous carbon dioxide is the major constituent of the atmosphere. There is very little water vapor in the martian air, and because there are no living organisms on the surface, there is little oxygen. Carbon monoxide, nitric oxide, nitrogen, argon, neon, ozone, krypton and xenon are also present in traces.

The martian day is very similar in length to the Earth's, 24 hours 37 minutes, and the obliquity of 23° 59′ compares with 23° 27′ for the Earth. Because of its greater distance from the Sun and its thinner atmosphere, however, the surface temperature is much lower, scarcely rising to freezing point on warm summer afternoons. When the atmosphere was thicker, the greenhouse effect would have raised the temperature, and the evidence of surface water action makes this more credible. At that time, conditions might have been favorable for the development of primitive life. If so, there is no evidence that it has survived the 100-million-year drought.

Before spacecraft examined Mars, the term "canals" was often used to describe the faint, linear markings suspected, from Earth-based observations, to dissect the red plains. These are now known to be illusory, and the entertaining theory that intelligent life constructed canals to irrigate the planet can be abandoned.

Mars has two satellites, PHOBOS and DEIMOS.

Martian Clouds

Like Earth's

Although there is very little water vapor on Mars, clouds may form whenever the atmosphere is cooled drastically, as in crossing a high mountain. These photographs show two different types of cloud on Mars, but they by no means exhaust the varieties.

The photograph (p.194) shows a region of streaky and diffuse clouds very reminiscent of cirrus and cumulus clouds frequently seen by passengers in aircraft. The close similarity of terrestrial and martian clouds indicates that they form in much the same way. These clouds move around with the wind.

On p.195 is a cloud formation less common on Earth: wave clouds. Wind is blowing from left to right across a martian crater whose rim rises high above the surrounding plains. Air is forced to rise as it passes over the crater rim. It subsequently falls too far, rises again, and continues in a succession of up and down movements as it recedes from the crater. The air is behaving like water in a stream flowing over a submerged rock but, unlike water air may maintain the pattern for 20 or more waves. The crest of each wave is cold enough for clouds to form; the clouds evaporate at the troughs. These crests rise 20 or 30 km above the planet's surface. Wave clouds occasionally form downwind of the Andes and other terrestrial mountains, but at most six waves have been seen on Earth.

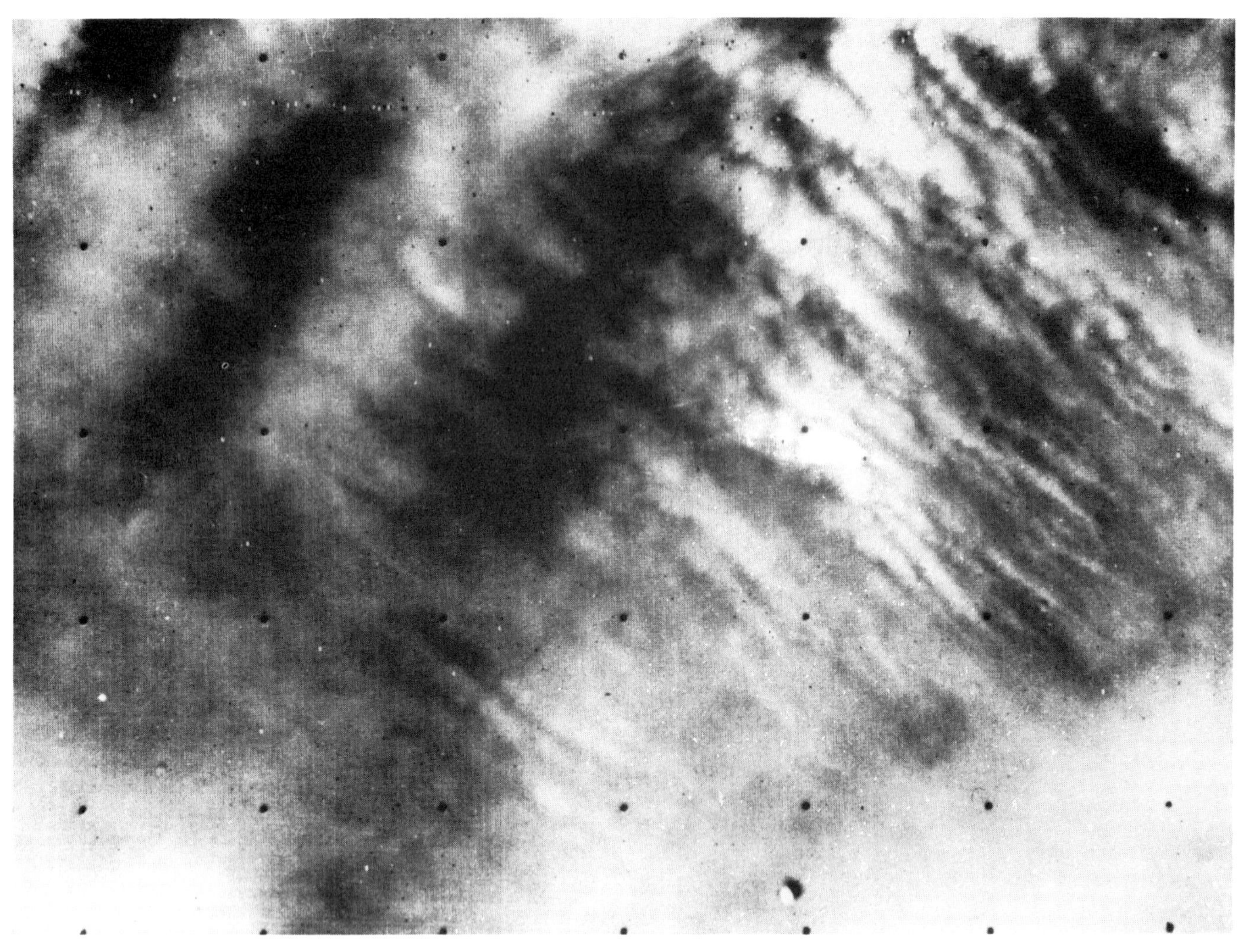

Cirrus cloud on Mars *Mariner IX*

Near the martian poles the air is colder, and carbon dioxide clouds can form. In autumn, around the edges of the polar ice caps, ground fogs of carbon dioxide occur. These closely resemble terrestrial ground fog except that the air temperature within them is about –127°C.

An Impact Crater

Melting the crust of Mars

On p.195 is a beautiful example of an impact crater typical of craters over much of the surface not only of Mars but of all smaller bodies in the solar system, like MERCURY, MOON, PHOBOS, VENUS, some ASTEROIDS etc.

In the distant past, probably a few billion years ago, a large lump of rock that had been quietly orbiting the Sun collided with Mars. The original rock would have been about one-tenth the diameter of the crater or less—a tiny body only a kilometer or so across. It hit Mars at a speed of perhaps 10 or 20 km/sec, or about 50,000 km/hour. It is hard for Man to conceive of the force and devastation of such an impact, even though it was a very minor one in the history of Mars. Certainly Man has been unable to generate an explosion capable of producing a crater like this.

Such was the impact that the small body would have been almost totally obliterated. The martian crust also suffered greatly. Much of the region within the crater was instantly melted, and rock strata further from the center were tilted upwards to produce the almost circular outer walls. As when water drips into a puddle, the molten rock at the center splashed upwards. Before this splash subsided, the rock cooled and solidified, leaving a central peak within the crater. Considerably more of the molten material was thrown out of the crater, and it produced a compact, surrounding ring of debris. The debris has a soft, rippled appearance indicating that much of the material was molten when it landed. On planets smaller than Mars, where the force of gravity is less, the splash pattern around craters extends farther.

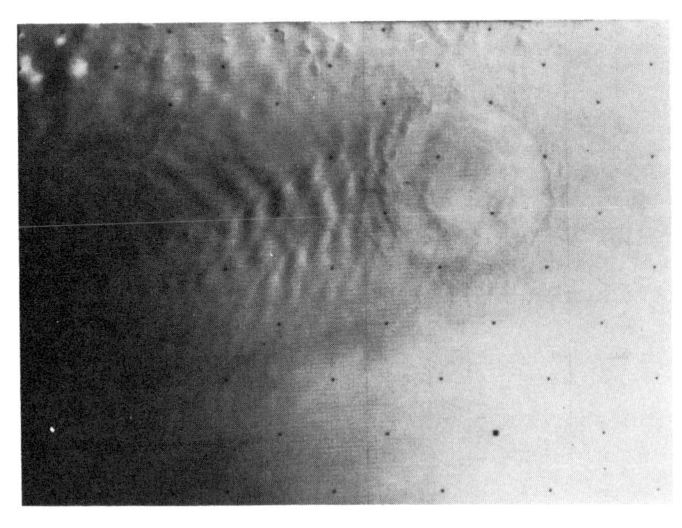

Wave cloud on Mars *Mariner IX*

Syrtis Major

Windblown streaks

The darkest feature on Mars is the Syrtis Major. This, indeed, was the first marking on the planet ever recorded by astronomers, and can be recognized on a sketch by the Dutch scientist Christiaan Huygens made in 1659. We now know that the Syrtis Major is not a single dark patch, but a series of streaks of dark material all running parallel to one another, and all originating from small craters. It seems certain that strong winds have produced these streaks, particularly since the Syrtis Major occupies a sloping part of the planet where the atmosphere would be expected to flow downhill.

The material that produces the streaks is sand: long dunes are known to form downwind of obstacles in terrestrial deserts. The martian sand is dark, but

Crater on Mars *Mariner IX*

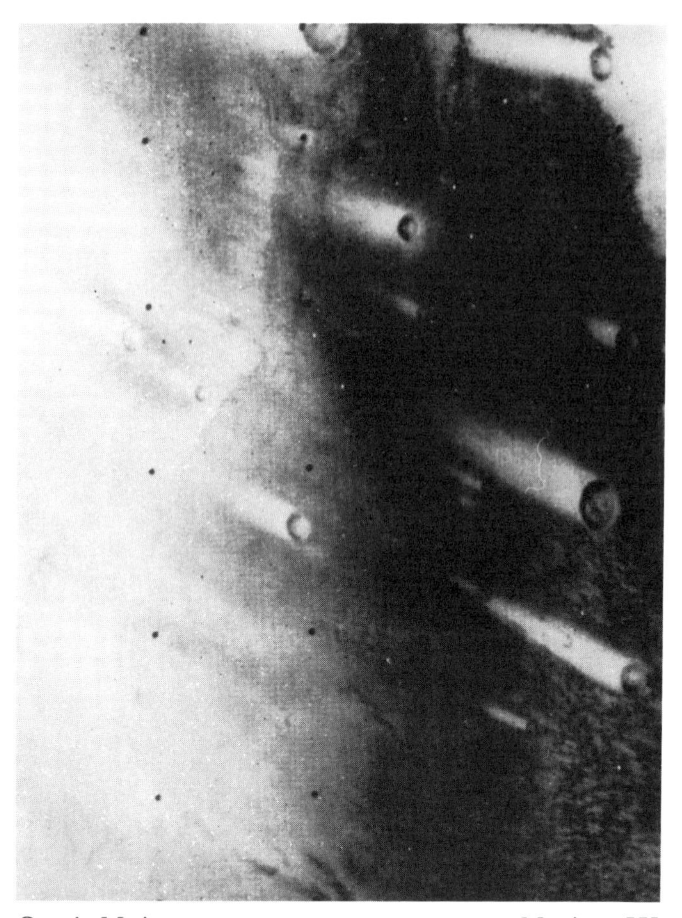

Syrtis Major *Mariner IX*

exactly why this should be so is unclear. The situation is made more complex by the occurrence of bright streaks in some parts. This photograph shows a set of bright streaks at the eastern edge of the Syrtis Major, where the highest wind speeds are expected.

Possibly the difference between light and dark streaks is that of a variety of particle sizes in the sand. Finer powders tend to be brighter than coarse sands. Alternatively, the material may have a different mineralogy in the two distinct types of streak. Sands rich in silica (quartz) are generally brighter than those with little silica content.

Olympus Mons

Martian volcano

The two largest mountains on Earth are volcanoes. Their names are Mauna Loa in Hawaii and Mount Teide in the Canary Islands. Their bases, on the floors of the Pacific and Atlantic Oceans respectively, straddle about 200 km. Both rise some 9 km, or about 30,000 feet, from the surrounding submarine plain.

The volcanoes of Mars dwarf those of the Earth into insignificance. Olympus Mons (Mount Olympus) is the largest of these, measuring 600 km across its base and 21 km high (69000 ft). The bulk of Olympus Mons exceeds 20 times that of Mauna Loa or Teide. This seems all the more impressive when it is remembered that Mars is a small planet, only one-ninth the mass of the Earth. All the volcanoes of Mars are dormant or extinct, and have been inactive for a considerable time. Some researchers date the last volcanic eruptions to a million years ago, but many believe this figure is a hundredfold underestimated.

This composite photograph shows the entire volcano from directly above (compare CERRO LA MALINCHE). At the summit of Olympus Mons is a collection of craters, or calderas, very like those found on terrestrial volcanoes, but measuring 90 km across. The slopes below the summit are made up predominantly of lava flows. The lava which poured out of Olympus Mons was very fluid and spread readily down the shallow gradients, accumulating to form the main body of the mountain. The low viscosity is indicative of a type of rock known as basalt.

More detailed photographs show sinuous channels trending down the slopes of Olympus Mons. These are believed to be collapsed lava tubes. Terrestrial basalt usually forms lava tubes when the flow of molten rock has dwindled to its last trickle, and most of the flow has solidified.

Around the base of Olympus Mons is a roughly circular ring of cliffs which were clearly produced by erosion. However, basalt flows do not erode in this fashion. Thus the outer portions of Olympus Mons must be composed of some other material. The most likely material is compacted ash, and if this interpretation is correct the ash was probably transported the vast distances involved within a cloud of hot gas. Such a cloud is known as a nuée ardente, and examples have occurred on Earth. The erosion agent would have been wind. Martian winds are less erosive than Earth's—the atmosphere is thinner, so the winds flow faster but with less ability to carry heavy particles. In the past, however, there may have been periods when the martian atmosphere was denser, and erosion may have been more severe.

Tharsis Shield

Fractured terrain

The majority of the martian volcanoes, including Olympus Mons, lie on a dome-shaped swelling known as the Tharsis Shield. This huge blister has been forced to swell from the planet's surface by interior pressures; the same pressures caused the volcanoes to burst through the blister's skin in a few weak places.

A planet's surface can withstand only a limited amount of bending. The dome which rose in Tharsis forced Mars' crust to its limit. A series of cracks

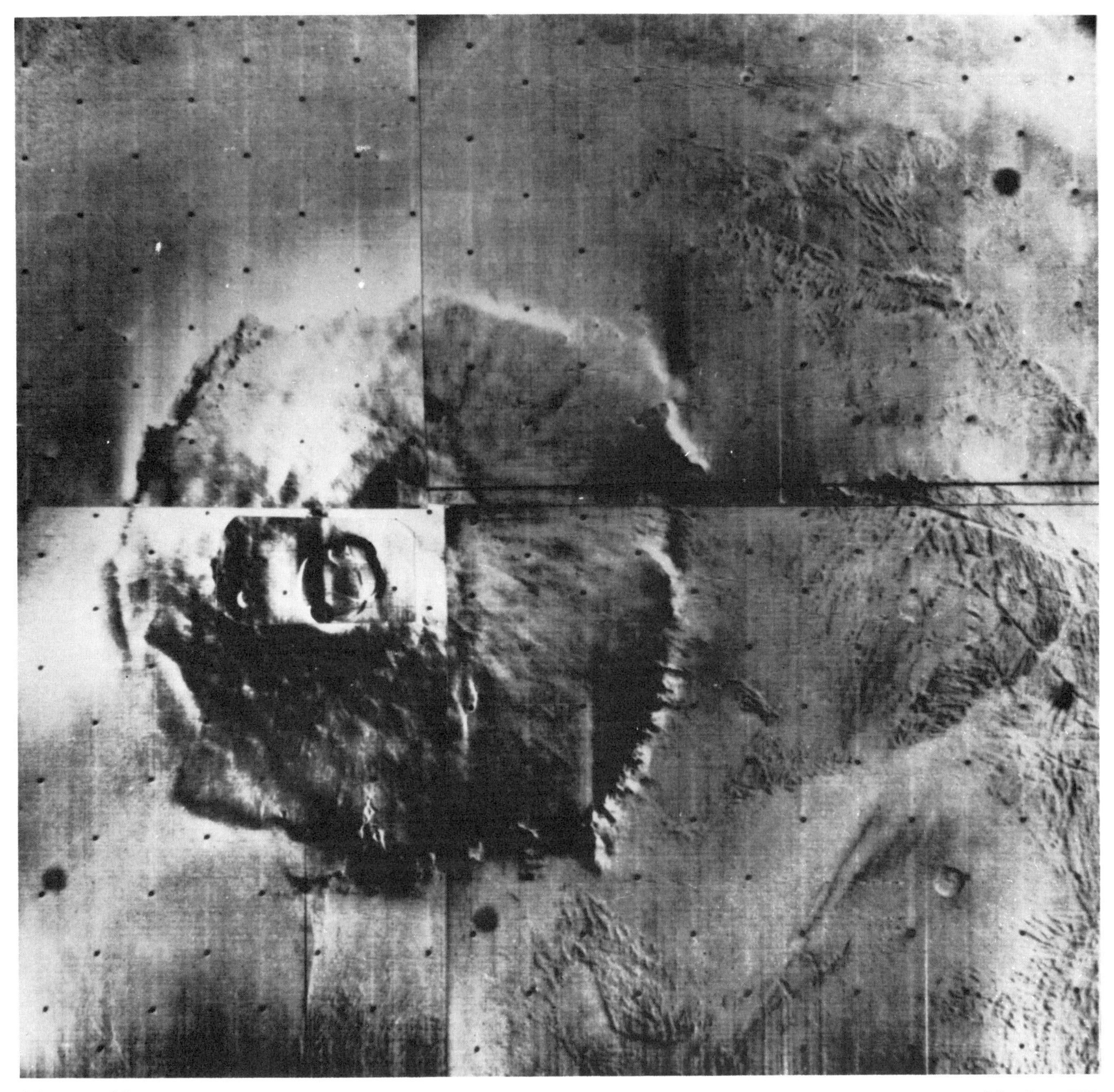

Olympus Mons *Mariner IX*

formed around the Tharsis dome, radiating directly away from the center of the upheaval. These cracks produced the fractured terrain pictured here. The Tharsis dome lies about 200 km distant, towards the lower left, and the majority of cracks trend from lower left to upper right.

Within each fracture, the planet's crust has slumped downwards. Each one takes the form of a broad, shallow gully up to a few hundred meters deep. Some extend almost 1000 km. The largest fracture, however, considerably exceeds these figures, and this is the subject of a separate photograph on p. 199.

Valles Marineris

Martian rift valley

This mosaic of photographs covers the largest single feature on Mars. Terrestrial observers had reported a faint, linear marking (one of the so-called canals) in this region and had named it Coprates. Unlike all other linear features seen telescopically, Coprates corresponds to a real martian feature. It is a portion of what we now call the Valles Marineris.

Valles Marineris is one of the fractures running radially out from the THARSIS SHIELD. It is dis-

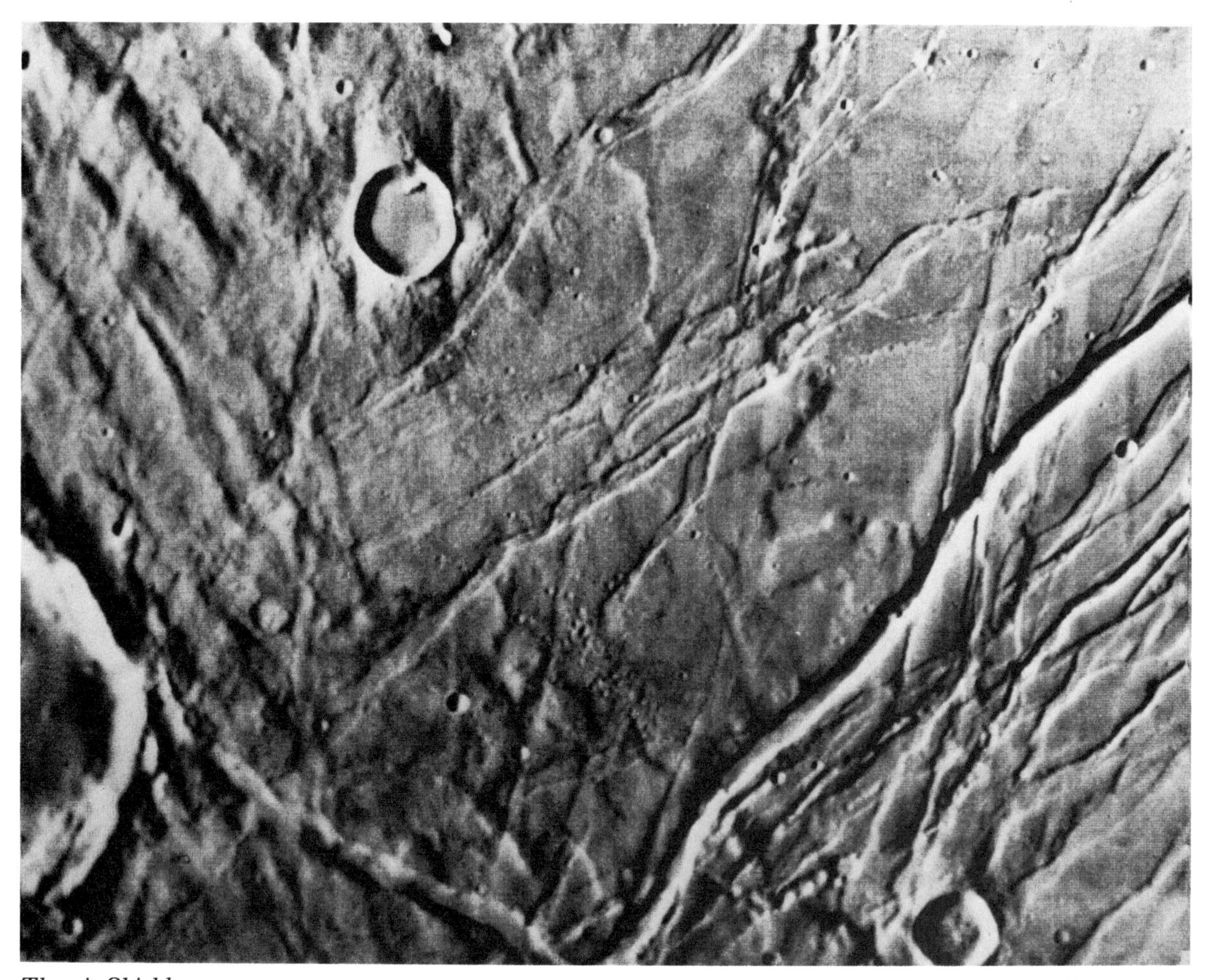

Tharsis Shield *Mariner IX*

tinguished from others by its dimensions. With a length of 4000 km, the valley stretches one-sixth of the way round Mars. At its widest point the rims are 20 km apart, and the floor is 6 km below the surrounding plateau, almost 20,000 feet.

Two parallel faults developed in the crust of Mars and between them the land collapsed inwards. This was about a billion years ago. Similar phenomena have occurred on Earth; the most impressive of these is the East African rift system, which includes the Red Sea. Typical dimensions of the East African system are: length 3000 km, width 300 km, depth 2 km. Extraterrestrial examples are found on the Moon (ARIADAEUS RILLE) and the SURFACE OF VENUS.

In the martian case, the collapse of the canyon floor by no means ended the formation of Valles Marineris. Erosion has subsequently widened the walls and carved out tributary canyons. Landslides have scalloped the rims, sending thousands of tons of rock crashing into the canyon floor. In one place half a crater remains on the rim, the other half having tumbled into the depths. In all probability, during the wetter period of Mars' history rivers flowed through the canyon, and its tributaries, and waterfalls may have cascaded down from the cliffs. However, any erosion by river action has been relatively little.

Elysium

Lava channel

Some of the martian volcanoes lie apart from the Tharsis region. This photograph is a bird's eye view of the summit crater of a volcano in the area known as Elysium. Like Olympus Mons, this volcano was built up from successive outpourings of molten lava.

Running down the volcano from the summit crater is a prominent channel. It might be thought that this was produced by water flowing out of the crater. But water channels always widen as they descend, whereas

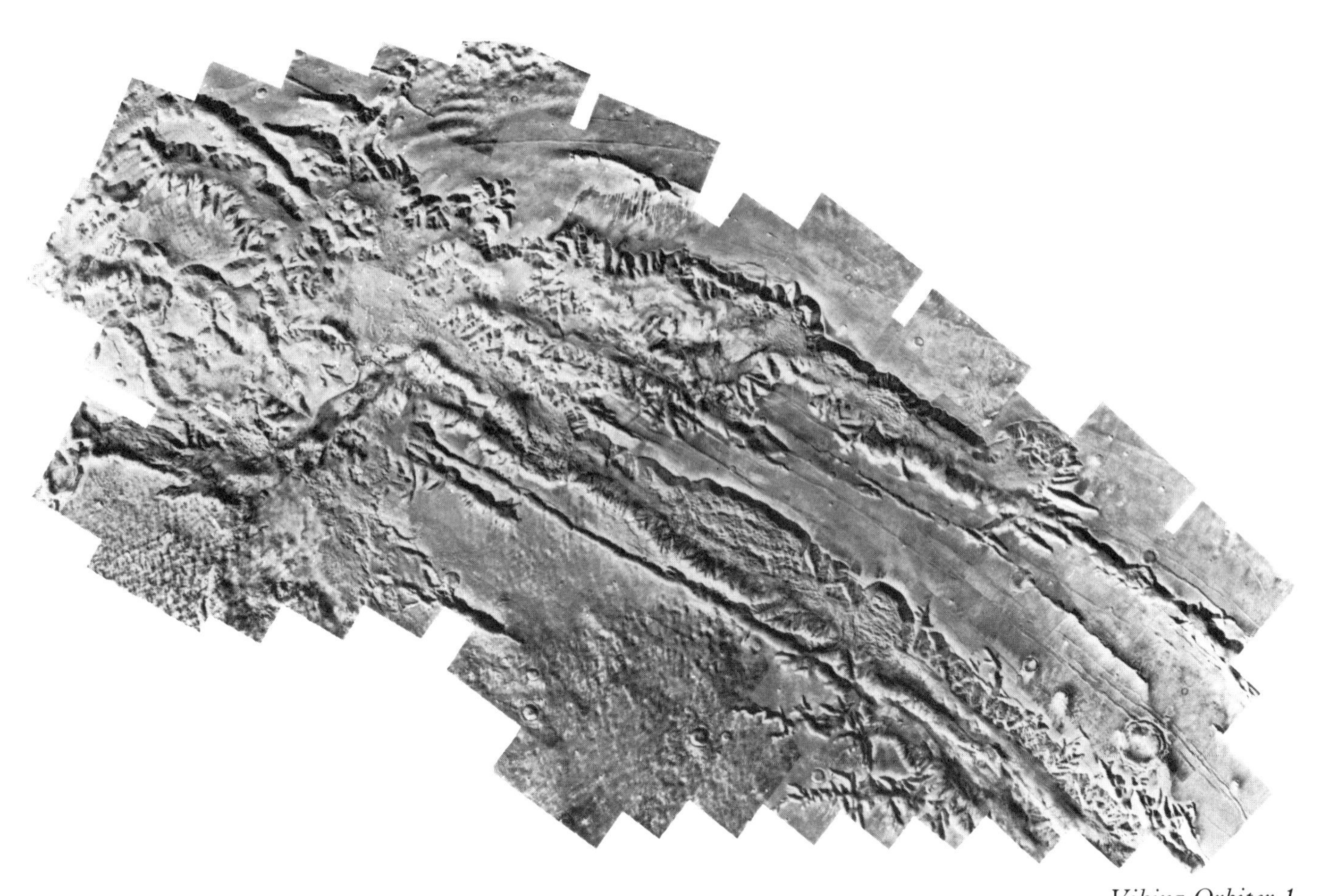

Valles Marineris *Viking Orbiter 1*

Elysium *Mariner IX*

this channel narrows away from the crater. This simple piece of evidence indicates that the channel was formed by molten lava. As lava flows down a slope, some of the material progressively cools and is left behind. Hence the amount of flowing lava

decreases downhill, and the channel it produces narrows.

Other very fine lava channels can also be seen farther down the slopes of this volcano. Some of them open out into crater-like depressions.

Flow Patterns on Mars

Evidence for flowing water

Something has evidently flowed across the surface of the planet in this region. The prominent craters, which appear to have been produced by meteoric bombardment, have raised outer rims probably a few hundred meters high. These have been sufficient to resist the flow, and the teardrop-shaped patterns of high ground which formed around and downstream of these are typical of the patterns one finds at ebb tide on beaches the world over. The number of small craters caused by random meteoric bombardment, is much the same both on and off these raised portions, indicating that no such floods have occurred for a very long time. One crater in fact straddles the edge of the right hand formation. The high ground is all that remains in this region of a layer of rock several hundred meters thick. Upstream of this region the layer is preserved intact.

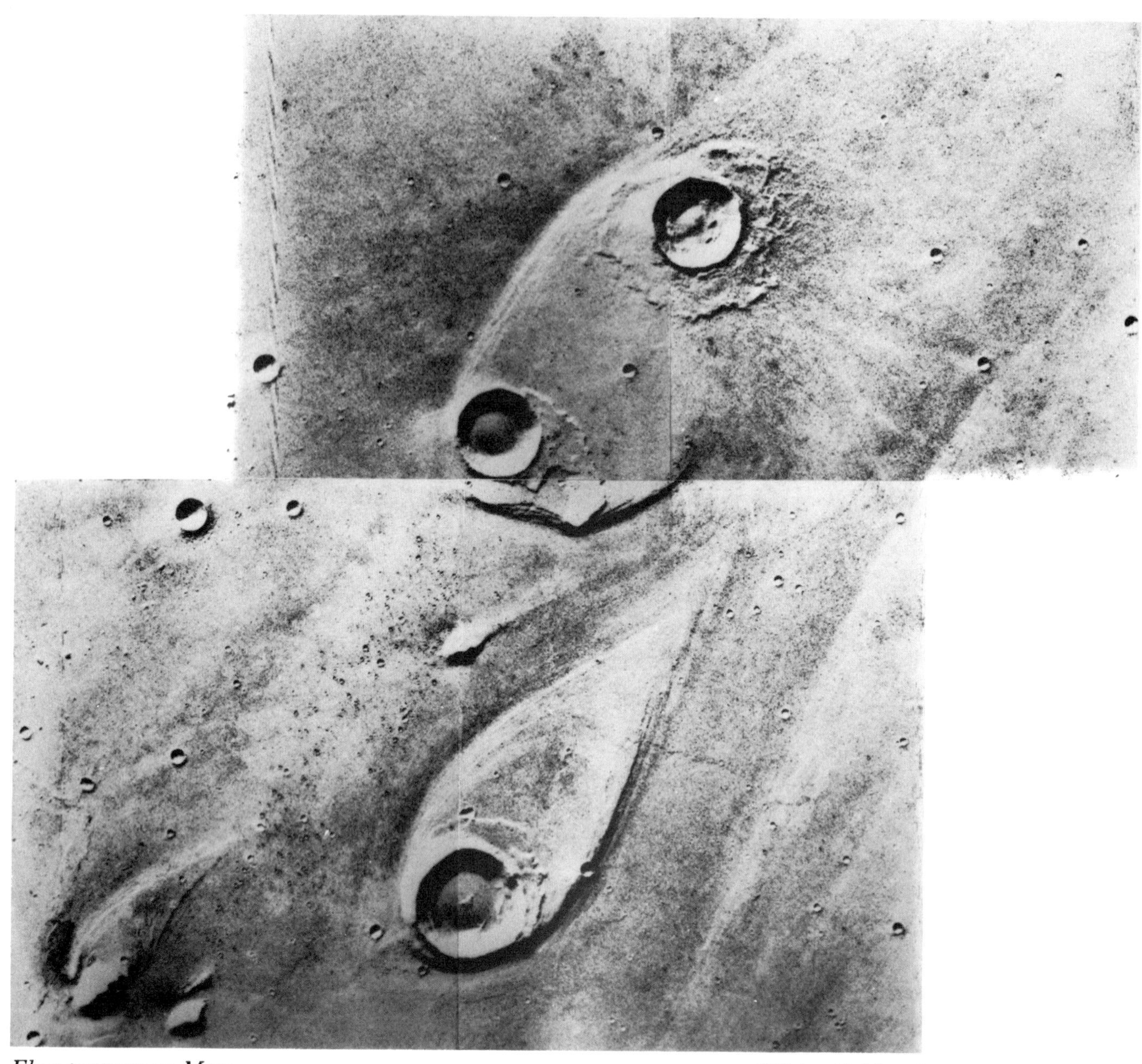

Flow patterns on Mars *Viking Orbiter 1*

It is natural to think that water produced these patterns. Liquid water cannot exist on Mars now, for the atmospheric pressure is so low that any introduced would very rapidly evaporate. However, the atmosphere was much denser and warmer in the past, particularly during the eras when giant volcanoes threw off larger quantities of gas. The former existence of flowing water is very credible.

Water need not be invoked: other substances which might produce patterns of this sort as they flowed past include lava, ice, and a fluid comprising a mixture of hot volcanic gas and fine particles of dust. That water was responsible in this case is deduced because of the mounting evidence that other formations on the planet were eroded by that liquid.

Fretted Terrain

Wind-eroded layered ash

Much of this photograph is occupied by a type of surface known as fretted terrain. It comprises raised, flat-topped mountains dotted upon a level plain. These mountains, each about 10 km across, resemble mesas on Earth. Mesas arise in regions where sedimentary rock abounds. If the sedimentary layers have not been disturbed, erosion can produce portions of landscape at discrete levels, corresponding to the tops of particularly resistant layers. This results in a level plain with flat-topped plateaux. In terrestrial examples, erosion of the upper layers is usually by water. At an advanced stage of evolution, isolated mountains

Fretted terrain *Mariner IX*

develop, and on aerial photographs these would closely resemble this martian scene.

But layered rocks on Earth were laid down under vast oceans. Does this imply that parts of Mars once lay beneath oceans? Astronomers think not, preferring to attribute the layering to deposits of windblown dust or volcanic ash. The erosion is also believed to have a different origin, for it is doubtful that sufficient water has flowed over this portion of Mars' surface to remove so much material. As in the polar regions, ice may have been trapped in the layers of sand. Evaporation of ice at the edges of the mesas would progressively cause these rocks to collapse. Wind could then remove the crumbled debris, exposing more ice-bearing layers, which would in turn erode back farther.

Laminated Terrain

Terraces

This photograph shows laminated terrain on Mars. This is moderately common at high latitudes on the planet, particularly in the southern hemisphere where it is found quite close to the polar cap. It comprises a series of steps or layers, typically 50 meters (150 ft) thick, which together form a wide staircase of terraces.

Planetary scientists are confident that Mars never experienced oceans and land movements capable of depositing strata as on Earth. Thus, whereas exposed strata forming terraces are quite common in the terrestrial landscape (see, for example, p. 189), on Mars their presence is enigmatic. The only mechanism capable of generating layers is dust storms, which could spasmodically deposit sand or volcanic ash. However, such strata would not in themselves be sufficiently cohesive to remain in layers while being weathered into laminated terrain.

Several theories can account for the form of laminated terrain. Perhaps the most widely held is that the dust layers settle with water and carbon dioxide (martian air) trapped within them. At the latitudes where laminated terrain is found, the carbon dioxide would alternately freeze and evaporate, and this could be the erosion agent which wears back each layer. The trapped water ice would remain frozen, since it melts at a higher temperature, and would give each stratum the strength to remain prominent.

An alternative theory is that the layers represent the limits of flow of former carbon dioxide glaciers which advanced from the polar caps. If this is correct, they might map out the peregrinations of the polar caps over the last few million years. It seems likely that the positions of the martian poles meander, by at least a few kilometers and possibly much more, as do the Earth's poles.

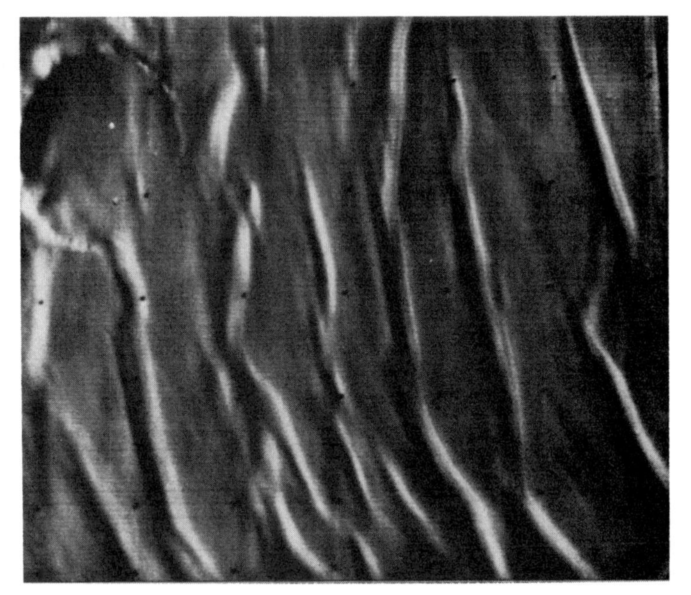

Laminated terrain *Mariner IX*

Arandas

Thaw after impact

The ejecta blanket around the martian crater Arandas is clearly different in appearance from that which surrounds normal impact craters. A comparison may be made with the photograph on p. 195. There are radial streaks and some secondary impact craters, indicating that Arandas is of impact origin, but the primary component of the ejecta blanket is smoothly

Arandas *Mariner IX*

rippled, and ends in a rounded scarp slope.

Astronomers believe that this offers further evidence of the existence of ice trapped in the martian soil. If the impacting body had landed in a region of frozen ground—permafrost—the energy released in the impact would have thawed the ground. Fluid, muddy terrain would have been formed over an extensive region, and waves of mud might be expected to have spread outwards until they refroze. Such flows are called lahars. The scarp marks the flow limit, where the mud froze.

At the site of the impact, the ground was excavated to bedrock. The walls and central mountain of Arandas are made of the bedrock, and thus were not destroyed by the mud flow.

Chaotic Terrain

Cave in

At the center of this mosaic of photographs is a chaotic region. Here a canyon has formed, and its floor is littered with the debris of its birth. Huge pyramids of rock lie in jumbled ranks across the canyon, a scene well-described by the official geological term of "chaotic terrain".

At left, the canyon extends off the photograph. Here its form is more regular, and is that of a normal river valley, steep-walled and flat-floored. In Mars' wetter past, it appears that a mighty river rose in the chaotic terrain, flowed down its canyon, and discharged into the shallow seas which probably then inundated the lower-lying terrain.

On Earth, rivers do not rise in chaotic terrain. Indeed, landforms of this kind are hard to come by, especially on so large a scale. Geologists therefore had some difficulty in accounting for this rather common martian phenomenon. The present consensus of opinion is that chaotic terrain was produced by permafrost.

It seems that long ago the martian surface cooled below freezing point, but that at the time the surface rocks contained considerable quantities of water. On cooling, the water became frozen so that ice was trapped in the rocks and soil. Such frozen ground exists in the tundra of Canada and northern Eurasia. Although the top few meters melt during the summer to form thick mud, some distance underground the ice does not thaw. That portion of the ground which does

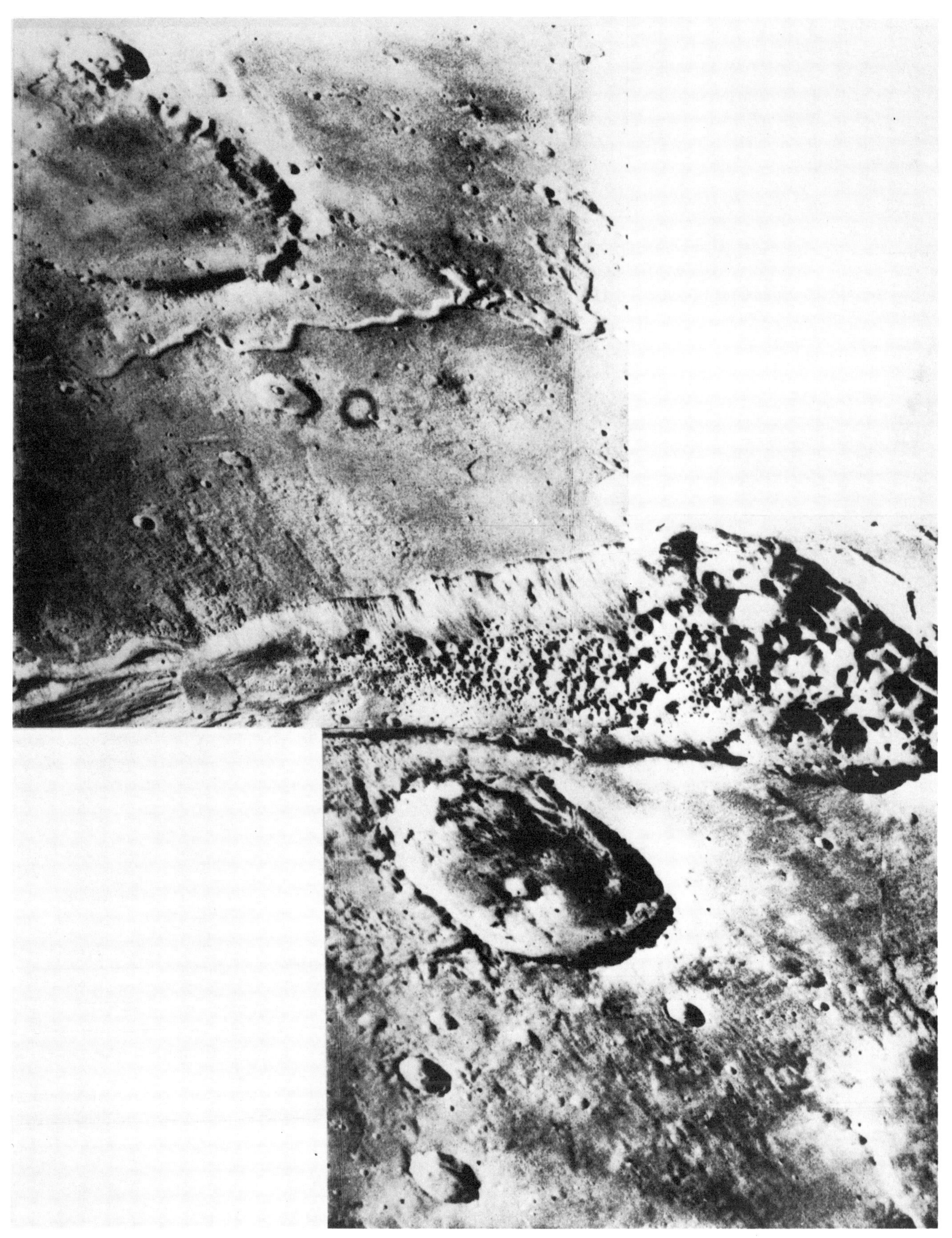

Chaotic terrain *Viking Orbiter 1*

not take part in the annual thawing and refreezing is called permafrost.

On Earth, regions of permafrost have survived undisturbed for millions of years, and the same would be true of Mars. However, during a warmer spell the martian permafrost would have thawed, at least in the equatorial regions. In highland regions where the soil contained much ice, enough water escaped from the ground to feed rivers. These flowed to the lowlands, carving canyons and causing other erosion. Meanwhile, the permafrost region was undermined as ice melted and migrated away. Subsurface caverns formed and enlarged, until finally the surface caved in. The chaotic terrain which was thus produced is evidence of the greatest natural mining disaster known.

The Polar Caps

Dry ice and water

Mars, like our own planet, has white deposits at both its poles (Plate 27, inset). A detail of the martian north polar cap is shown in the black and white photograph. Here, layers of the white material are seen overlying sandy terrain.

It has been known for some time that the polar caps change with the seasons. Sir William Herschel found them to enlarge during the martian winter and to shrink when summer came. The south polar cap is reduced to a very small region during the southern summer, whereas a sizeable fraction of the planet's northern hemisphere is covered by a permanent cap. Since the summer temperatures at the martian poles are considerably below the freezing point of water, the polar caps cannot be composed of water ice, as they are on Earth, otherwise they would not change significantly with the varying season. Instead, dry ice—solid carbon dioxide—is likely. Recent observations by orbiting space probes have confirmed that dry ice accounts for that portion of the polar caps which waxes and wanes with the season. The surprising fact which the satellite observations revealed was the presence of water ice too. The permanent polar caps, as seen on this photograph, are almost pure water ice.

As on Earth, the polar ice caps account for a great amount of water. Astronomers have performed the "thought experiment" of imagining what would happen if the polar caps melted. On Earth, no major change would ensue, the main result being an increase in the ocean depth and the inundation of some land. On Mars, however, dramatic events would occur. The first water released would quickly evaporate. But as water vapor entered the atmosphere, it would set up the greenhouse effect, preventing the escape of heat from Mars. Hence the air temperature would rise. Eventually liquid water would be able to exist, and the melting polar caps would flood desert basins, producing shallow seas. Clouds would rise from the seas and drop rain on the high land. Rivers would flow.

Over the last few pages, evidence has been presented that rivers once flowed on Mars. This was attributed to a wetter period when volcanoes enriched the atmosphere. But could volcanic eruptions have introduced sufficient gases to account for the ample evidence of water erosion? A daring alternative is that once the polar caps melted: is this credible?

To melt the polar caps requires either an increase in the Sun's radiation or a change in Mars' obliquity, so that the poles were turned farther towards the Sun in summer. Geologists find no evidence of great increases in the Sun's past output as recorded in the features of the Earth's surface. To change the tilt of a planet's axis seems, at first sight, impossible. However, the tilt is determined by the shape of the planet. Recent calculations suggest that if Mars were slightly closer to spherical, its obliquity would change thus allowing the poles to receive much more sunlight. The relevant change of shape would just about be accomplished if the Tharsis dome were removed. Put another way, before volcanic activity raised the massive Tharsis blister in Mars' northern hemisphere, the poles probably tilted more towards the Sun, and may have fully melted each summer. This theory is currently finding support among some astronomers, for it neatly accounts for many of the recent observations of Mars. But it must be regarded only as a theory: there is no proof of it so far.

Even if we are able to dissociate from our minds a planet covered with oceans and plants, we are conditioned to expect a polar ice cap to border on a frozen landscape of level, featureless plains and hard, unyielding soil. It comes as a great surprise to learn that Mars' polar caps are ringed by sandy deserts. This is particularly true in the northern hemisphere. The entire north polar cap is girdled by a vast field of sand dunes, the most extensive such field in the solar system. As winter approaches, mists of carbon dioxide fill the valleys between the dunes, and gradually a hoar frost of dry ice condenses on their slopes and summits. For half the year they are buried beneath the dry ice cap. Only during the summer can the chill martian winds shape them.

The dunes are large, perhaps up to 100 meters or more in height. As on Earth, dunes of this size will be modified only slowly by wind action. No appreciable change has yet been seen in these martian dunes. At the boundary between ice cap and sand, intricate fronts and tongues of ice have formed. Some can be seen on this photograph. It is difficult to imagine how this region would appear from the planet's surface, but in all probability the permanent polar caps do not simply thin down to a few centimeters of broken ice, but

Polar caps of Mars *Viking Orbiter 2*

present a spectacular front. Ice cliffs and stubby glaciers may abut against drifting sand dunes, perhaps a scene of beauty and contrast unparalleled on our planet.

The Surface of Mars

Desert

Before spacecraft ventured to Mars, there were many predictions about the planet's surface properties. Even discounting the more bizarre schemes in which intelligent life sustained itself by irrigation, their number was matched by their variety. Some astronomers argued that the darker regions were areas of vegetation, living on moist highlands which overlooked red sandy deserts. Others maintained a more staid approach, claiming that life cannot be supported on Mars and that the darker regions were merely darker rock exposed. In the lighter areas, which the Viking landers investigated, three experiments were performed to determine whether any terrestrial-like life existed. Ambiguous evidence of biochemical activity has left most exobiologists doubtful that life existed in the soil examined. However, that vegetation is present in the darker areas is not entirely ruled out even now, for we have examined the surface only in the red portions.

The first detailed photographs of the planet enabled astronomers to understand at last the topography and geology of Mars. The findings, which have been summarized in the preceding pages, enabled scientists better to predict what would be found when, in 1976, spacecraft finally landed on the surface. It is an interesting thought experiment to predict the surface characteristics of a planet which once, as now Earth does, experienced erupting volcanoes, swift-flowing rivers, quakes and hurricanes; but which for 100 million years has witnessed none of these.

These photographs and Plate 27 provide the answer: desert is all that remains. Attacked by sunlight and desiccating wind, the softer rocks have crumbled to sand and the windblown sand has etched the more durable rocks to rounded boulders and pebbles. Sand and rocks alone survive.

If once great cities stood here, they have crumbled to unrecognizable shapes. If trees bowed before moist zephyrs, they have returned to the dust whence they

Surface of Mars

rose. If aircraft landed here, they too have vanished, or been buried beneath unknown depths of sand and rocks. There is no life here. The soil is richly capable of supporting life as we know it, given water. But there is no more water than the thinnest of hoar frosts just before dawn. Nor is there evidence that life flourished when water was more abundant.

Suppose the cycle of climate were to bring a hundred-million-year drought to Earth. Spacecraft despatched to its surface would probably record much the same scene as this view of Mars. The greatest mechanical triumphs of mankind would vanish on that time scale. To appreciate this one need only delve into modern archaeological literature, or see firsthand the unexcavated remains of the great civilizations of a few millennia ago.

Surface of Mars

Viking Lander

Rocks and sand. Almost without exception these are tinged a deep red, very like the deserts of central Australia and some lesser-known parts of the Earth. Ayers Rock, that bold, eroded chunk rising from the Australian desert, would not appear out of place on Mars. As on Ayers Rock, the red coloration is probably only skin deep, a coating known on Earth as desert varnish, and mineralogically termed limonite. Limonite is an oxide of iron, but not the better-known oxide, rust. The limonite is so widespread that, even from Earth, Mars appears red to the unaided eye. The presence of limonite was, in fact, predicted by some astronomers two decades before the martian surface was sampled, purely on the grounds of the planet's color.

Viking Lander

The martian sky is also tinged with limonite. The air is very thin, and therefore little sunlight is scattered to produce a blue sky similar to that on Earth. Most of the sky color near the horizon is due to suspended dust particles, and is therefore pink. The martian air is 100 times murkier than the terrestrial atmosphere. Towards sunset, as the Sun's rays penetrate more atmosphere, the proportion of blue light increases. Dramatic sunsets on Mars are blue.

Phobos

Captured satellite

Phobos, the inner of the two tiny satellites of Mars, is believed to be typical of all small bodies in the solar system: irregular and heavily cratered. Cratering is produced by impacts. Somewhere in Phobos' long history it was jostled by even smaller bodies which collided with it and chipped out the many pits and craters seen on this Mariner photograph. This probably occurred when Phobos was just one of the innumerable asteroids circling the Sun a little beyond the orbit of Mars. On one of its circuits, Phobos passed sufficiently close to Mars to be captured by the

Phobos *Viking Orbiter 2*

gravitational attraction of the planet. From that time it has orbited Mars as a satellite.

Despite its unpropitious start, Phobos is now one of the more distinctive satellites in the solar system. It lies closer to its parent planet than any other known satellite: a mere 9350 km from Mars' center. In order to balance the strong gravitational pull of Mars, Phobos must orbit very rapidly. It requires only 7.6 hours for each orbit. To an observer on Mars, Phobos would appear to travel backwards across the sky, rising in the west and setting in the east twice every martian day of 24 hours 37 minutes.

The proximity of Phobos to Mars causes it great stress. The face nearer the planet is attracted more strongly than the other face, and tries to orbit more rapidly. It is believed by some astronomers that the prolonged stress has ruptured faults through the satellite which manifest themselves on the surface as the long striations seen on this photograph. These striations are typically 100 to 200 meters wide and 5 to 10 meters deep.

Phobos measures 27 by 21 by 19 km, yet its largest crater, Stickney, is 10 km across. The collision which produced Stickney might also have fractured Phobos internally, and another school of thought attributes the striations to this event. Stickney itself is not seen on the photograph.

Phobos' density is about twice that of water, and its surface is extremely dark. The darkness and density suggest that it is composed of what is known as carbonaceous material, like minor planets such as Ceres (p. 235), and if this is correct, it should have condensed much farther from the Sun than the orbit of Mars. How Phobos was captured by Mars is therefore unclear.

Deimos

Outer martian satellite

The outer satellite of Mars is known as Deimos. It is a body closely resembling Phobos, but is rather smaller, measuring 15 by 12 by 11 km. Like Phobos its surface is pitted by craters which testify to the large number of small objects which it once encountered. This photograph shows in remarkable detail a small portion of Deimos' surface. The smallest craters recorded are only a few tens of meters across. Only one other satellite has been photographed in greater detail—Earth's own Moon. There are no linear striations on Deimos, as on Phobos. Since Deimos orbits 23,490 km above the surface of Mars, the rupturing effect of the planet's gravitational pull is less. Moreover there

Deimos *Viking Orbiter 2*

Earth's surface (Hawaiian Islands)

Earth's surface (Grand Canyon)

Mars (inset) and its surface

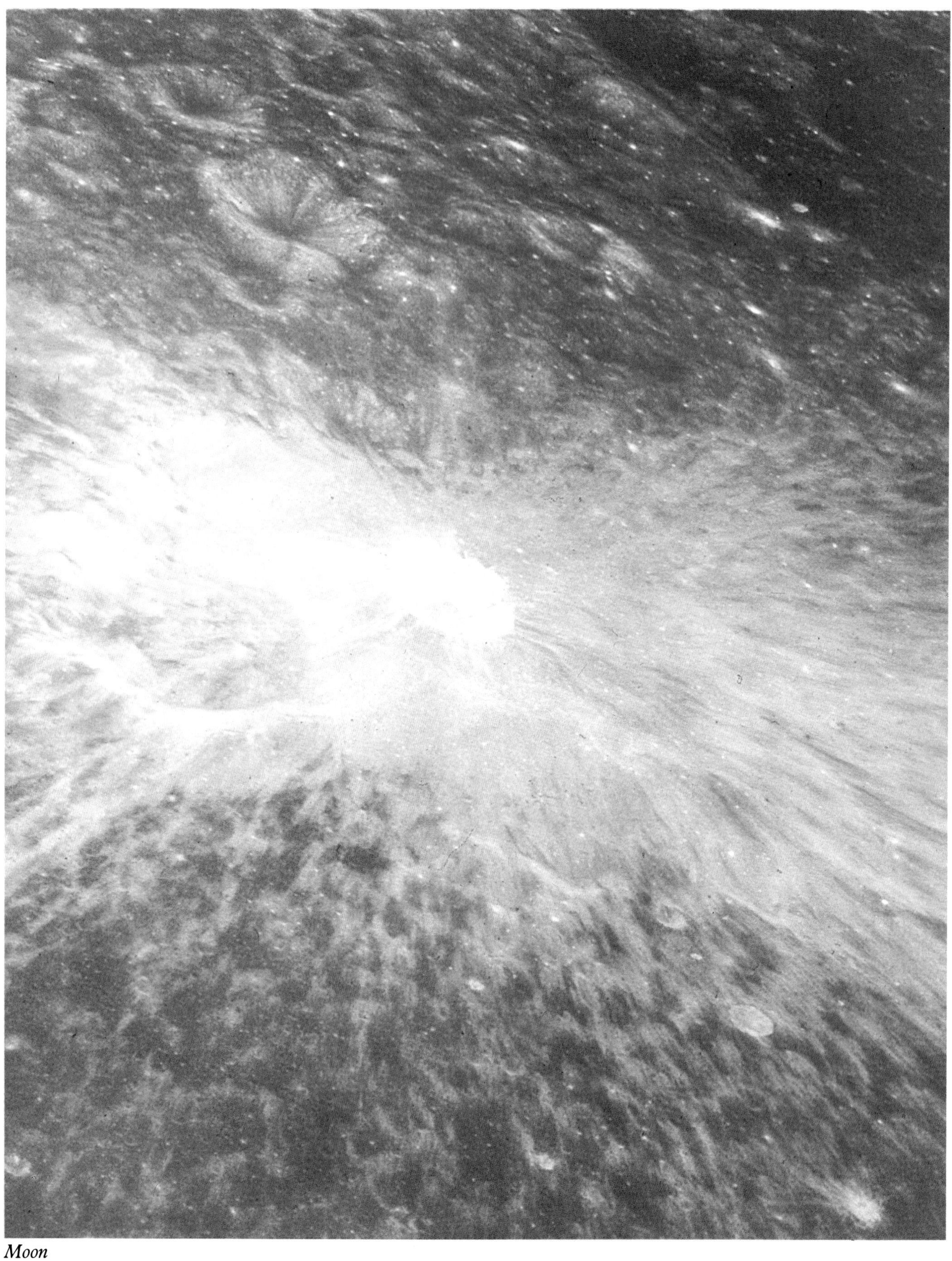

Moon

are no craters as large as Stickney on Deimos. Striations might not therefore be expected.

An experiment carried aboard a Mariner spacecraft measured the polarization of the light reflected off Deimos and Phobos. Anyone who has worn polarizing sunglasses will find that light is more highly polarized when reflected off rock than off soil. By measurement of the polarization and brightness, scientists were able to determine that Deimos and Phobos are covered with a thin layer of dust. Such a layer is called a regolith, and is produced by the continuous rain of tiny particles which bombard the surfaces of the satellites. With every impact a small amount of dust is produced, and at least 90% of this will be thrown up with sufficient velocity to escape the satellite. But the small portion which remains accumulates to a thin film possibly as little as 1 mm thick.

Deimos, like Phobos, is thought to be a captured asteroid. That the captures occurred long ago is shown by the fact that the slowly acting tidal forces of gravity have caused both satellites to turn one face predominantly towards Mars.

Mercury

Innermost planet

Of all the major solar system bodies, Mercury is closest to the Sun. This simple fact gives the planet a number of distinctive characteristics, some obvious, some obscure.

It might, for example, be thought that Mercury would be the hottest planet. This is not the case. The average temperature at the surface of Mercury is much lower than that of Venus, and even its maximum falls 50°C short of the normal Venus temperature. Conversely, it is no surprise that Mercury travels faster round the Sun that any other planet, for the laws of gravity dictate that this should be so. An obscure consequence of Mercury's proximity to the Sun is that it is the densest planet in the solar system, mean density 5.4 times that of water.

The temperature is easily explained. Mercury has almost no atmosphere to act as a greenhouse. The surface temperature by day is dictated purely by the Sun's radiation, and reaches 430°C. Mercury's day is 176 terrestrial days long. During the 88-day night, the surface cools dramatically, reaching −180°C. This enormous range in temperature must cause erosion of any exposed rock, a circumstance which must contribute to the layer of dust which cloaks the surface.

Mercury's high density is a product of the local composition of the solar nebula from which it condensed. Theory predicts that at that distance from the Sun, a planet should contain a sizeable quantity of iron, nickel and similar metals. Observations indicate that the nickel-iron has formed a core which accounts for 80% of Mercury's mass. Because iron has a higher density than the silica-rich rocks of the mantle, the average density of the planet is raised above that of the Earth, Venus and Mars which have smaller cores and larger mantles. Some iron remains in the crust, however, and is at least in part responsible for the planet's slightly reddish tinge.

Mercury's thin atmosphere is a consequence of the planet's low surface gravity. What atmosphere exists is composed of traces of hydrogen and helium deposited on its surface by the solar wind. Solar material itself streams from the Sun in an outflowing wind, which sweeps through the solar system. Mercury intercepts the wind and refreshes its sparse atmosphere from the Sun, until it is lost again from the planet, unretained because of the planet's weak gravity.

At first glance our photograph might be thought to portray the Moon. The surface of Mercury certainly resembles that of the Moon, with a mixture of craters and plains. The craters are mostly quite old, but some relatively recent specimens can be identified by the splash pattern of bright rays. Rays from some craters stretch for 1000 km around the planet. Ray craters are brighter and slightly bluer than their surroundings, but over the next few billion years they will fade and redden eroded by the 600°C daily temperature cycle.

A Mercurian Crater

Impact of a giant meteor

The most important factor contributing to the surface characteristics of Mercury is the absence of an atmosphere. On the Earth and Mars, and presumably on Venus too, the presence of an atmosphere causes, or has caused, tremendous erosion. Rivers, glaciers, winds, tides and in the Earth's case life, have greatly modified the surface on a large scale: all of these are manifestations of an atmosphere. On Mercury, erosion proceeds at a much slower rate. The bombardment of the solar wind, the effects of the enormous temperature range, and an ever-present rain of high-speed dust grains, combine to wear down the exposed rocks to a fine powder. But the rate of erosion is slow, and features persist from early stages in the history of the solar system.

A second result of the sparse atmosphere is that Mercury's surface is vulnerable to attack by all the rocks and specks of dust which get in the planet's way. Moreover, at Mercury's distance from the Sun, where gravity forces everything to move rapidly, the impacts are much more violent than on Earth or Mars. Most infalling objects hit the planet's surface at about 30 km/sec, but speeds four times this are possible. A car traveling at 120 km/hour can do a lot of damage when

Face of Mercury *Mariner X*

Crater on Mercury *Mariner X*

it collides with a stationary object; a piece of rock the size of a mountain traveling at 120 km/second wreaks considerably more damage.

Such an event produced the large crater on this photograph. From rim to rim the crater falls just short of 100 km diameter. Its floor lies considerably below the surrounding plain, as can be seen by the different shadow lengths from the opposite walls. A central mountain group, formed of material which was thrust up from the interior following the removal of the pressure of the overlying material, is particularly prominent; it rises almost as high from the crater floor as do the walls, something like 5 km.

In the excavation of so large a crater, a vast amount of rock was removed. The majority of this landed outside the crater. Up to about 50 km beyond the rim the rock was still molten when it landed. This accounts for the soft, washed appearance. Farther out, the ejected rock had time to solidify into boulders before landing. These boulders, on impact, themselves produced smaller craters, known as secondaries. The field of secondary craters covers an area much greater than that of this photograph.

Other craters on this photograph, particularly the three prominent examples near the bottom, were also produced by impacting bodies. Craters of this size are usually bowl-shaped and lack central peaks. The splash pattern around them is much smaller. At least the upper of these three was formed more recently than the main crater, for its splash pattern has obliterated the field of secondary craters.

The date of these impacts cannot readily be determined, but it probably lies between three and four billion years ago.

Plains and Craters

Volcanism aided by impacts

On this photograph can be seen the two dominant mercurian land forms: craters and plains. Of these, the cratered uplands are the older. They were produced

Plains and craters *Mariner X*

about four billion years ago during a period in the solar system when great amounts of planetary debris were being scooped up by the newly formed planets. The density of craters in these highlands is so great that astronomers believe them to be saturated. If this is correct, the addition of further impact craters would make no difference to the basic appearance of the terrain, for the new craters and their ejecta blankets would cover up previous specimens.

In the lower right of the picture is part of a smooth plain. This plain occupies a circular basin about 450 km in diameter which was itself produced by an impact. The impact probably predates most of the craters on the highlands, for if it were younger, the rock necessarily ejected by so major an event would have nearly obliterated them. Why, then, is the basin's floor not also peppered with craters? Clearly, something has happened to it in the four billion years since the era of massive cratering.

Evidence that the basin was flooded by lava is provided by the remnant rims of craters which protrude from its floor. These must be the highest rims of a population of craters which was completely engulfed. The smooth plains are therefore Mercury's equivalents of the great lava field of Mars. They are slightly lighter and redder than the cratered uplands.

The flooded plains are the only evidence of volcanic activity on Mercury. Being a smaller body than the planets considered earlier in this book, Mercury has less internal heat to power its volcanism. There are no certain volcanic mountains or craters. The lava which filled this basin must have been extruded from cracks and fissures in the ground, and these are now buried beneath it. They, in turn, were probably generated by the impact itself, which weakened the crust. Possibly, in the absence of large impacts to facilitate its journey, no lava would have flowed on Mercury's face.

The Caloris Basin

Largest crater on Mercury

The largest known crater on Mercury is the Caloris basin. The feature has never been photographed in its entirety, and this view shows only a portion of its floor, a portion with great character. The whole crater is

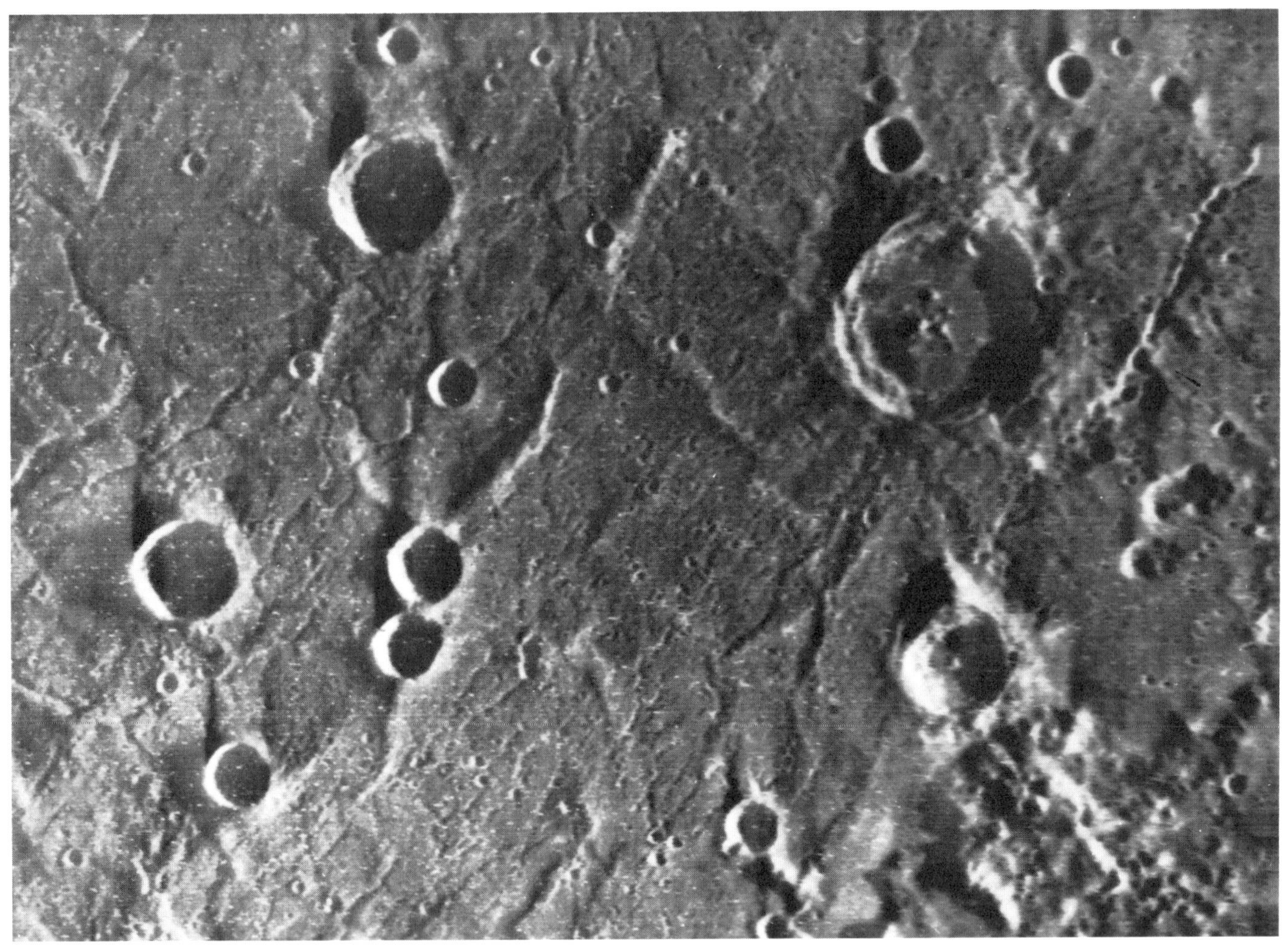

Caloris basin *Mariner X*

1300 km in diameter, more than one-quarter of the diameter of the planet. The crater floor reaches to 9 km below the mean surface of the planet. This figure sounds impressive, and is similar to the greatest ocean depth on the much larger planet Earth, but an observer standing within the Caloris basin would be quite unaware of the degree of excavation. Due to the curvature of the planet's surface, the Caloris basin is domed in the middle, and one rim is invisible from the opposite rim. A large body, probably over 100 km in diameter, collided with Mercury to produce the Caloris basin, and the impact had repercussions over the whole planet, as subsequent photographs will reveal.

It is rather difficult to define the rim of the Caloris basin. Along most of the circumference there is no major mountain range such as surrounds smaller craters. Instead, there is an abrupt change of terrain which outlines a roughly circular area of the quoted diameter. The boundary runs from the upper right corner of this photograph to right center of the lower edge. The lower right corner shows a region of mountainous terrain. The hills range from 1 to 2 km high, and form a girdle about 200 km wide most of the way round the Caloris basin. These mountains were almost certainly thrown up by underground pressure following the Caloris impact. At the center of the right hand edge of this photograph, a gap in the mountainous terrain is seen. The gap seems to have been filled in by molten rock, and the amorphous craters lacking rims within this smooth area resemble the collapsed calderas found in volcanic regions. These may, in fact, be the only large volcanic craters on Mercury.

Beyond the mountainous girdle to the Caloris basin are other features clearly associated with the impact. These include valleys which radiate from the basin and which were gouged out by low-flying debris from the impact. The valleys are up to 1000 km long. They slice through craters, demonstrating that the Caloris impact occurred after most of the other craters of Mercury had been formed. In other parts of the planet, similar but shorter valleys exist, and probably radiate from smaller basins on the side of Mercury which has not been photographed in detail.

The floor of the Caloris basin is a textured region of smooth plains, meandering ridges and cracks, and is dotted with younger craters. Some ridges rise to 700 meters above the plains. The ridges and cracks tend to describe crude circles around the center of the Caloris basin, which lies off the left edge of the photograph. Such patterns are typical of the cooling of molten rock, and after the impact which produced the Caloris basin, much molten rock must have fallen back in and flowed slowly towards the center, solidifying on its journey.

However, astronomers find difficulty in believing that an area as large as one million sq. km could be filled by the molten debris of the impact. Instead, therefore, it is proposed that the basin floor was filled shortly after its formation by outpourings of underground lava. This lava, as in the smaller basins, probably flowed out from cracks and fissures opened up by the impact. The difference in appearance between the floor of the Caloris basin and the majority of smooth plains can be attributed to a shallower layer of lava extruded onto the steeply sloping floor of the Caloris basin.

Smooth Plains

Solidified lava

On a planet as textured and heavily cratered as Mercury, it is unusual to find quite so smooth a plain as this one. The photograph covers an area 50 by 85 km, and the largest craters on it are shallow basins only four or five kilometers in diameter. It is a region typical of the so-called smooth plains material which is found in irregular patches between about 1000 and 1500 km from the center of the Caloris impact. However, smooth plains are not unique to Caloris, and similar patches are found elsewhere on Mercury.

Clearly the older crater fields have here been buried, and molten rock is the most obvious agent responsible. Some of the molten rock could have been ejected by the Caloris impact, but the estimated volume of this material, 50 million cubic kilometers, is too great to attribute to ejection alone. The smooth plains are therefore most probably solidified oceans of lava.

The bright spot just to left of center is the halo of debris surrounding a small and very young impact crater. A million years or so hence, the brighter material will have darkened and will be imperceptible from the rest of the plain.

The Caloris Antipodes

Weird terrain

A small portion of Mercury's surface is occupied by a land form of the type photographed here. This has

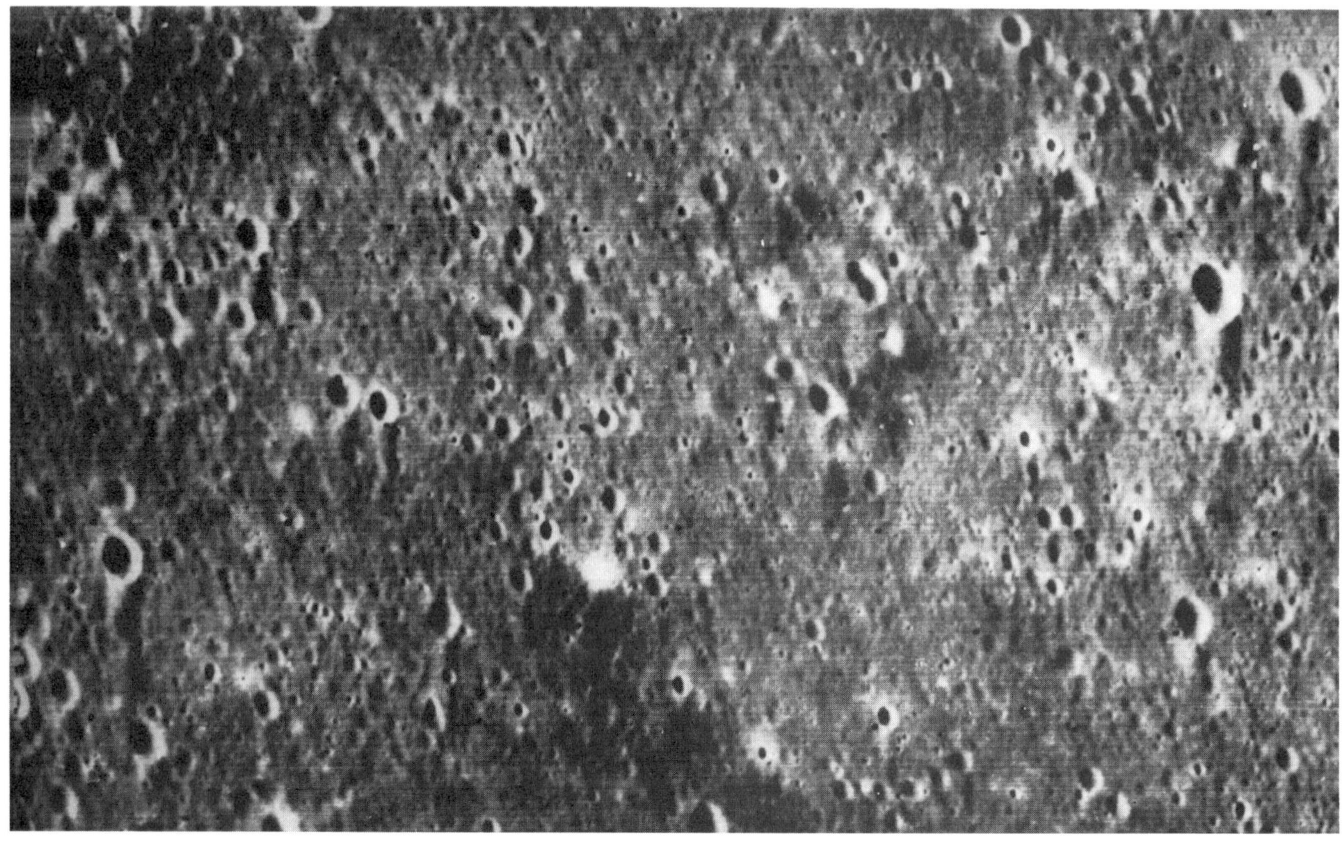

Smooth plains *Mariner X*

Weird terrain *Mariner X*

been aptly christened "weird terrain," and it is found nowhere else in the solar system. Jumbled blocks and mountains are scattered over the area, in chains and in isolated stands. What forces could generate such scenery?

A clue to the origin of weird terrain is afforded by its geographical location, at the antipodes of the Caloris basin. Planetary scientists believe that weird terrain was generated by the Caloris impact, and that therefore the effects of that impact were truly global. A credible mechanism is provided by seismic waves.

On Earth, seismic waves accompany any violent event—the eruption of a volcano, an earthquake, the detonation of an atomic bomb, and so on. Like ripples across a lake, these seismic waves travel through the curved surface of our planet, carrying their message of distant events. Similar phenomena must occur on Mercury, and though we have no information on how seismic waves propagate on that planet, it would be no surprise to learn that they do so at least as efficiently as on Earth.

When the Caloris basin was formed, waves set off around Mercury in all directions. Because the Caloris event was a particularly big one, the waves were very energetic. They traveled round the planet until they met at the diametrically opposite point. Here the waves combined, their energies adding to produce a quake of unimaginable proportions. The result was to raise the blocks and mountains of the weird terrain.

The Santa Maria Rupes

Rounded ridge

Even within the cratered highlands there are small regions on which the density of large craters is below average. This photograph shows one such region. It is far from a smooth plateau, however, for the number of tiny craters, with diameters less than 5 km, is very high. These are almost certainly secondary craterlets produced by large boulders thrown out from bigger impacts which lie all about. Such portions of plateaux are probably remnants of the very first solid crust which formed on Mercury, immediately before the spray of colliding bodies so drastically reshaped the planet's topography. The largest crater on this view, at upper right, is 40 km in diameter.

Crossing the plateau is the wavy line of the Santa Maria Rupes. *Rupes* is the Latin word describing a low ridge, and this particular specimen was named after Columbus' flagship. Ridges such as this are technically described as lobate scarps, which signifies that they are steeper on one side than the other and have rounded crests. Several are known on Mercury, with lengths up to 500 km. The Santa Maria Rupes is about 3 km high.

The ridges meander through plains, plateaux and craters alike, and are therefore geologically young. Such formations offer evidence that the crust of Mercury has undertaken some excursions. In this case, the portion at left of the photograph appears to have pushed towards the upper right, squeezing out the rupes in the process.

Strangely, although several of these compressional ridges have been found on Mercury, each indicating a local shrinkage of the crust, there are no corresponding geological features indicative of stretching. Does this mean that Mercury is shrinking like an old apple? Astronomers are as yet unsure, and an answer must await adequate photography of the other face of the planet.

Santa Maria Rupes *Mariner X*

The Moon

Earth's satellite

Our Earth, in common with most other planets, does not travel alone around the Sun. It is attended by a single satellite which we call the Moon. The Earth makes up for its lack of number of satellites by the Moon's size. The average retinue in the solar system is rather more than four per planet, each with a mass which is thousandths that of the planet. The Moon, however, weighs almost one-eightieth as much as the Earth. This distinguishes it from all other solar system satellites, except the newly-discovered companion to Pluto, which are minuscule compared to their parent planets. For this reason the Moon is sometimes considered to be a separate planet which was captured by the Earth after its formation. However, the laws of physics make such a capture quite difficult, and astronomers are not in agreement about how and where the Moon originated.

With a diameter of 3500 km, the Moon is smaller than any planet except the remote Pluto. It has been unable to generate enough heat from the decay of its radioactive materials to power prolonged volcanic activity, although it has experienced an era of volcanism. Today it is a dead, rocky planet with only a very small core accounting for about 4% of its mass. The Moon is too small to retain an atmosphere.

The Moon rotates on its axis at the same rate that it orbits the Earth. This is called captured rotation, and results in the Moon always turning the same face towards the Earth. Examples of captured rotation exist in other satellites. Because of the large mass of the Moon relative to its parent planet, the situation is more remarkable than in other examples. It appears to have arisen because the Moon is far from spherical, but has a pronounced bulge on the side facing the Earth. It was the gravitational attraction of the Earth on the Moon's bulge which gave rise to the captured rotation.

As this photograph shows, the Moon's surface comprises two distinct types of terrain: bright, rugged, cratered areas and dark, smooth plains. In this respect it resembles Mars and Mercury, but lacks the red coloration of these two planets. Indeed, the Moon has one of the least colored surfaces in the solar system. Moreover, even the brighter portions of its surface are extremely dark, reflecting no more than one-tenth of the sunlight which falls on them.

The cratered highlands have been shaped almost exclusively by meteoric bombardment. The craters remain as fossils of this bombardment, for erosion on the Moon is almost nonexistent, save for that produced by the bombardment itself. Recent impact craters, particularly those under about 100 million years old, are surrounded by a splash pattern of bright rays, which slowly become obliterated.

In the early days of the telescope, the darker portions of the Moon's surface were thought to be oceans, and were called *maria*, from the Latin for seas (singular: *mare*, pronounced mare). There is no water on the Moon, but the maria do represent ancient expanses of liquid. They were formed by molten lava filling shallow impact craters. However, it is true that most of the maria have been covered by a succession of lava flows occurring at different times. This indicates that some volcanism existed after the big impacts. Volcanic activity is thought to have died out three billion years ago.

Very few large craters pepper the maria despite their great antiquity. This suggests that the solar system has been swept clean of most of the rocky debris which existed at the time that the Moon and planets developed their solid surfaces. Astronomers see this as evidence that the present planets were built up by the accumulations of many small rocks and dust grains which had condensed from the solar nebula. A Darwinian survival-of-the-fittest law would have dictated events, large objects becoming larger by engulfing smaller ones. On this hypothesis we might envisage the Earth and Moon competing for the same material, growing to their present dimensions and then confronting one another. The confrontation was resolved not by a collision, but by the development of a planet-satellite relationship. A collision between the Earth and Moon would probably have been the largest single event within the formation of the solar system, and might have been more destructive than constructive.

A curious fact, not fully explained, is that there are scarcely any maria on the side of the Moon which is permanently turned away from the Earth. However, there are on that face a number of bright basins of comparable size to the maria. These are called thalassoids. Thus the paucity of maria on the far side does not indicate a favored side for impacts, but rather a favored side for subsequent volcanism. Since the Earth exerts greater tidal forces on the nearside than on the far side of the Moon, we have a mechanism favoring volcanic activity on the face we view. The stronger tidal forces could open deeper the faults and cracks in the Moon's surface through which volcanic activity would occur. In order for such a mechanism to have functioned, the Moon must have turned the same face to the Earth for as long as three billion years. This is not too surprising, however: there is evidence that the Moon was captured by the Earth at least 4.2 billion years ago.

In addition to the lava flows on the maria, there are other lunar formations indicative of a volcanic origin. All volcanism probably died away about three billion years ago. Thus all the Moon's rocks are older than

Moon AAT

virtually every known Earth rock. On the other hand, it is a little surprising that volcanism persisted longer on the Moon than on the larger Mercury. This probably reflects the slightly different chemistry of the two. The only evidence of any volcanic activity continuing to the present day is small moonquakes. Some of these are caused by impacts of the minor debris still circulating in the solar system. One large impact occurred on the far side of the Moon on 17 July 1972, for example. Other quakes are of internal origin; these generally occur within three days of the Moon's monthly closest approach to the Earth, when the gravitational stresses are most intense. There are in addition occasional reports of local patches of red coloration. These may represent the release of pockets of gas from beneath the surface. Such gases would quickly disperse and escape into space.

Astronomers have been able to understand the internal structure of the Moon only with the help of spacecraft. Particularly important in this regard has been the placing of seismometers on the surface of the Moon. Geologists can interpret the records of moonquakes from different parts of the Moon in the same way that they do on Earth, and hence probe the material through which the seismic waves traveled. The first results were surprising. Whereas on Earth all seismic waves die away very quickly, on the Moon they persist for several hours. In effect, the Moon rings like a bell when hit by an impacting body. This shows that most of the Moon's interior is solid and rigid, a confirmation that no significant volcanism can now exist.

The crust of the Moon is 65 km thick, about twice that of the Earth. Below that is a rather solid mantle which accounts for most of the body of the Moon. The core, of 500 km radius, is only partly melted, and is the seat of many of the moonquakes. There is some difference of opinion as to whether the core is composed of iron or simple rocks.

Because of the continued bombardment of the surface by solar system debris, it has not survived as solid rock. Instead, a fractured, compacted type of material known as breccia has been generated. This broken material is remarkably deep—up to 25 km in places. A loose, dusty soil covers this, with a maximum thickness of 100 meters.

A Bright Ray Crater

Rays thrown out by impact

Plate 28 shows a small, young impact crater on the far side of the Moon. It is typical of such craters not only on the Moon, but on Mercury and Mars too. Here a rocky morsel of the solar system met its end by slamming into the far larger body. The speed of impact was probably 10 or 20 km/sec, about 500 times faster than a family car can travel; the object weighed a trillion times as much as most family cars. The crash was, therefore, spectacular. So much energy was released that the rocky lump together with a sizeable area of the Moon's surface were instantly melted, and the force of the impact, which was, in effect, an explosion, blasted a hole in the rocks of the Moon. We call such holes craters.

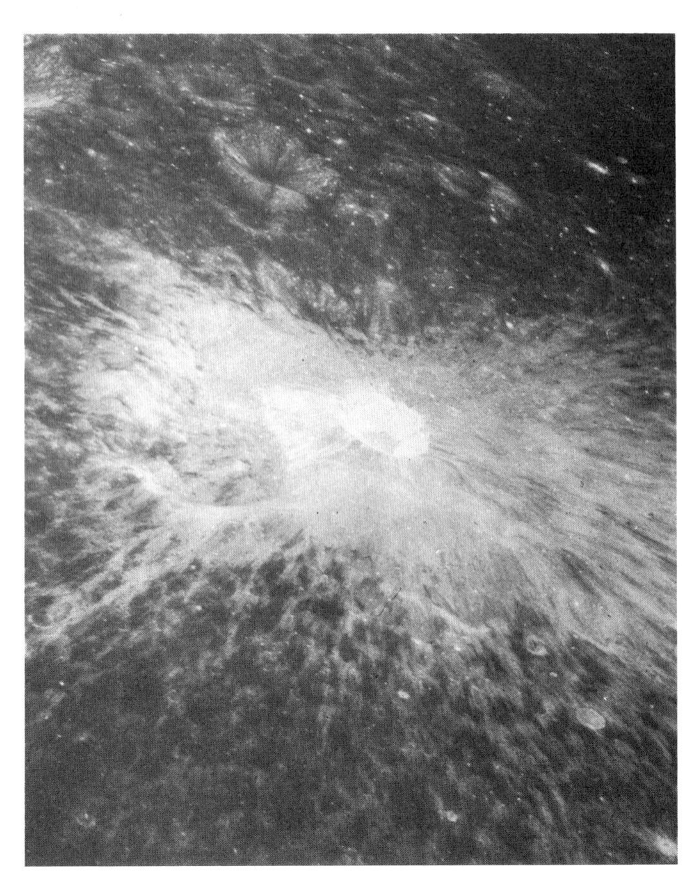

Ray crater *Apollo 10*

On good photographs, impact craters are easily recognized. They are almost perfectly circular, except in rare instances where the impacting body approached on a very low-angle trajectory. Their interiors lie well below the surrounding area, and their walls rise steeply on the inside but more gently on the outside. Small impact craters are bowl-shaped, while larger ones have flattish floors with central mountains. Very large impact craters may have two or even three concentric rims. In most craters the inner walls are so steep as to be unstable: landslides are often found within them, and many crater walls have slumped to form terraces.

Outside impact craters lies evidence of the material thrown out by the explosion. Close to the crater walls, molten rock was deposited as an undulating, smooth terrain. Farther out the ejected material cooled in flight and was solid when it landed. Formations resembling sand dunes were produced, and the larger lumps on landing themselves produced small craters which are known as secondary-impact craters. In certain preferred directions, rows of secondary-impact craters have formed long, bright rays. Because of the low gravity, ejecta patterns are well developed, and in some cases rays completely encircle the Moon.

Most impact craters lie in the brighter, highland portions of the Moon. This does not indicate a deliberate choice made by the infalling bodies, of course. Rather, the darker areas have been flooded by molten rock which obscured the craters.

The bright rocks of the highlands, and particularly of the far side of the Moon, are the oldest known, and probably were the first solid crust to form on the satellite, about 4.5 billion years ago. They cover 70% of the Moon's surface. The rocks are rich in minerals of sodium and potassium: like common salt, such minerals tend to be bright. However, even these minerals are darker than most rocks on the Earth's surface. The rocks are called anorthosites by geologists, and are very coarse and granular, indicating that they cooled slowly.

Because of the intense bombardment these rocks have received, they have become badly shattered. Fragments of broken rock subsequently became cemented together by later impacts to form patchy boulders and outcrops known as breccias.

Aristarchus and its Environs

Brightest spot on the Moon

Most prominent on Plate 29 is the crater Aristarchus, the brightest spot on our side of the Moon. Like most bright craters, it is a young specimen and was formed by impact. When dinosaurs roamed the Earth, a few hundred million years back, a lump of rock probably 10 to 20 km across plowed into the Oceanus Procellarum and generated Aristarchus. A hundred, perhaps 1000 cubic kilometers of rock were excavated by the resulting explosion, and flung over vast stretches of the Moon's surface. The bright ray pattern surrounding Aristarchus can be seen towards the right-hand side on the whole Moon photograph, p. 217. On the photograph above, some of the ejecta pattern can be seen surrounding Aristarchus.

To the right of Aristarchus, on both this and the whole Moon photograph, is Herodotus, a crater of similar size. It is much older, and almost certainly an impact crater whose floor subsequently became covered with lava. Behind Herodotus a low cliff meanders across the mare: its origin is unknown.

In front of Aristarchus and Herodotus is a strange formation known as Schröter's Valley. A deep canyon has been excavated by some mechanism, and stretches in an angular curve from a crater-like head to a shallow foot off the picture at right. Its origin is still debated, but there is general agreement that it can be only of internal origin. The most widely accepted theory describes Schröter's Valley as a gigantic fault, possibly occasioned by one of the impacts nearby, but more likely reflecting an old weakness. The ground has collapsed inwards within the fault, producing a valley.

In the floor of Schröter's Valley is a narrow, meandering feature exactly resembling a mature river valley. It wanders along the full length of the valley, weaving from side to side in a series of graceful curves. So clearly does it mimic terrestrial rivers that its discovery promoted a wave of enthusiasm for theories of the Moon's evolution which incorporated floods and inundations. It was soon pointed out, however, that molten rock can flow in much the same way. Today there is ample evidence that water has not flowed on the Moon; lava is the fashionable explanation. This formation in Schröter's Valley is called a sinuous rille. Many are known, and there is strong evidence of the association of most of them with lava flows.

Kopff

A volcanic crater

This 45-km crater on Mare Orientale seems to be an example of a volcanic crater on the Moon. Many others may exist, but lie in crowded regions where they are less easily studied.

For several centuries, controversy raged among astronomers concerning the origin of lunar craters. Volcanism and impacts were the popular mechanisms, and much dogma was expounded in support of one or other. Gradually, as better and better photographs became available, astronomers began to accept that both mechanisms had been active on the Moon, as on Earth. It now seems fairly certain that the great majority of lunar craters are of impact origin, that some of these have been modified by subsequent volcanism, and that a small proportion are of purely volcanic origin. Three basic criteria are used to recognize volcanic lunar craters—shape, wall structure, and ejecta. All are evident in the crater Kopff.

Volcanic craters usually have polygonal outlines. Their walls are formed from a succession of straight sections and are sometimes discontinuous. The straight sections usually align with the tectonic grid pattern of the Moon. This is a global network of linear features—portions of crater walls, rilles, valleys, mountain ridges, etc. The grid comprises three principal directions, NW-SE, NE-SW and a weaker N-S pattern. It represents trends of weakness through the lunar crust, directions where it is easier for valleys to form or crater walls to rise. The grid lines are

Aristarchus *Apollo 11*

manifested on scales from a few centimeters to hundreds of kilometers.

The walls of volcanic craters were extruded along the tectonic grid lines, or in short curved arcs. Each section is a volcanic ridge, and the gradients on the inner and outer sides are generally very similar. Equally, the walls are of the same height on each side: in other words the crater floor is not much depressed below its environs. Finally, the walls of volcanic craters are not usually terraced.

The formation of volcanic craters appears to have been sufficiently violent that some material was ejected in the process. However, the ejection pattern is quite unlike that surrounding impact craters. There are no secondary craters, no patterns of bright rays. Rather, the ejecta blanket is smooth where molten rock has landed and cooled. One imagines that much the same effect would have been produced if several million bucketfuls of black paint had been tossed off the crater rim.

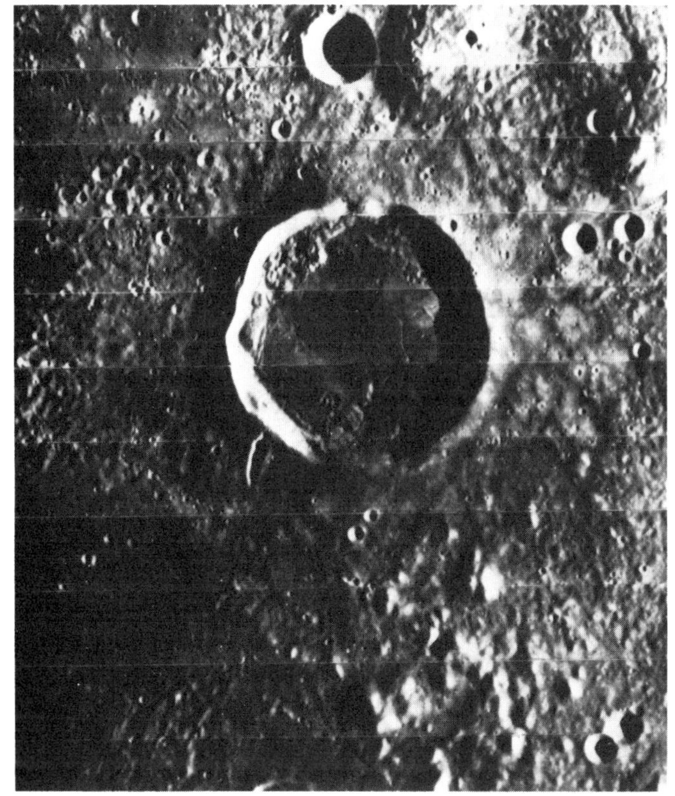

Kopff *Orbiter IV*

Dark lava fills the floor of Kopff, and some narrow cracks and shallow depressions can be seen in it. These, too, point to a volcanic origin, for they are cooling cracks and collapsed calderas. The origin of calderas is easily described. Below the surface a cavity developed, where molten rock drained back into the Moon at the end of the volcanic activity. The local surface could not support itself and thus subsided. Volcanic calderas are common on other planets especially EARTH (see p. 187).

Mare Imbrium

Filled impact crater

The dark, circular feature which occupies almost the whole of this view is Mare Imbrium, the Moon's largest crater. Its walls are outlined by bright mountain ranges which have been obliterated at right, but are continuous elsewhere. A large bay to the bottom of the mare, known as Sinus Iridum, is a smaller crater merged with Mare Imbrium; both are flooded with dark mare material. Compared to the picture on p. 217, Mare Imbrium appears more circular. This is because the photograph has been adjusted to give the view we would have from directly above.

The Mare Imbrium crater has a diameter of almost 1100 km. It is undoubtedly of impact origin. Astronomers can estimate the size of body which produced it, and have derived a mass of 10^{19} kilograms. This corresponds to an object about 150 km in diameter. Fortunately, few objects of this size now exist in the solar system, and none of these can now collide with the Moon, or more important, its neighbor the Earth. In order to find sufficient objects in the solar system to generate several craters as large as Mare Imbrium, we must go back some four billion years. At that time there were frequent collisions, and most of the Moon's maria were formed by impacts occurring within a period of 400 million years centered on that epoch. The majority of smaller craters also date from that period. It seems that the solar nebula contained a great many bodies in the 1 to 100 km diameter range. These were absorbed by the larger planets, as the 150 km body which generated Mare Imbrium became part of the Moon. Some astronomers have argued, however, that most of the impacting bodies were not members of the solar system, and that instead the Sun passed through a region of the Galaxy containing many such bodies.

Immediately after its formation, Mare Imbrium would have been about 200 km deep at its lowest point. Today its floor is almost level with the surrounding plains. The dark mare material which filled it could not account for so great a depth; indeed there is probably no greater depth than 1 km of this material. Instead, the wound was healed from within by the entire Moon changing shape. Geologists are familiar with the elastic properties of rocks, a process known as isostasy. Large portions of the Earth's surface are today moving to compensate for past changes. For example, much of Canada is gradually rising; this reverses the subsidence of its rocks under the great weight of ice that covered them during the last ice age. In just the same way, the floor of Mare Imbrium slowly rebounded to its original level. This process took several hundred million years. We know that the Moon is now too rigid for isostasy to occur, and conclude that four billion years ago it was hotter and therefore more malleable.

An impact as large as the Imbrium event produced greater effect than a vast crater. Much of the material excavated by the impact was thrown out across the rest of the Moon's surface. Nowhere escaped the inundation, and in many places hundreds of meters of molten or crushed rock landed. In some parts the debris landed in several successive waves, having followed a variety of trajectories on its journey.

Some of the rock molten by the impact would have flowed back into the crater. As much as a few hundred meters of rock could have been deposited, scientists calculate. But the dark mare material which currently occupies the Imbrium basin has been reliably dated from 3.3 to 3.8 billion years old. Nor is the spread of ages an indication of errors in the measurements.

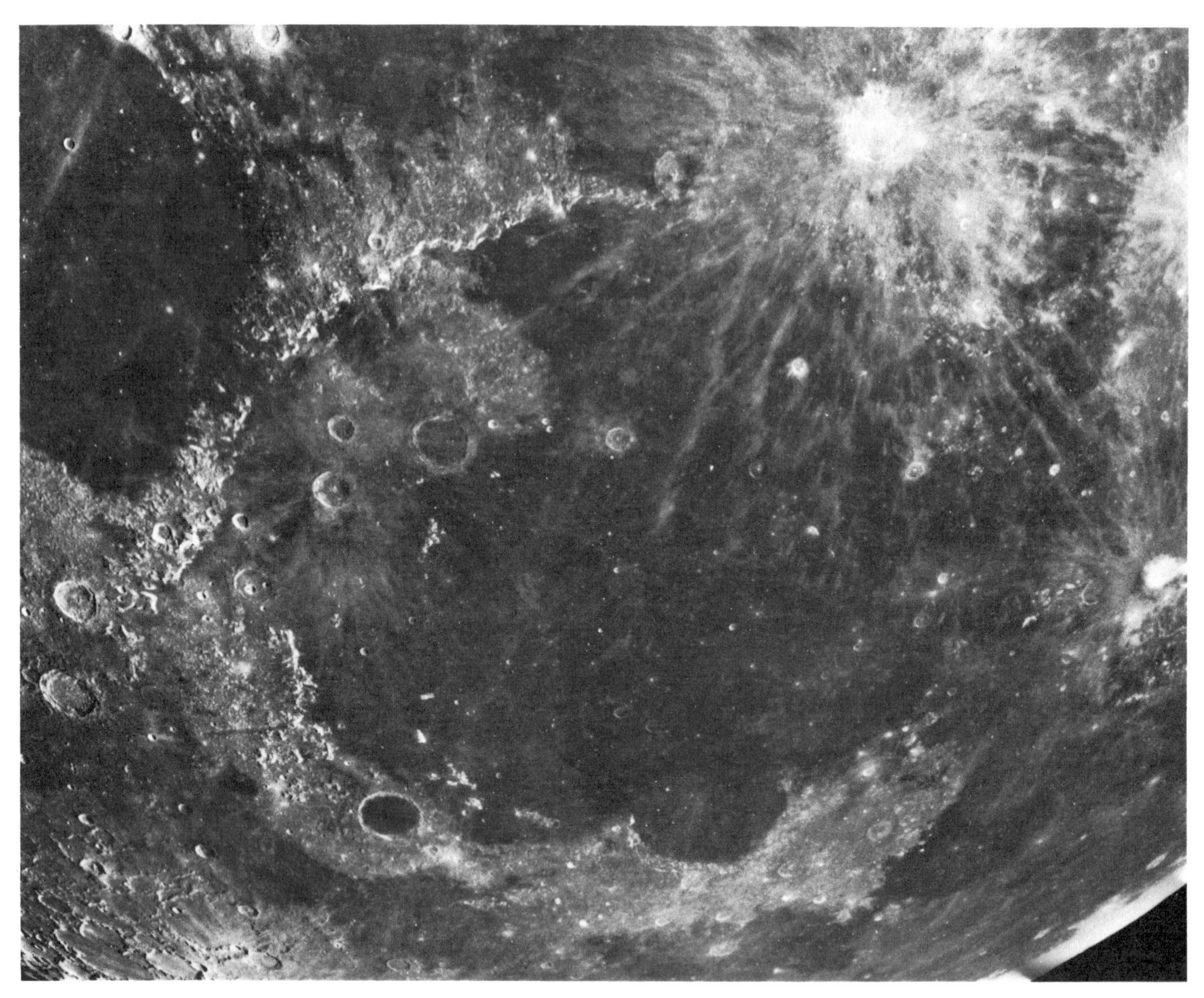

Mare Imbrium AAT

Mare Imbrium was filled by dark material in several successive episodes, the last occurring 700 million years after the impact. The rock which has filled the Imbrium basin and other maria is basalt, a rock commonly extruded by terrestrial volcanoes. Volcanism therefore concluded the formation of Mare Imbrium.

Some basalt flows were darker than others. The edges of successive flows can be picked out by the slight change in color or darkness. One fairly prominent flow edge forms a meandering line starting at the rim of Mare Imbrium, just to the left of Sinus Iridum, and ending near the mare's center. The basalt flows completely engulfed the giant crater's rim at right of this picture, and must there be a kilometer deep. They join an extensive area of basalt known as Oceanus Procellarum.

By the time of the last lava flows, the Moon had cooled so much that it was no longer malleable. Isostasy has not occurred in the intervening 3.3 billion years. Since the basalt is denser than most lunar rocks, Mare Imbrium is out of balance with the rest of the Moon, being top-heavy. This is manifested in a locally higher gravitational field, a fact which was discovered by its effect on the first lunar-orbiting satellites. Such local effects are known as gravitational anomalies, or mascons (mass concentrations). There are smaller gravitational anomalies on Earth, caused not by the absence of isostasy, but by internal activity proceeding faster than the crust can accommodate.

Lava Flows on Mare Imbrium

KREEP basalts

The surface of Mare Imbrium, as illustrated on the previous photograph, has been covered by successive flows of molten basalt. Some of these flows are seen in

Lava flows on the Moon *Apollo 15*

greater detail in this photograph.

Molten rock—lava—appears to have been extruded from the low, meandering formations known as wrinkle ridges. Only a few tens of meters in altitude, wrinkle ridges abound on the maria but are seen only when a very low sun angle casts abnormally long shadows. They are narrow formations; one or two kilometers wide, but may traverse distances measured in hundreds of kilometers. In places, wrinkle ridges become sinuous rilles. The classic sinuous rille was described in connection with the crater ARISTARCHUS, on p. 219. However, sinuous rilles occur on many maria too. Most are about 200 meters wide. The association of sinuous rilles, wrinkle ridges and lava flows makes it virtually certain that all are of volcanic origin. It is widely believed that wrinkle ridges are hollow lava tubes, and that in places where their roofs have caved in, sinuous rilles result.

The basalt must have come welling up through underground cracks as a hot fluid. Immediately on reaching the surface it encountered a vacuum. The trapped gases bubbled out of the lava, turning it frothy. Frothy basalt would have spread rapidly, for its consistency is about the same as engine oil at room temperature. It flowed down the gentle slopes of the mare basin until it cooled and solidified. In all, about 1000 cubic kilometers of basalt were extruded onto Mare Imbrium.

The basalt flows of the maria are dark and colorless. This is dictated by their chemistry. Most of the basalts are rich in iron and titanium, elements forming dark minerals. There is more iron and titanium than in almost all terrestrial rocks. On Earth, these metals diffused into the core, leaving other minerals to form a crust. On the Moon, this process of fractionation was much less active, so iron and titanium remained near the surface. Iron, of course, tends to rust on Earth. The absence of red coloration therefore indicates that there never has been much surface water on the Moon. Water *is* present, but as a constituent of many of the mineral crystals. This water was not released, contrary to the terrestrial example, because the Moon never became hot enough.

Some of the basalts have a different chemistry, being rich in potassium (chemical symbol K), rare earth elements, and phosphorus (P). These are known by the acronym KREEP basalts. Again they differ from any known terrestrial rock. Yet despite the different *chemical* compositions of Earth and Moon rocks, only three *minerals* have been found on the Moon which are unknown on Earth.

Although these basalts flowed 3.3 billion years ago, there are few impact craters on them. A photograph such as this does reveal many small craters a kilometer or less in diameter, but larger specimens are very rare. Moreover, there are more volcanic than impact craters on this photograph. The paucity of impact craters is in great contrast to terrain which was not inundated, and is therefore 4.5 billion years old. The different rates of crater generation in the first quarter and subsequent three quarters of the Moon's history is clearly demonstrated.

Mare Orientale

Most recent major impact

Only slightly smaller than Mare Imbrium is the three-ringed basin of Mare Orientale. From the vantage point of Earth, Orientale lies mostly on the far side of the Moon, and only a small portion of it protrudes, very foreshortened, around the limb. Thus before the advent of lunar-orbiting satellites, astronomers had no inkling of what a remarkable formation it would turn out to be.

Unlike Mare Imbrium, and indeed all the giant impact craters on our side of the Moon, Mare Orientale has not been filled by dark mare material, save for a few portions which may even be rock which the impact itself melted. Moreover, it is the most recent of the major impacts, with an age of 3850 million years. This date places it at the end of the great era of bombardment. There has been relatively little damage by subsequent impacts: we see in Mare Orientale the fossil of an explosive event unequaled in human experience.

Most obvious of Orientale's attributes is the triple mountain ring. The inner, 480 km in diameter, is unnamed, being barely visible from Earth. Next comes a 620-km ring, called the Rook Mountains (officially Montes Rook). A small portion of these was seen and named by Earth-based observers. Farther out, and more easily seen from the Earth, are the Montes Cordillera, with a diameter of 930 km. The origin of these three rings is still debated by astronomers and geologists. Two schools of thought prevail. The mountain rings may be the shock waves of the impact, frozen, as it were, in mid-flight. An impact of this magnitude would melt or soften the lunar crust, and shortly after the crust must have cooled again. At the time of solidification there might have been three shock waves traveling outwards. The alternative model argues that the mountain rings were formed after the impact by readjustments of the crust. Quite possibly neither hypothesis is correct.

Between the inner two rings is an annulus of smooth, low hills, extensively cracked. The small picture shows a portion of these in enlargement. The smoothness of this underlying terrain suggests that it was melted by the explosive release of energy which accompanied the impact. A surface cooling from the molten state is likely to crack open as it shrinks,

Aristarchus

Surface of Moon

Surface of Moon

PLATE 31

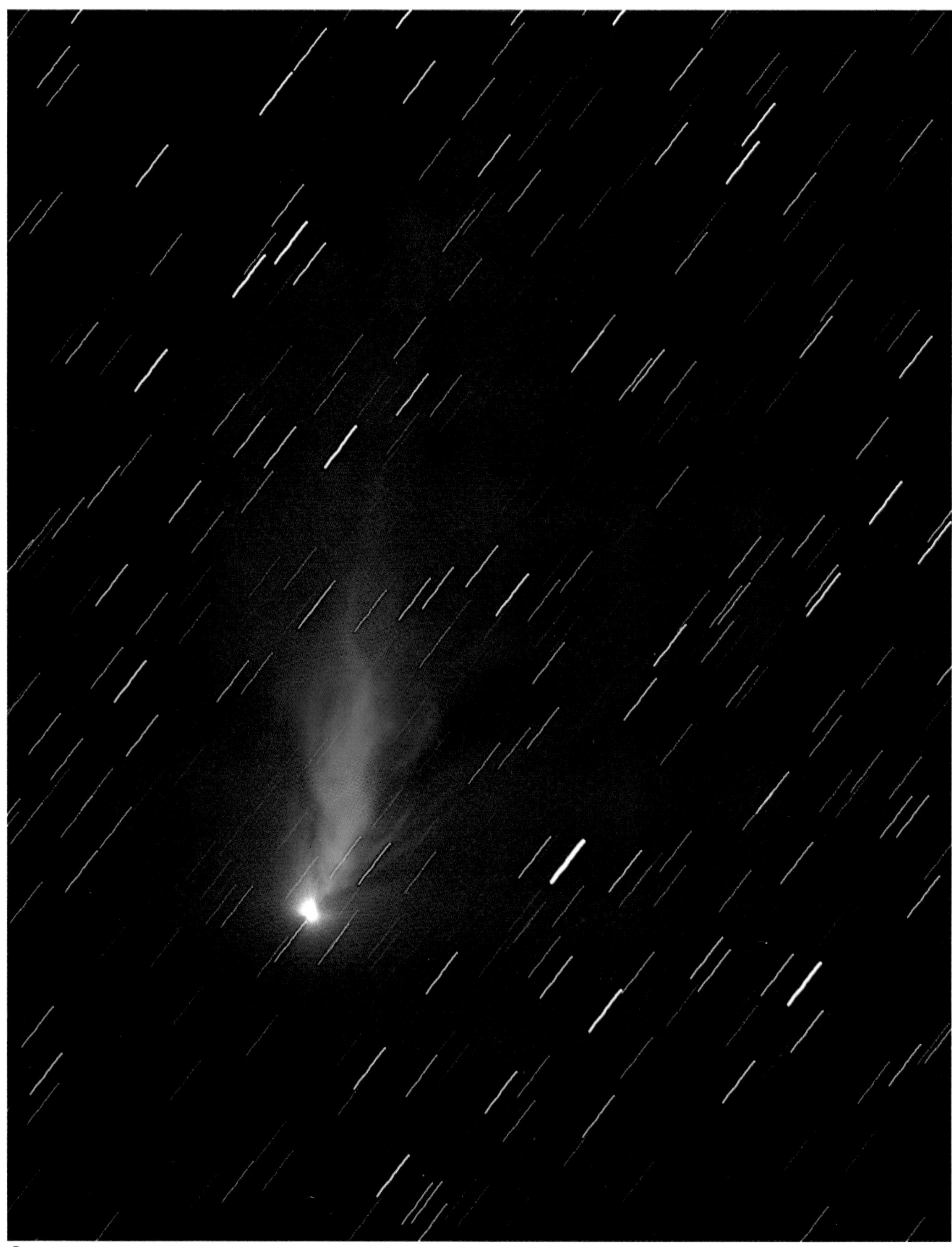

Comet Humason

Mare Orientale *Orbiter IV*

especially if the melting happened violently.

Within the inner ring lies a plain which has been covered by a thin veneer of basalt. The volume of basalt released into the Orientale basin is much less than that released into Mare Imbrium and similar craters. Probably this reflects the later date of the Orientale event. Other evidence of volcanism is present in the Orientale basin, in particular the crater Kopff pictured on p. 221.

Beyond the Montes Cordillera are many streaks radiating directly away from Mare Orientale's center. Some of these are deposited material: in effect, dunes. Generally, however, the Orientale event produced many fewer ejecta than did the rather larger Imbrium impact. Others of the radial formations are grooves which may have been made by low-flying debris traveling at speed. If many of the ejecta traveled in low trajectories, their absence over many parts of the Moon is readily understood.

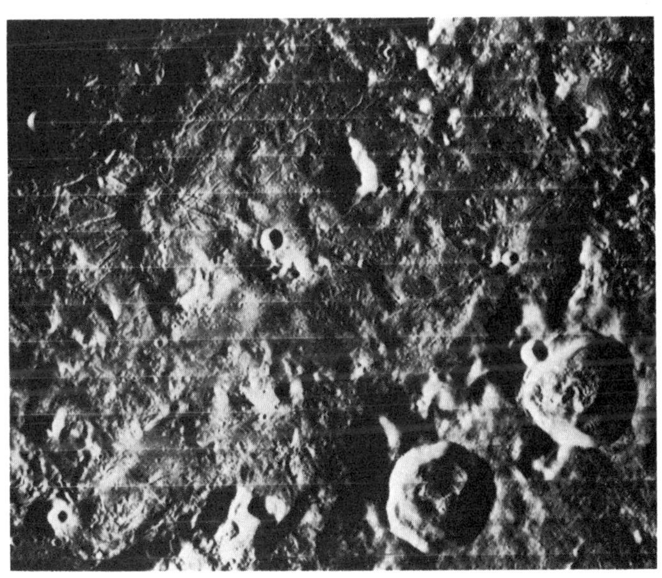

Mare Orientale *Orbiter IV*

The Ariadaeus Rille

Lunar rift valley

These two views show the Ariadaeus rille, sometimes known by its Latin name Rima Ariadaeus. Rilles are common on the Moon, and many were known before the era of spacecraft. From the distance of the Earth they appear as meandering lines of shadow, and in the larger examples a sunlit, bright edge parallels the shadow line. Terrestrial observers interpreted these as steep-sided, V-shaped valleys and therefore termed them clefts. Photographs such as these, taken from a closer location, show the formations to have wide, flat floors between their steep walls; the term rille is therefore now preferred. The total width of this rille is 5 km, and its depth is close on 1 km.

The view along the Ariadaeus rille, at left, shows the gentle meanders very clearly. It also reveals the constancy of the rille's width from its point of origin near the double crater Ariadaeus (left of center) almost to the horizon. In the photograph at right, the rille crosses a variety of different terrains – smooth mare at left and mountain ridges at right. Clearly the rille is unperturbed by mountains, for it continues its course across them with no more than a modest increase in its width. From this we learn that the Ariadaeus rille was formed after the impact which created these mountains and, indeed, after the flooding of the local mare. Rilles are among the youngest formations on the Moon.

What forces produce them? There seems little choice of mechanisms. Clearly the ground slumped inwards between two roughly parallel walls. Hundreds of square kilometers of lunar landscape do not fall in without some good reason. Beneath each rille there must have been a weakness – maybe even a hole. The hole in this case took the form of a long, thin line across the Moon's surface. Such formations on Earth are known as faults. A fault can be thought of as a split in the Moon's crust. On Earth the constant seething of the crust usually promotes vertical motions across

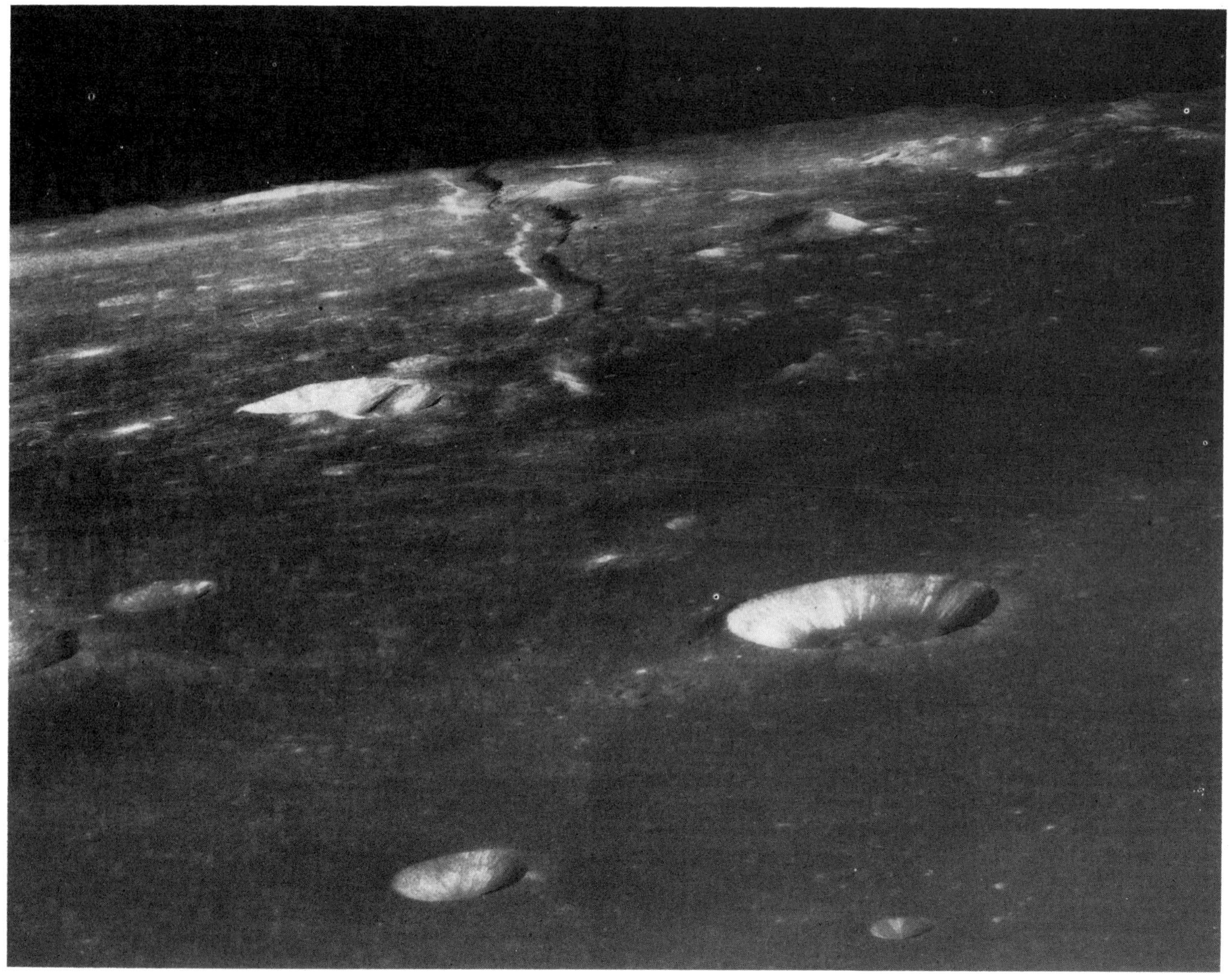

Ariadaeus rille *Apollo 10*

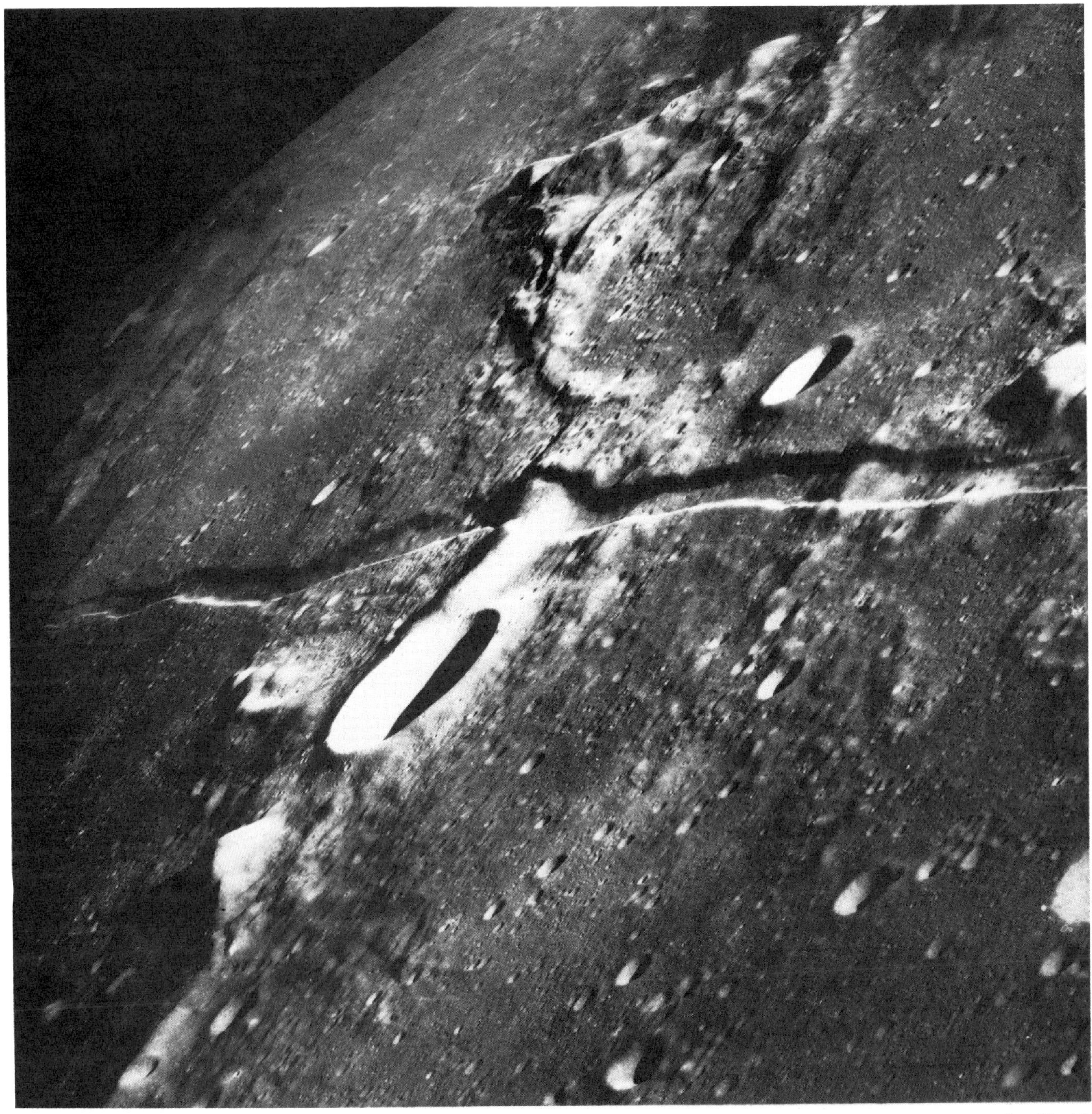

Ariadaeus rille *Apollo 10*

faults, one portion of land being forced to rise or fall relative to the other. The Moon's quieter history has not induced vertical motions except in very rare cases, like the Ariadaeus rille. On Earth rilles would be called rift valleys or graben.

Triesnecker

Crater near a rille system

Triesnecker was an Austrian astronomer of the 18th century whose name is perpetuated in this 28-km crater rather near the apparent center of the Moon's disk as seen from Earth. This photograph clearly demonstrates that Triesnecker is an impact crater, with its almost circular outline, terraced walls and outer splash pattern. The field of secondary craterlets produced by the main impact covers most of the region as far as the distant mountains. Astronomers do not know the age of the Triesnecker impact, but would generally agree that a figure around 2 to 2.5 billion years is about right.

Triesnecker *Apollo 10*

Although Triesnecker is of impact origin, it appears to be the focus of a complex system of rilles. As shown earlier, rilles are of internal origin, being manifestations of underground faults. Does the association of crater and rilles indicate that the impact fractured the Moon's crust? This seems unlikely because few rilles are seen in association with other and larger impact craters. Moreover, it is certain that an impact of this size could not fracture a crust 65 km thick over so extensive an area.

A crucial test is whether the splash pattern of Triesnecker overlies the rilles. This photograph is barely adequate to allow a definitive answer, but the indications are that it does. Near the crater, the walls of the rilles are softened in appearance by material thrown out in the Triesnecker impact. Thus the siting of Triesnecker at the focus of these radiating rilles is entirely fortuitous.

Hygines

Row of volcanic craters

This striking feature is Hyginus and its crater chain. Hyginus itself is the larger double crater which sits at the bend of the formation. The remainder of the chain has variously been described as a rille or cleft. These descriptions were introduced a century or more ago when telescopes were incapable of showing the true nature of the feature. As better optics were developed, astronomers realized that the valley comprises a series of crater-like depressions merging into and overlapping one another and linked by short sections of a rille.

Obviously such a distribution of craters could not occur by the chance arrival of lumps of rock from space. And astronomers can show by calculations that a bevy of such rocks could not long travel in the necessary arrangement, tempting though it is to argue

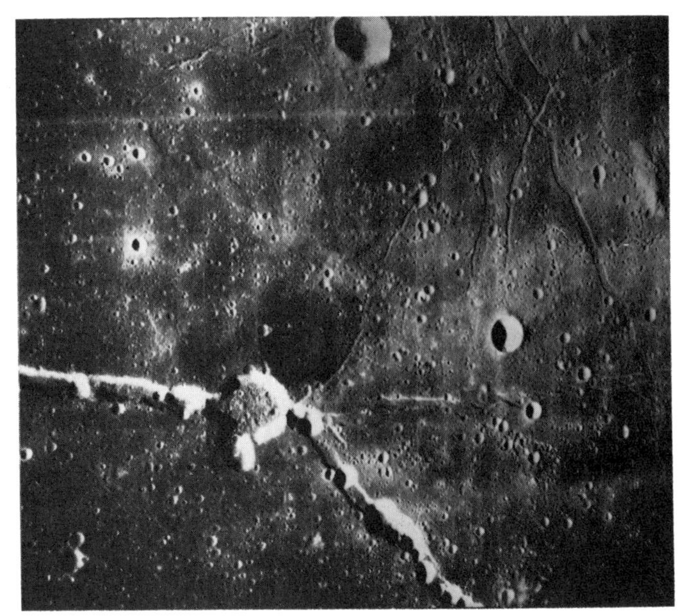

Hyginus *Orbiter V*

that a V-formation of arrivals produced a bent crater chain. In fact, the Hyginus feature is undoubtedly of volcanic origin. Crater chains are known in volcanic regions on Earth, several craters frequently having formed along a subterranean line of weakness. In most terrestrial examples the craters do not merge, and Hyginus may be the best example of a merged crater chain in the solar system.

On the floor of Hyginus is a number of domelike swellings. The most prominent lies just to the left of the crater's center. Domes occur in many places on the Moon and are also of volcanic origin. Quite possibly many are hollow and will one day collapse to form shallow craters.

The diameter of Hyginus is about 6 km, and the crater chain has a total length of some 200 km. The network of rilles at upper right connects with the Triesnecker system shown on the previous page.

Vitello

Rolling boulders

Details of the outer slopes of the crater Vitello are seen on this photograph. The hillside slopes gently down from top to bottom of the picture, and down the gradient have rolled two large boulders. Their tracks in the soft soil are clearly seen and look fresh. One might be tempted to think that both made their journey only a few hours before the photograph was taken. Indeed they may. But it is equally likely that the boulders tumbled a million years or more ago, so slow is the pace of erosion on the Moon. A million years hence the tracks will probably appear much as they do today.

Why did the boulders roll? The answer may seem obvious, yet the Moon is so devoid of eroding agents that any such change in its appearance merits attention. At their birth, possibly coeval with the formation of Vitello, the boulders were probably buried in the dusty soil, together with many smaller rocks. Gradually they were exhumed. Moonquakes were responsible. Every few years a moonquake would occur of sufficient intensity to make the big boulders shudder. With each shudder, a few grains of the dusty soil would fall beneath the boulder. Imperceptibly, dust migrated from around and above the boulder to beneath it: in other words the boulder rose to the surface. Once the boulder was at the surface, the process continued until it was totally exhumed. Quakes then caused the soil to slide gradually downhill, undermining the boulder. Eventually came a quake which removed the last supporting grains of dust. The boulder toppled.

Impacts by dust grains of the solar system also cause erosion, but only of surface formations. Had the boulders sat on the surface since their formation, the constant whittling action of these minuscule impacts would have smoothed their contours, or even worn them to dust. We can see, from the shadows they cast and from the tracks they made in the soil, that these boulders are angular. This lends support to the arguments that they have spent most of their lives underground.

These boulders on Vitello are among the largest known on the Moon. This specimen is as big as a house.

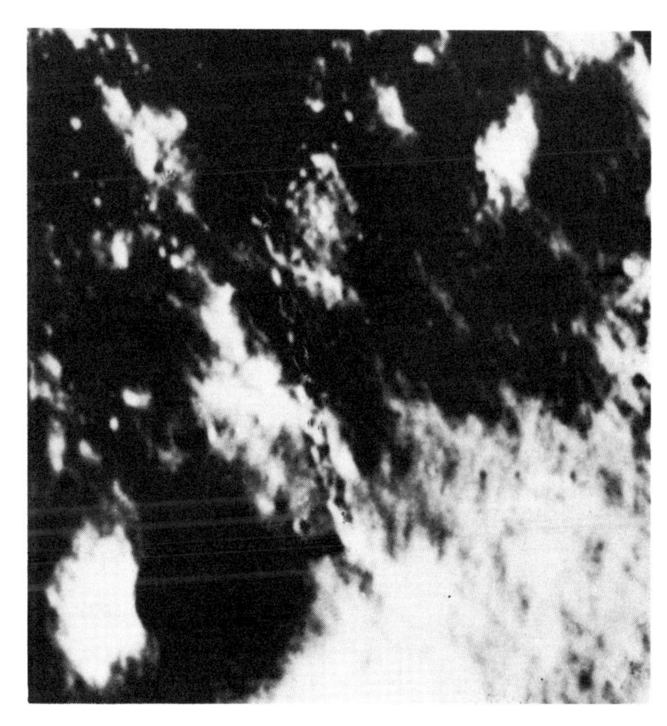

Vitello *Orbiter IV*

Silver Spur *Apollo 15*

Silver Spur

Mountain face

On an airless planet distances can be unexpectedly deceptive. The mountain face in this photograph, Silver Spur, looks to be only a kilometer or so distant. Even under the reduced gravity of the Moon, it would require a considerable effort to walk to it, for Silver Spur is 20 km away from the camera.

With a total height of 800 meters (2600 ft), Silver Spur is a low mountain by lunar standards: the Himalayas are dwarfed by several of the ranges in the lunar highlands. Indeed, Silver Spur is only one of the foothills to the Apennine Mountains, the range which forms the rim of Mare Imbrium, at upper left of the photograph on p. 222.

Most striking in the case of Silver Spur is the layering, or stratification. The strata are each about 100 meters thick, and all slope gently down towards the left. Stratified rocks on Earth usually indicate a sedimentary origin. Most of the exposed rocks on the Earth's surface were laid down under water, as sand or lime or chalk, and subsequently raised to the surface. On the Moon, however, no oceans have ever existed, and sedimentary rocks are unknown. All Moon rocks are igneous. To explain the layering of these rocks one must appeal to one of two mechanisms. Ejecta from the giant mare impacts have deposited layers up to 100 meters deep, exactly as seen in Silver Spur. However, the regularity of these layers seems to argue against such an interpretation. It is difficult to envisage a series of impacts in far-flung regions of the Moon all depositing just the same depth of material at this spot. Instead, the second mechanism seems more credible, namely lava flows.

The six to eight flows which accumulated to form

Silver Spur must have predated the Imbrium impact. When the Apennine Mountains were formed, a portion of the former crust was pushed up, creating the small ridge which we now call Silver Spur.

The Surface of the Moon

Airless

The blackness of the sky and hence of all shadows; an almost complete absence of color in Plate 30: these are the clues to the location of this photograph. It could be nowhere but the Moon.

On an entirely airless world such as the Moon, the sky can only be black. Only dust grains or gas molecules can scatter sunlight and so spread light from the Sun across the sky. On the Moon there is neither. The Sun is a cruelly bright disk in the star-studded blackness. Its strong light raises the Moon's surface to a temperature slightly above the boiling point of water. By night, with no atmosphere to insulate the ground, heat escapes rapidly and the land cools to the same temperature as liquid air. The night, it should be remembered, is 14 terrestrial days long.

The night-time temperature is determined by the physical properties of the ground. Where solid rock is

Surface of the Moon *Apollo 11*

exposed, the lowest temperatures found are around −50°C. A dusty soil cools 100°C and more below this value. The explanation of this phenomenon is that rocks conduct the Sun's daytime warmth deep into their interiors and have an inner heat store to last them through the night. Dusty soil does not conduct heat well and therefore has no reserves. As long ago as 1948, the Dutch astronomer A.J. Wesselink realized that the measured night-time temperature of the Moon was indicative of a dust-covered surface. Only in 1969 was this first revealed directly.

This photograph shows clearly the typical landscape. The soil is fine, but its surface is heavily textured by the innumerable tiny impacts it has received. Craters abound, from the smallest features discernible to the 40-meter depression which forms the foreground. This large crater is not easily recognized, for its contours are very gentle. The foreground shadow is cast by a slight eminence on its near rim; the rim beyond is more clearly seen, for its outer ramparts face the Sun and are brighter than any other part of the scene. Only half of the crater is covered by this photograph.

In addition to the dusty soil, the presence of numerous rocks and stones is apparent. These, too, range in size—from the limit of detail to boulders several meters across in the far distance. A number of rocks, some nearly a meter across, have rolled down into the center of the large crater and are seen at lower right of the photograph.

Rocks on the Moon have two basic origins. Some are the remnants of the solid surfaces which formed as much as 4.5 billion years ago, or, according to location, by mare flooding as recently as three billion years back. The continuous battering by impacting debris that these surfaces have endured has broken down the toughest rocks to pebbles and boulders. Some of the stones are conglomerates, however—fragments of stone and soil fused together into a chaotic mixture. Conglomerates are produced when impacting material partially melts the local surface and ejects blobs of it. These blobs cool into boulders even before they land, and produce secondary-impact craters. Most of the small craters on this photograph are secondary-impact craters, and many of the rocks which sit within them are themselves the ejecta of distant and much more violent events.

Crater

Lying in an ejecta blanket

Even within the dark maria, different locations on the Moon's surface exhibit different characteristics. This view should be compared to the previous photograph. The first obvious difference is the depth of the main crater. Here the gradients of the crater wall are steep. They probably lie at the "angle of repose" of the local soil, the steepest slope that will not collapse, for there is evidence that material often slides down the crater wall.

A second difference is the greater number of rocks and stones on the surface. Moreover, these rocks generally do not lie in shallow craters, but are embedded in the soil. They are therefore not projectiles which have recently arrived from distant impacts. Rather, they were buried at formation. Migration of the soil, caused by moonquakes, has exposed them. The formation of the nearby crater may also have helped to expose the rocks by stripping off the topsoil.

If the rocks were laid down within the local soil, we might expect both to have been deposited at the same time, in the form of a major ejecta blanket. We know that up to 100 meters of rocks and soil can be deposited by a single mare-forming impact. Geologists viewing this photograph therefore seek evidence that deposition of material has occurred. Such evidence is present, though hard to recognize. On the crater wall, just above the cross on the upper left, is an outcrop of crumbling rock which takes the form of three or four parallel bands. This stratification would have been produced by successive depositions of ejecta. Probably a single impact produced these strata, the material arriving here in several waves.

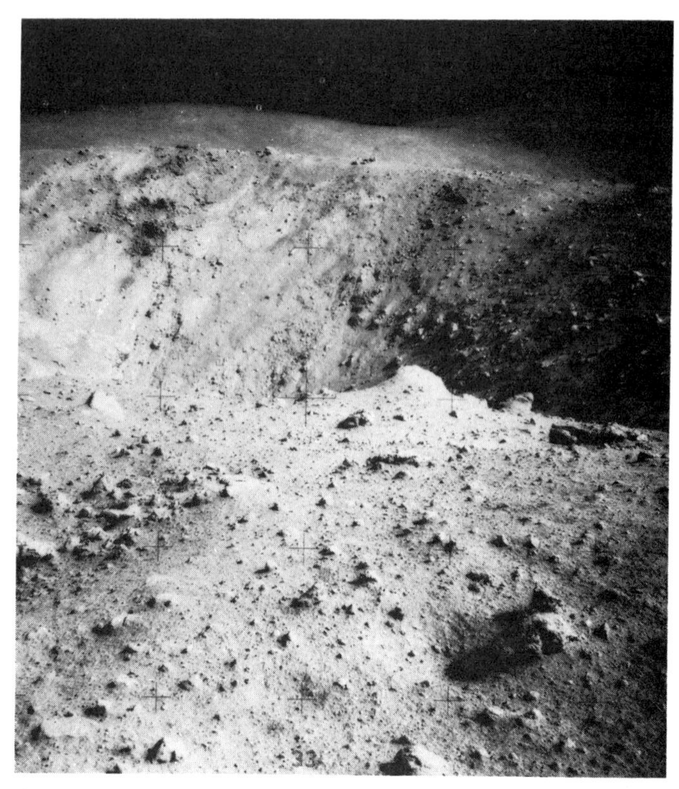

Crater *Apollo 16*

Boulders

Whittling away to dust

These photographs show typical boulders on the Moon's surface. A figure standing beside the larger specimen gives scale to the scene (Plate 31).

The lower boulder is of basalt, a dull gray rock. Its rocks formed when lava flooded a mare basin. A later impact fragmented the rock, and this boulder was one of many which resulted. Probably it lay partly buried for a time, and was exposed as the soil moved away. Since that time the boulder has suffered the predations of impinging meteoric grains from space, and slowly is being whittled down to dust. Some of the soil around the boulder is material sandpapered off by these impacts. Note the angular fragments of paler rock caught up in the basalt as it cooled.

In the upper photograph, the boulder was exhumed from a mountain slope behind the camera. It rolled down the hill, fractured, and came to rest as a split rock at its present location. The rock type is anorthosite, a rock common in the bright highlands of the Moon. There are also fragments of pale turquoise breccias imbedded in the boulder.

When these and other rocks were formed, the Moon had quite an intense magnetic field. Astronomers are still puzzled by this fact, for the lack of a large iron core to the Moon makes it difficult to maintain a permanent magnetic field. As the rocks cooled and solidified, the magnetic field was trapped, weakly. These boulders, therefore, are slightly magnetic.

Boulder on the Moon *Apollo 17*

Boulder on the Moon *Apollo 14*

Minor Members of the Solar System

Asteroids

Smaller planets

The solar system is by no means restricted to the planets and their satellites. A very great number of smaller bodies also orbit the Sun, and over 2000 of these are sufficiently large and bright to have been catalogued. These are called minor planets or asteroids. The etymology of this latter word indicates the principal characteristics of these bodies: they are starlike. Only the very largest of the asteroids can be distinguished from stars through ground-based telescopes.

Asteroids reveal themselves by their motion against the background stars, as did the most distant planets. Our photograph shows the motion of three asteroids during a 40-minute exposure taken while the telescope tracked the stars.

Ceres was the first asteroid to be found, on 1 Jan 1801, the first day of the 19th century. Measurements of the motion of Ceres soon showed its orbit to be reasonably circular and to lie between those of Mars and Jupiter. Ceres was thought to be the planet which contemporary theories predicted should fill this gap. However, discoveries of other asteroids soon followed, and it rapidly became apparent that no single planet occupied this niche in the solar system.

The asteroids are so numerous that a search for them will never be completed. At first, names were given to them. These were usually female names from mythology; later, everyday female names were used, but after about the first 1000 discoveries this was abandoned in favor of a system based on the date of discovery, though names may still be given.

Asteroid trails UKSTU

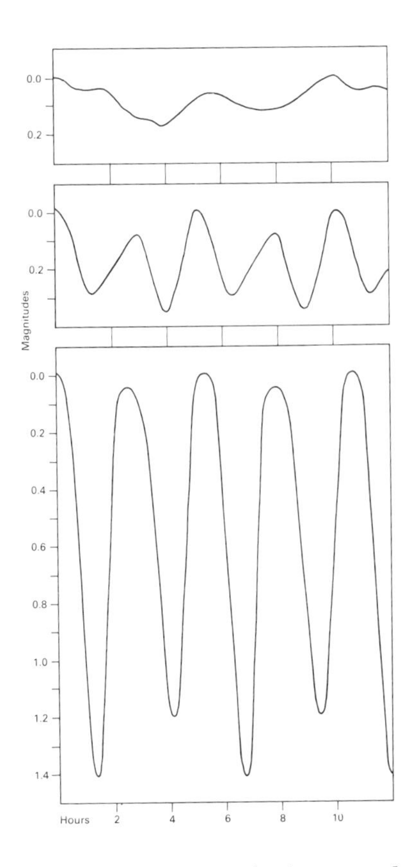

Light curves of Luisa (top), Antigone and Eros (bottom)

Most of the asteroids reside between Mars and Jupiter, but some range more widely in elliptical orbits. Asteroids are known which pass closer to the Sun than Mercury, while others exceed the distance of Saturn. A small number is trapped by the combined gravitational fields of the Sun and Jupiter, and have settled into two stable positions exactly 60° ahead of or behind the planet in the same orbit. These asteroids are usually referred to as the Trojan group, and have been given the names of Trojan warriors.

A popular theory of a few decades ago held that a planet once orbited the Sun at about the average distance of the asteroids, and either exploded or was somehow destroyed; the fragments of this hypothetical body formed the multitude of asteroids. Such a body need not have been large: the estimated total mass of the asteroids is 1/2500 that of the Earth.

Theories of the formation of the solar system have been improving of late. It is currently fashionable to believe that the asteroids condensed more or less in their present configuration from the gas which surrounded the Sun after its formation rather than first coalescing to form a planet. Recent observations indicate that there are at least two distinct chemical populations of asteroids, which are distinguished also by their distance from the Sun. Since different minerals would be expected to have condensed from the gas at different distances from the young Sun, these observations favor the current theories.

The new observations are measurements of the diameters of the asteroids. About the turn of the century, attempts were made to measure directly the disks of the four brightest asteroids, Ceres, Pallas, Juno and Vesta. No other asteroids were satisfactorily measured. Asteroid diameters are now determined by two techniques, one involving the infrared emission and the other the polarization of reflected sunlight. Both methods agree satisfactorily, and indicate that the old optical measurements were too small by 30-40%. Ceres has been shown to be the largest, with a diameter of 1000 km.

From a knowledge of the diameter and apparent brightness, it is a simple matter to calculate the albedo of an asteroid, that is, the fraction of incident sunlight reflected. It is found that two distinct types occur. The majority of asteroids are very dark, and some are the darkest known objects in the solar system. These are believed to have surfaces rich in carbon, and are known as carbonaceous. A smaller proportion have high albedos, reflecting up to 40% of incident sunlight. These generally lie nearer to the Sun and are thought to have surfaces rich in silicate minerals, as do the inner planets.

Most of the asteroids are only a few kilometers across, and will be recorded only as points of light by any terrestrial telescope. Although we cannot see any features on their surfaces, we can determine their rotation periods and learn something of their shapes. This is accomplished by measuring their brightness variations.

The light curves show how the brightness of typical asteroids varies with time. In most cases the variations are small and must be measured carefully. Since the asteroids shine only by reflecting sunlight, the brightness variations can be produced only by changes of the reflectance of the asteroid.

The reflectance is determined by two factors: the apparent size of the body and its albedo. The albedo can obviously change if the asteroid has dark and light patches, like the Moon. Surprisingly, the brightness variations in asteroids seem mainly to be produced by changes in apparent size rather than by portions of different albedo rotating into view. This does not mean that asteroids physically change size, for they are solid lumps of rock. Rather, they are not spherical objects. As an irregular object rotates, its projected size can change, just as this book appears much smaller when viewed on your bookshelf from its spine than when laid flat on a table.

Computers can be programmed to reproduce the observed light curves by calculating the brightness variations of spinning shapes. In this way it can be demonstrated that some asteroids are long and thin whereas others are almost perfectly spherical. Generally, the smaller asteroids are more irregular. Large asteroids are not exclusively regular shapes, however. Vesta, the fourth asteroid to be discovered and one of the largest, is shaped like an American football and has, in addition, a dimple near its southern end. This dimple may be a large crater.

Most asteroids rotate several times in an Earth day. The commonest values for rotation periods are about six hours. Some spin in only two hours: these are generally very small. The longest period found to date is 32 hours, for asteroid number 654, Zelinda.

For Ceres and a couple of the other large asteroids, it is possible to calculate the mass. This is accomplished by careful analysis of the motions of some smaller asteroids, and hence of the gravitational effect of Ceres on them. These calculations lead to a mass for Ceres of 0.0002 times that of the Earth, and, using the new diameter, to a density of 2.3 times that of water. This can be compared to the density of the Earth, which is 5.5, and the Moon, 3.3.

Chiron

Farthest asteroid

On 1 November 1977, Palomar Observatory astronomer Charles Kowal discovered the object which is now named Chiron. This photograph reproduces the plate on which he identified the object: Chiron is arrowed. During the exposure Chiron moved very much less than the asteroids also seen on the photograph. This immediately suggested to Kowal that the object lies farther from the Sun than most asteroids.

Subsequent measurements have confirmed this suggestion. Chiron currently lies between the orbits of Saturn and Uranus, and spends most of its time there. Only for a few years every five decades does it move to within the orbit of Saturn. The estimated diameter of Chiron is about 200 km, comparable with many of the asteroids.

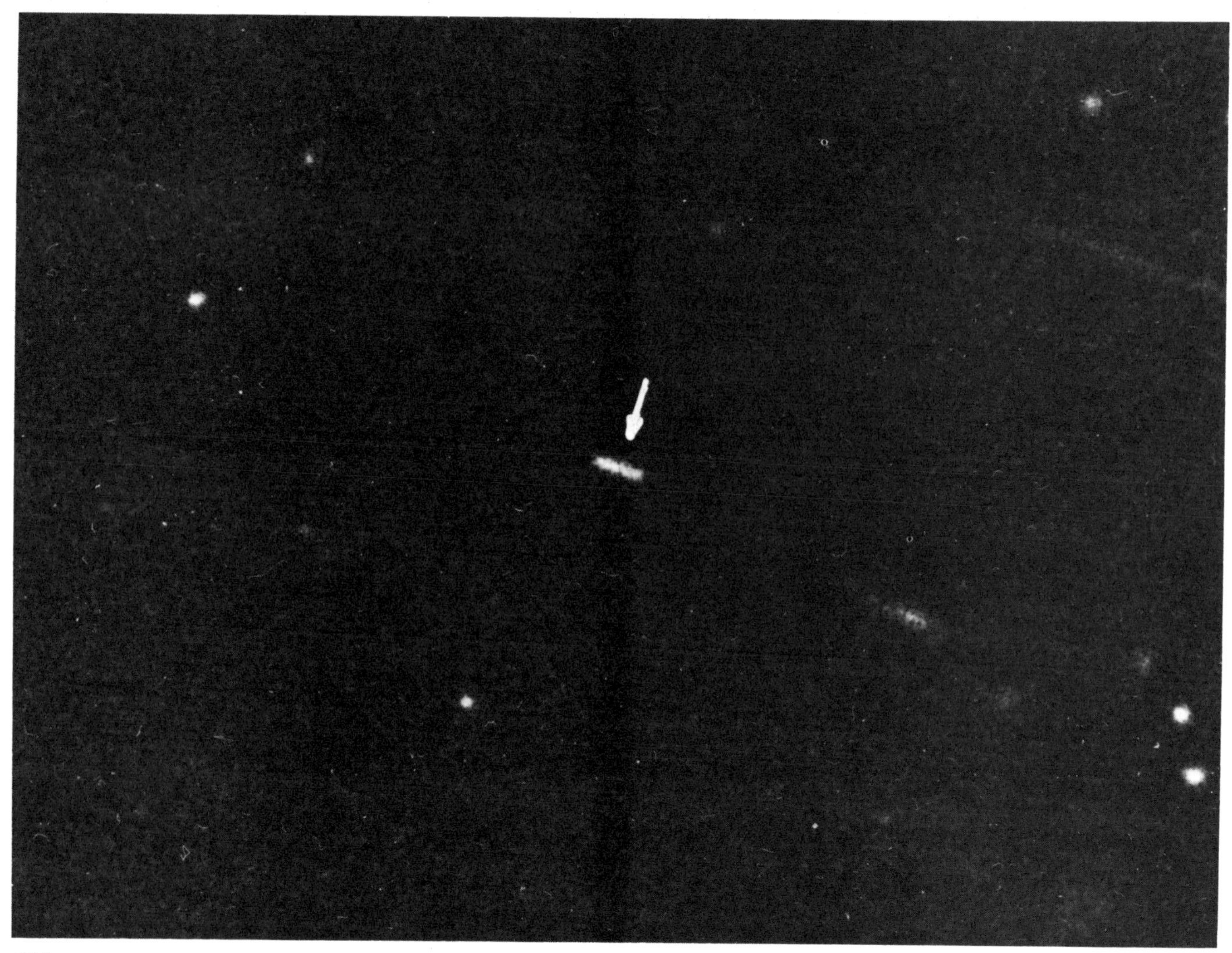

Chiron *Kowal*

Astronomical opinion is currently divided about the significance of Chiron. Some astronomers believe that it is the first to be discovered of a major swarm of asteroids lying between Saturn and Uranus. The alternative suggestion is that Chiron is one of a small group of asteroids which originated between Mars and Jupiter but were subsequently perturbed out to greater distances.

If Chiron was formed near the outer edge of the asteroid zone, it would sooner or later have passed close to Jupiter. Once within 50 million km of Jupiter its orbit would be drastically altered. That this can happen is evidenced by the assortment of captured asteroids which now orbit Jupiter as satellites. The perturbation could have thrown Chiron farther from the Sun, into the region between Jupiter and Saturn. Several asteroids inhabit this region of the solar system. If, by chance, Chiron's new orbit carried it close to Saturn, it could have been perturbed once more into its present orbit. About 1% of asteroids originating between Mars and Jupiter could eventually be thrown out as far as Chiron's present orbit.

By calculating the past motions of Chiron and Saturn, astronomers have found that the second perturbation might have occurred as recently as 1664. In that year Chiron passed within 15 million km of Saturn. If this is the case, Chiron has completed only six circuits of the Sun in its present orbit.

Comets

Captured by the Sun

The unaided human eye can see in the night sky fixed stars, wandering planets, a few bright galaxies, the Milky Way and one or two nebulae. Every few years a completely different type of object appears in the sky: a comet. About once in a lifetime comes a comet which outshines everything but the Sun and Moon, which stretches most of the way across the sky, which is visible in broad daylight for a week or so, and which fades to invisibility a few weeks later.

Comets appear fuzzy and nebulous. They have bright, round heads from which stretch sweeping tails. Despite their appearance they move imperceptibly across the sky, though swiftly compared with planets; sometimes their tails follow, sometimes they lead. The tails do not indicate the direction of the comet's motion: instead they point roughly away from the Sun. Comets come from afar, quickly pass the Sun, and return to distant regions.

Every comet is different. On deep photographs all are beautiful (Plate 32). Our photographs give some idea of the structure and variety they assume. The appearance of every comet changes with time. As it nears the Sun, the comet brightens and its tail lengthens; the opposite happens during its subsequent retreat from the Sun. When near the Sun, the head may disrupt into several components, some of which disperse and fade.

Throughout recorded history man has regarded comets with awe. Their ephemeral nature and great brilliance naturally command his attention: their intangibility ruffles his subconscious. Until quite recently the appearance of a comet in the sky was treated as a portent, ominously prophesying some "disaster" in the lives of men.

Today, astronomers can tell us much of the nature of comets. No longer are these objects regarded with trepidation and misgiving. Yet no one can predict the appearance in the sky of more than one or two percent of bright comets. While astronomers are not entirely in agreement about the origin of comets, the following scheme finds much favor in astronomical circles, and has the advantage of providing an explanation for most of their properties.

In its circuit of the Galaxy, the Sun and its retinue of planets occasionally intercepts a cloud of gas and dust. When this occurs, the Sun plows through the cloud leaving a wake behind it. Because particles in the cloud have been attracted by the Sun, they fall in towards its line of motion. Thus the stream of the wake converges on a point some way behind the Sun. Here the gas and dust grains coalesce to form a tenuous, nascent comet, weighing typically one-hundred-millionth as much as the Earth.

The young comet falls towards the Sun, pulled by the gravitational attraction of that body. It is unlikely to hit the Sun; more reasonably it will assume a very narrow, cigar-shaped orbit which carries it close to the Sun, then back to its starting point. The orbit repeats, with a period perhaps measured in tens of thousands of years. On each orbit the comet will spend only a few months close to the Sun.

When near the Sun, a comet is blasted by the solar radiation. Some of its dust and gas is driven out to form a long tail. Both the original head and the tail shine, partly because the dust grains reflect sunlight and partly because molecules of gaseous carbon and cyanogen are excited, causing them to glow. Comets which pass within the orbit of Mercury generally shine brightly enough to be easily seen by the unaided eye. Two distinct types of tail can often be seen, a smooth dust tail and a knotty, structured gaseous tail.

Because the Sun drives material out of the nucleus and sweeps it away along the tail, a comet cannot survive indefinitely. It is estimated that a few hundred passes would be enough to destroy most comets, giving them a life expectancy of several million years. Within the last few million years, therefore, the theory suggests that the Sun must have moved through a dark gas cloud.

In addition to orbiting the Sun, comets must negotiate the planetary system. If a comet passes close to a planet, particularly Jupiter or Saturn, it will be deflected. Such a deflection could eject the comet completely from the solar system. More likely it would force the comet into a much more circular orbit relatively close to the Sun. Comets are known which orbit wholly within the distance of the outer planets. These are called periodic comets because they pass close to the Sun every few years. In truth, virtually all comets are periodic even though many have been seen to make only one pass of the Sun.

The periodic comets spend all their time comparatively near the Sun, and are subject to its disruptive radiation and gravitational pull. They cannot survive long, perhaps only 1000 years. Virtually all the periodic comets are insignificant objects because they have been driven early into senility. Most can be photographed only by large telescopes. About 100 are known, and many of these have become obviously fainter and more insignificant with each orbit they have been observed to complete.

Comet Arend Roland

A tail towards the Sun

Every amateur astronomer knows that comets' tails generally point away from the Sun. But the comet of 1744 had six tails spread out in a broad fan. Not all of these could have pointed directly away from the Sun, so the popular notion is not entirely true.

The comet which most obviously contradicted the popular notion was Comet Arend Roland, which succeeded in displaying a tail which pointed directly *towards* the Sun. That bald statement must be qualified by admitting that it also had a perfectly normal tail, pointing roughly away fro the Sun. There have been other comets with similar Sun-pointing tails—they have been called "bearded" comets—but Arend Roland was the brightest and best-seen for many years, being observed by millions of people throughout the northern hemisphere in April 1957.

The phenomenon of comet tails can only be understood by regarding them not as solid appendages, but as streams of gas and dust being continually emitted from the comet's nucleus. The entire length of the tail represents material emitted over a period of many days.

Several forces can act on the material, with different effects for gas and for dust. To begin with there is the orbital motion of the comet's nucleus around the Sun. Then there is the velocity of ejection from the nucleus, which can be in the form of definite puffs of material, resulting in knots or separate streams of material. Additionally, any chunks of frozen material emitted by the nucleus will evaporate a stream of gas on their sunward side only, providing a jet action. In the case of the gas emitted from the nucleus, the solar wind plays a part in causing it to stream away from the Sun, perhaps in a curve as a result of the original motion.

Each particle of dust emitted by the nucleus begins its own independent orbit of the Sun, following a path dictated by its initial motion, direction and time of ejection. The particles are subject to various non-gravitational forces such as the pressure of radiation from the Sun. These effects combine to produce tails which are swept out along the line of the orbit of the comet, separate streamers being created by large individual ejections of material. These streamers by no means point directly away from the Sun, because the comet's own motion smears them out.

Complicating these various effects is the angle at which we observe the orbit of any particular comet. We rarely see them face on, as a wide parabola around the Sun. At some point the Earth, in its own orbit, passes through the plane of the comet's orbit. In the case of Arend Roland, this happened when the comet was within the Earth's orbit, and not far from its closest point to the Sun. This gave the clue to the appearance of the Sun-pointing tail, or anti-tail, which appeared

Comet Arend Roland *Lick*

strongest when the Earth was exactly in the plane of the comet's orbit. The hazy, dusty tail is really a broad fan of particles swept out by the various effects mentioned earlier, spread along the comet's orbit. The comet's head is moving away from us, in the direction that the main tail points. Material emitted several days previously is still strewn along the path of the comet, and forms the anti-tail, 56 million km long but only 15,000 km wide.

As we are exactly in the plane of the orbit the concentration of particles is high enough to show it up, just as, in reverse, even a sheet of transparent glass appears black when seen edge on. As the Earth left the plane of the comet's orbit, so the anti-tail broadened to one side and faded.

The photograph shown here was taken by an amateur astronomer, R. L. Waterfield, from his observatory at Ascot, England. He tracked the comet for 36 minutes for this exposure, resulting in the star images drawing out into short streaks.

Halley's Comet

The once-in-a-lifetime visitor

Most famous of all comets is Halley's, on p. 240 shown at its last appearance, in 1910. The blotchy portion of the tail is due to dust grains which had been ejected from the nucleus in a series of hiccups; the long, smooth tail is purely gaseous.

Most comets except Halley's bear the names of their discoverers. But Halley's comet has been "discovered" thousands of times. Instead, it is named in honor of Edmond Halley, Britain's second Astronomer Royal, who demonstrated in 1682 that it was the same comet coming round periodically which precipitated the discoveries. As far as we can tell, Halley's comet has been appearing in our sky every 76 years for several millennia at least. That it was known long before the invention of the telescope indicates that on many occasions it was visible to the unaided eye. The year 1910 was no exception, and this photograph shows it to have been an impressive sight.

Halley's comet is unique in our solar system. It is the only comet with a period of about a century. Comets of much longer period appear so infrequently that they have been recorded in adequate detail only once: their next return cannot be predicted reliably. Comets of significantly shorter period than Halley's spend so much of their time near the Sun that they are wasted by solar radiation and thus become insignificant objects visible only in large telescopes.

The next appearance of Halley's comet will be in 1986. Alas it will be a poor showing. On that occasion, the comet makes its closest approach to the Sun, and thus becomes intrinsically most luminous, only when it lies some distance away from Earth. It will appear brightest in February of 1986, but will require binoculars to be seen at all.

Comet Humason

A different tail

To most people, a comet has a long sweeping tail, stretching across the sky. Halley's comet fits this picture perfectly, as the photograph on the next page shows. But Comet Humason showed a very different type of tail, as this view shows, and the difference between them reveals interesting features about the way in which comet tails form.

The Sun plays a crucial part in the formation of a comet's tail. When the comet is close to the Sun, its nucleus is heated strongly. This results in the production of gas from the ices which make up the nucleus, and also in the release of dust mixed in with the ice. If it were not for the glowing of the gas and the reflectivity of the dust, the nucleus of a comet would be very dim indeed, rather like most asteroids. So most comets can only be seen when they are close to the Sun.

Comet Humason *Hale*

Halley's comet *Lick*

Occasionally, though, there are exceptions, and Comet Humason is one of these. Although it did not come close to the Sun on its only appearance, in 1961-62, its nucleus gave rise to quite large quantities of gas but very little dust. In absolute terms it was several times brighter than Halley's comet, yet it remained fairly faint because of its great distance, beyond the orbit of Mars.

The color photograph shows the result. From the nucleus streams a blue wisp of gas, rather like the trail of smoke from a cigarette in a still room. At such a distance from the Sun, there was very little strength to the solar wind, that stream of particles from the Sun which sweeps most comets' tails out into dramatic streamers. So the tail is weak and unspectacular. Had the comet been in an orbit which brought it closer to the Sun, it would probably have been one of the most spectacular comets in history.

Comet Tritton 1978

Faintest comet

While bright comets, of which Halley's comet is the best known, are spectacular and memorable, they are by no means representative of those which make close passes to the inner solar system. Most comets are faint, insignificant objects, and we know from statistical reasoning that the majority pass undiscovered.

The small percentage that are noted receive very little attention. Mostly they are discovered on routine photographs taken for another purpose. A few more photographs will usually be taken, in order to measure accurate apparent positions from which the orbit can be determined. The orbit, and to a lesser degree the brightness, specify a comet quite precisely, and these data are filed. The files of comet orbits reveal that some comets exist in families. Although originating at very great distances from the Sun, they may approach along essentially the same trajectory. Members of comet families are typically separated in time by a century or so. We are not witnessing a return of the same comet, since the orbital period would be many thousands of years. Indeed, some families appear to have open orbits, and in this case each comet makes only a single pass of the Sun.

This photograph reproduces the discovery plate of comet Tritton. During the course of the 80-minute exposure the comet moved, about 50 arc seconds, producing a fuzzy streak.

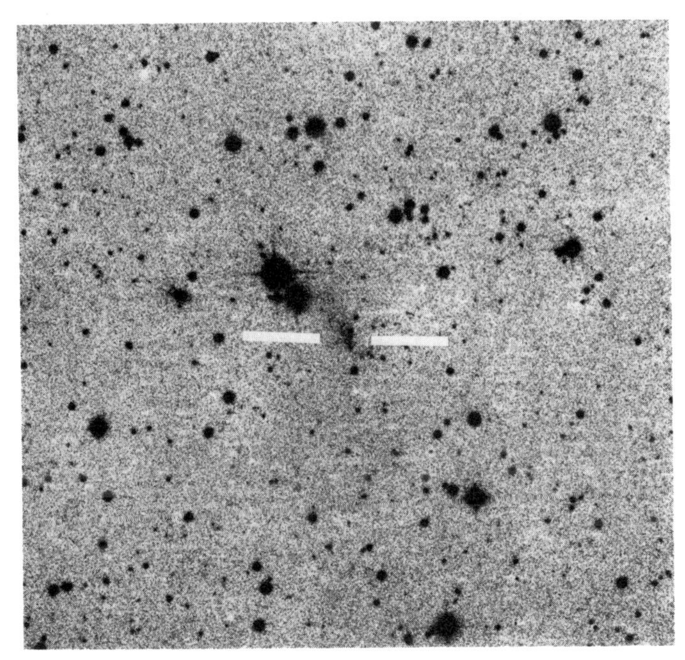

Comet Tritton UKSTU

Fireball

Brightest meteor

The largest bodies orbiting the Sun are the planets, but these represent only the upper end of a distribution of objects. In decreasing order of size, the solar host comprises planets, satellites, asteroids, comets, meteoric particles and dust grains.

Solar system objects under about 1-km diameter cannot be seen directly; these meteoric particles and dust grains, known collectively as meteoroids, are much too faint even when passing close to the Earth. Before the era of artificial satellites, their existence was known only because they occasionally collide with the Earth.

If we inhabited an airless planet, we would experience a continuous rain of these comparatively tiny solar system objects, amounting to several hundred tons per day over the whole surface of the planet. The atmosphere effectively prevents all but the larger specimens from reaching the ground. The dust grains

Daylight fireball *Olsen*

enter the atmosphere at velocities measured in tens of thousands of km/hour. At such speeds the air offers considerable friction, and in traveling through it the particles heat up, become liquefied and evaporate away. Only bodies about a meter or more across stand much chance of reaching the Earth's surface. Those that do are called meteorites. About two per day are thought to land, but this figure may be revised following recent research on meteorites frozen into Antarctic ice.

The molecules of the atmosphere also suffer from the collision. They are heated and ionized so that, like the gaseous nebulae pictured earlier in this book, they glow. Meteors, often mistakenly called shooting stars, are the result of this interaction: the streak of light, or train, is the temporary nebula produced by a tiny grain of dust on its destructive journey.

Several times per year on average a large meteoroid collides with the Earth and penetrates to ground level as a sizeable fragment of rock. Its path through the atmosphere is dramatic. It may have a train brighter than the Full Moon, and nearly as wide, traveling rapidly across the sky. The train may persist, slowly fading to invisibility over a period of several minutes. Sometimes the meteoroid will fragment into many pieces, each generating its own luminous path.

Extremely bright meteors are known as fireballs, or bolides. The photo on p. 241 captured one spectacular example. In the afternoon of 10 August 1972 a fireball, easily bright enough to be seen in daylight, appeared. It moved south over Montana, Wyoming and Utah, traveling 1500 km in 100 seconds, and was witnessed by a great many people. On our photograph can be seen the luminous train left by the meteorite, terminating near the horizon in the bright portion immediately surrounding the meteorite.

Daylight sightings of fireballs are, at best, rare. This example was particularly so, for unlike most of the rock fragments which produce fireballs, it neither burned up in the atmosphere nor fell to Earth. The trajectory on which this fireball approached the Earth was such that it merely skimmed across the very top of the atmosphere and then continued on its way. To this day it is pursuing its journey through the empty spaces of the solar system, merely atrophied by its encounter with our planet.

Even more unusual was the fact that this fireball was also seen from above. An artificial satellite happened to be looking at that particular portion of the planet at the time, measuring the Earth's infrared radiation. Because the meteorite heated up, it glowed strongly at infrared wavelengths, and was therefore recorded by the electronic equipment on board. Using in part the satellite observations, astronomers have been able to estimate the size of the body responsible. They find a diameter in the vicinity of 30 meters.

Before its path was intercepted by the Earth, this body traveled round an elongated orbit, spending most of its time beyond the orbit of Mars. This path must have been grossly altered by the Earth's gravitational field, and insufficient information is available to predict its present orbit around the Sun. Since, however, it appeared in the daylight sky, it must have passed between the Earth and the Sun. At that time, the Earth's gravitational pull was so intense that the fireball swung rapidly round our planet and was sent off much faster than it had arrived, in a direction roughly away from the Sun. Thus its new orbit must be much more elongated than its former.

The Leonid Meteor Shower

Stars fell on Alabama

The fireballs just described are still considerably larger than the majority of solid particles in the solar system. A host of grains of dust and dirt, most no bigger than the letters on this page, orbit the Sun. Their orbits are much less regular than those of the planets: they may follow elongated ellipses, climbing high above the plane of the orbits of the planets, and even orbiting the Sun in the opposite direction to the major bodies.

Because of their disorderly motions, these grains often find themselves on collision courses with the planets, as do the larger objects which produce fireballs. There has been frequent reference, in the pages describing planets, to the erosion produced by impacts of small and large particles. Our Earth, too, encounters this meteoric rain. Moderately large particles produce fireballs as they travel through the air; small particles give rise to the fainter but more frequent metors.

Meteors are streaks of light, some slow, some fast, typically 5° long as seen in the sky. They range in brightness from those comparable to the brightest stars down to those at the limit of the human eye and beyond. At the bright end of the range they become known as fireballs, but the difference between meteors and fireballs is purely semantic. Meteors are also known popularly as shooting stars; however, since they bear no relationship to stars, this is a misleading term which is avoided by astronomers.

On a dark night an observer will see an average of from six to eight meteors per hour. Each will last for about half a second, during which time it will travel roughly 10 km. To produce a 5° track through the sky, therefore, it must be about 100 km distant. This figure was first determined by triangulation, using the sighting of the same meteor from two different places on the Earth's surface. The rate of meteors increases towards dawn because at that time the observer stands

Milon

Leonids

on the leading face of the Earth as it orbits the Sun. Before midnight, meteors must overtake the Earth to collide with it.

At certain times of year the rate of meteors is greatly increased. Moreover, the meteors mostly appear to move across the sky from a fixed point, the radiant. Such events are known as meteor showers. The most reliable meteor shower for northern hemisphere observers is the Perseids. At its maximum, usually around August 12, about one meteor a minute can be seen. The radiant lies in the constellation Perseus.

The particles of a meteor shower orbit the Sun together. In the case of the Perseids, the whole of the orbit is strewn with meteoric bodies, so that once per year, when the Earth plows through it, a shower will occur. All the meteors hit the Earth's atmosphere in the same direction, but the effect of perspective is to make them appear to radiate from a point in the sky. In the same way, when driving through a rain or snow storm, the drops or flakes appear to radiate from a point in front of the car.

Meteors are closely related to comets, for the orbits of many meteor showers are identical with those of old comets. It seems that as a comet dies, its dust grains spread out around its orbit. This suggests that the grains of a young meteor shower may be concentrated within a small portion of the orbit. If so, the Earth would encounter a richer shower once in a while. An example of this is the Leonid meteors, which appear in mid-November. Most years the shower peaks at about ten meteors per hour. But every 33 or 34 years the rate is much higher, for the Earth then encounters the densely crowded nucleus of the dust cloud.

There is always a tendency to downplay old reports of remarkable events. Big displays of Leonid meteors were reported in 1799 (in Peru) and 1833 (in the U.S.A.; the popular song "Stars fell on Alabama" refers to this event). 1866, too, yielded quite a major shower. Astronomers therefore predicted an impressive display in 1900. The shower proved disappointing in that year, as it did in 1933. Many people began to suspect that the earlier reports were exaggerated.

The photo on p.243 is proof that the 19th-century reports were well-founded. On 17 November 1966, for just one hour, the Earth once again intercepted the Leonid shower's heart. Over the southwestern U.S. the shower was seen by many observers, all of whom were presented with the spectacle of their lifetime. Instead of its usual ten meteors per hour, the 1966 Leonid shower put on a display of 40 meteors *per second.* At any instant 20 or 30 meteors were visible.

Photographs record only the very brightest meteors. Nevertheless, in this 3½-minute exposure, Dennis Milon captured 61 meteors. During the exposure, stars trailed a little way across the picture from left to right, producing the parallel trails. The paths of the meteors are clearly seen to radiate from a point near the center of the photograph, and two meteors which approached the camera head-on left bright spots of light very close to the radiant. Eyewitnesses were impressed by the perspective: when facing the radiant, the observer seemed to be traveling through a tunnel, with bright spots on the walls streaming past. More vividly than any experiment, this Leonid display demonstrated the Earth's motion through space.

Hopefully, the 1966 observations will enable astronomers to predict future Leonid displays more precisely. Calculations for 1999 or 2000 have not yet been made. If these predict that the Earth will once again pass through the heart of the Leonid shower, the event will be one of the finest spectacles of this century.

Meteorites

Visitors from space

If we wish to study the Moon or planets, we must mount an expensive expedition to visit them. If we wish to study the small bodies of the solar system, we need do no more than wait. Every year several lumps of solar system debris penetrate the Earth's atmospheric shield and plummet to the ground. Once on the planet's surface they become known as meteorites.

Meteorites come in various shapes and sizes, from pebbles to boulders weighing many tonnes. Larger specimens do, of course, arrive occasionally, but these tend to be dashed to shreds on impact, producing instead craters like the RICHAT FORMATION (p. 188).

Our illustration is of the largest-known meteorite, which was found in 1920 near Grootfontein, in Namibia. It weighs over 60 tons.

Meteorites in general occur in two distinct chemical compositions, stony and iron. Stony meteorites are composed of various rocks and minerals common on Earth, though in somewhat unusual mixtures. For this reason small examples are not easily recognized. Iron meteorites, as their name implies, are composed almost entirely of iron. They are rarer than stony meteorites even though they survive better their passage through the Earth's atmosphere.

Pure iron did not condense within the solar nebula. Iron separates out within bodies large enough to have molten interiors; there is an iron core to the Earth and other of the smaller planets. Iron meteorites must have been formed when a body large enough to have an iron core broke up.

By analyzing the exact chemistry of the known iron meteorites, astronomers have determined that there were at least a dozen parent bodies, all of diameter 200 km or more. These asteroidal objects probably fragmented by colliding with one another. Similarly, the

Grootfontein meteorite *British Museum*

known stony meteorites originated in a dozen or so bodies, most of which were probably the outer mantles to the iron cores.

Tektites

Glassy visitors from space

Meteorites are found in small numbers all over the Earth. Tektites, by contrast, occur in large numbers over restricted parts of the planet. Meteorites are dense rocky or metallic objects which can weigh many kilograms. Tektites are small, glassy objects weighing only a few grams. Although relatively rare, meteorites have been seen to fall; the more numerous tektites have never been witnessed falling. Geological evidence indicates that tektites were laid down on Earth millions of years ago.

This photograph shows a selection of tektites from the field which covers Southeast Asia, Australia and part of the Indian Ocean. The variety of their shapes is evident, and more bizarre forms have been found. The pitting on their surfaces is weathering by the acid soil in which they were discovered; virgin tektites would almost certainly have had a glasslike polish. The smooth portion on the lower tektite is the interior of a large bubble which must have broken quite recently, for it is not pitted. These tektites are black, but some from other fields are olive green or brown.

When the rough exterior of a tektite is polished off the interior appears dark, transparent and normally brown or green. Tektites usually contain bubbles and streaks. Their chemical composition resembles volcanic glass (obsidian) but has unusually high quantities of silica and very little of the alkali metals usually present in volcanic glasses.

The first discoveries of tektites were made in the valley of the Moldau River in Czechoslovakia, and are known as moldavites. They lie a few meters underground. Other fields of tektites occur on the Ivory Coast, in Central Russia, and off the shores of the U.S. Few of this last field have been found. A single specimen was discovered on Martha's Vineyard, an island off Cape Cod, and a few occur in Georgia and Texas. These have been dated 30 million years old. The Australian—Southeast Asian field is the most recent, having fallen about 750,000 years ago, and appears to cover an area about 11,000 km across, much of it ocean.

What are tektites? Whence did they come? These questions have not been adequately answered. Their chemistry is unlike that of any known terrestrial rock,

Tektites DAA

meteorite or the sampled portion of any other planet. Some show evidence of having been molten twice, solidifying in between. The second melting occurred when the tektite passed through the Earth's atmosphere. But the first melting? No one can say when or where that occurred.

Some astronomers have argued that tektites come from the Earth itself, being thrown out by large impacts only to land some distance away, like the rays that surround craters on the other planets. Calculations of the force needed to eject them indicate that a crater 300 km in diameter would be formed by the impact. The absence of suitable craters has thrown this theory into disfavor. A more credible origin is the Moon. A lunar impact could eject a stream of blobs of rock so that they would fall onto a restricted portion of the Earth. A lunar volcanic eruption could also eject material. A cometary origin is preferred by other astronomers who point out that no Earth or Moon rock exactly matches the chemistry of tektites. None of these is entirely satisfactory; until the next tektite fall occurs we may not be able to decide among them.

The Zodiacal Light

Reflection nebula in the solar system

The zodiacal light appears as a faintly luminous, conical area in the western sky one or two hours after sunset, or in the east the same period before dawn. It tapers away from the Sun and is brightest at its base. But for the great brilliance of the Sun, and indeed of the twilight sky, we would be able to see the zodiacal light as an elliptical feature measuring something like 180° by 50° and having the Sun at its center. Instead, we are constrained to observe the very tips of the ellipse. Even this is visible to the unaided eye only against a very dark sky free from artificial lights.

The long axis of the zodiacal light lies along the ecliptic or zodiac, the path of the major solar system bodies across the sky, hence its name. An extremely faint extension, rarely seen, continues around the ecliptic to connect the two ends. This is the zodiacal band. At the point diametrically opposite the Sun, the band broadens into a diffuse patch known as the Gegenschein.

Dust grains lying in or near the plane of the solar system produce the zodiacal light by reflection of sunlight. Individually the microscopic grains are far too faint to record, but their numbers are so great, especially within the Earth's orbit, that their cumulative effect is quite marked. That the dust extends beyond the Earth's orbit is proved by the existence of the zodiacal band. The Gegenschein indicates that these grains reflect light particularly well back along its incident path: it is a giant halo surrounding the Earth's unseen shadow, akin to the patch of light that circles the head of an observer's shadow when seen against wet grass. There is no darkening of the center of the Gegenschein: this indicates that it is produced at such a distance that the Earth's shadow is very large and diffuse. The dust which produces the Gegenschein lies in the asteroid belt, out to 3.3 astronomical units or 500 million km from the Sun.

The zodiacal light is most easily seen from near or within the tropics. At greater latitudes in the northern hemisphere it is best observed on February evenings or before dawn in October. From the southern hemisphere the respective dates are August and April.

The zodiacal light is a reflection nebula around the Sun. Seen in this context it is a particularly meager example. Its extent and brightness are a very large factor down on the smallest and faintest reflection nebulae we can see surrounding other stars. Because of its proximity to us, however, it is the only reflection nebula in the sky visible to the unaided eye.

Zodiacal light *Blackwell*

Appendix: The Messier Catalogue

Messier number	Name	NGC number	Object type	Constellation	Page*
1	Crab Nebula	1952	supernova remnant	Taurus	150
2		7089	globular cluster	Aquarius	
3		5272	globular cluster	Canes Venatici	
4		6121	globular cluster	Scorpius	140
5		5904	globular cluster	Serpens	
6		6405	open cluster	Scorpius	
7		6475	open cluster	Scorpius	
8	Lagoon Nebula	6523	nebula	Sagittarius	108
9		6333	globular cluster	Ophiuchus	
10		6254	globular cluster	Ophiuchus	
11		6705	open cluster	Scutum	
12		6218	globular cluster	Ophiuchus	
13	Great Globular Cluster	6205	globular cluster	Hercules	137
14		6402	globular cluster	Ophiuchus	
15		7078	globular cluster	Pegasus	138
16		6611	open cluster	Serpens	
17	Omega Nebula	6618	nebula	Sagittarius	114
18		6613	open cluster	Sagittarius	
19		6273	globular cluster	Ophiuchus	
20	Trifid Nebula	6514	nebula	Sagittarius	110
21		6531	open cluster	Sagittarius	
22		6656	globular cluster	Sagittarius	
23		6494	open cluster	Sagittarius	
24		6603	star cloud	Sagittarius	76
25		I4725	open cluster	Sagittarius	
26		6694	open cluster	Scutum	
27	Dumbbell Nebula	6853	planetary nebula	Vulpecula	144
28		6626	globular cluster	Sagittarius	
29		6913	open cluster	Cygnus	
30		7099	globular cluster	Capricorn	
31	Andromeda Galaxy	224	spiral galaxy	Andromeda	52
32		221	elliptical galaxy	Andromeda	52
33		598	spiral galaxy	Triangulum	54
34		1039	open cluster	Perseus	
35		2168	open cluster	Gemini	
36		1960	open cluster	Auriga	
37		2099	open cluster	Auriga	
38		1912	open cluster	Auriga	
39		7092	open cluster	Cygnus	
40		—	two stars	Ursa Major	
41		2287	open cluster	Canis Major	
42	Orion Nebula	1976	nebula	Orion	102
43		1982	nebula	Orion	102
44	Praesepe	2632	open cluster	Cancer	123
45	Pleiades	—	open cluster	Taurus	121
46		2437	open cluster	Puppis	
47		2422	open cluster	Puppis	
48		2548	open cluster	Hydra	
49		4472	elliptical galaxy	Virgo	
50		2323	open cluster	Monoceros	
51	Whirlpool Galaxy	5194	spiral galaxy	Canes Venatici	29
52		7654	open cluster	Cassiopeia	
53		1024	globular cluster	Coma	
54		6715	globular cluster	Sagittarius	

Messier number	Name	NGC number	Object type	Constellation	Page*
55		6809	globular cluster	Sagittarius	
56		6779	globular cluster	Lyra	
57	Ring Nebula	6720	planetary nebula	Lyra	143
58		4579	spiral galaxy	Virgo	
59		4621	elliptical galaxy	Virgo	
60		4649	elliptical galaxy	Virgo	
61		4303	spiral galaxy	Virgo	
62		6266	globular cluster	Ophiuchus	
63		5055	spiral galaxy	Canes Venatici	
64	Blackeye Galaxy	4826	spiral galaxy	Coma	25
65		3623	sprial galaxy	Leo	
66		3627	spiral galaxy	Leo	
67		2682	open cluster	Cancer	
68		4590	globular cluster	Hydra	
69		6637	globular cluster	Sagittarius	
70		6681	globular cluster	Sagittarius	
71		6838	globular cluster	Sagittae	
72		6981	globular cluster	Aquarius	
73		6994	multiple star	Aquarius	75
74		628	spiral galaxy	Pisces	
75		6864	globular cluster	Sagittarius	
76		650	planetary nebula	Perseus	
77		1068	spiral galaxy	Cetus	
78		2068	nebula	Orion	
79		1904	globular cluster	Lepus	
80		6093	globular cluster	Scorpius	
81	Bode's Nebula	3031	spiral galaxy	Ursa Major	17 & 34
82		3034	irregular galaxy	Ursa Major	34
83		5236	spiral galaxy	Hydra	19
84		4374	elliptical galaxy	Virgo	
85		4382	spiral galaxy	Coma	
86		4406	elliptical galaxy	Virgo	
87	Virgo A	4486	elliptical galaxy	Virgo	36
88		4501	elliptical galaxy	Coma	
89		4552	elliptical galaxy	Virgo	
90		4569	spiral galaxy	Virgo	
91		4567	spiral galaxy	Coma	
92		6341	globular cluster	Hercules	
93		2447	open cluster	Puppis	
94		4736	spiral galaxy	Canes Venatici	
95		3351	spiral galaxy	Leo	
96		3368	spiral galaxy	Leo	
97	Owl Nebula	3587	planetary nebula	Ursa Major	144
98		4192	spiral galaxy	Coma	
99		4254	spiral galaxy	Coma	
100		4321	spiral galaxy	Coma	
101	Pinwheel Galaxy	5457	spiral galaxy	Ursa Major	20
102		5866	spiral galaxy	Draco	
102		581	open cluster	Cassiopeia	
104	Sombrero-hat Galaxy	4594	spiral galaxy	Virgo	16
105		3379	elliptical galaxy	Leo	
106		4258	spiral galaxy	Canes Venatici	
107		6171	globular cluster	Ophiuchus	
108		3556	spiral galaxy	Ursa Major	
109		3992	spiral galaxy	Ursa Major	

* Where page numbers appear in the right hand column, the object is to be found in the Catalogue of the Universe.

Note on the special photographic techniques used for some of the illustrations in this book

Those who look up to the sky from a city or suburb can never enjoy the full beauty of the dark night sky. Even the casual observer can see that the sky is not black. Dust particles in the air reflect the city lights as a diffuse glow which masks the glory of the Milky Way or Magellanic Clouds. The problem is one of contrast; the faint stars are still there but the eye cannot separate them from the uniform luminosity of the sky. If we could somehow turn up the contrast of the eye, many more stars would be seen against the man-made brilliance.

Most modern observatories are built in remote places well away from artificial lights. Unfortunately even at the best sites the night sky is never truly black. Energetic particles from the Sun and stars continuously enter our atmosphere and interact with it creating a faint glow. Added to this is sunlight reflected from dust in the solar system, sometimes visible as ZODIACAL LIGHT (p. 246). More diffuse light comes from dust in the Galaxy and from distant stars too faint to be resolved. The enormous light-gathering capacity of large telescopes collects this background illumination together with light from the objects under investigation, making it impossible to take photographs of faint objects without at the same time recording unwanted diffuse illumination on the plate. Exposures are limited not by the available telescope time, but by how much uniform fogging of the plate can be tolerated. In technical terms the signal from the faint object is lost in the "noise" from the sky. The problem is identical to that faced by the stargazing city dweller.

In recent years the Eastman Kodak Company have introduced new types of astronomical emulsions which dramatically improve the discrimination of faint objects against the night sky background. These materials, which combine fine grain with high contrast, have greatly advanced astronomical photography. With the new emulsions a large telescope such as the 3.9-meter Anglo-Australian Telescope is able to photograph objects as faint as the light of a candle burning at the distance of the Moon. Although much fainter than the night sky such an object is still visible on the developed plate as a slight increase in density above that produced by the sky.

A special method of increasing the contrast of processed plates even further has recently been used in the photographic laboratories of the Anglo-Australian Observatory and applied to many of the illustrations credited AAT or UKSTU. This book contains the first of the astronomical photographs which have been enhanced in this way to appear outside professional journals. Examples of the technique can be seen on p. 42 (CENTAURUS A) and p. 41 (FORNAX A) which show faint features previously unsuspected and quite unnoticed on the original plates.

The inherent high contrast of these new emulsions, so useful for the detection of faint objects, can be a problem where prints are needed. The more prominent astronomical objects often appear in prints in which detail in the brighter regions is completely missing. Simple enlarging or copying techniques are inadequate to display the full tonal range of the object hidden in the blackest parts of the negative. A technique, involving the use of a photographic mask, has been developed to overcome this. It is a kind of precise "dodging" often used by photographers. A blurred positive copy of the plate is made and then used in combination with the original to subtract gross density variations from the image. Only fine detail remains which can then be printed in the normal way. Many new pictures using the method are published here for the first time, such as THE LAGOON NEBULA (p. 108) and NGC 253 (p. 38).

Both these methods must be applied to original plates rather than copies. The authors are grateful to the staff of the 1.2-meter U.K. Schmidt Telescope for making many of their excellent plates available to us. The ready availability of Schmidt material and generosity of the many astronomers who allowed us to use plates taken on the 3.9-meter Anglo-Australian Telescope is shown by the large number of new southern hemisphere photographs which appear throughout this book.

Color photos credited AAT or UKSTU were made with Kodak Vericolor II film, pushed to double speed. This results in some loss of red sensitivity.

Acknowledgements

Pictures credited AAT were prepared by Malin from plates obtained by our colleagues using the 3.9-meter Anglo-Australian Telescope near Coonabarabran, New South Wales. Pictures are copyright © Anglo-Australian Telescope Board 1975, 1976, 1977, 1978.

Pictures credited UKSTU were prepared, mostly by Malin, from plates obtained with the 1.2-meter Schmidt Telescope near Coonabarabran operated by the U.K. Schmidt Telescope Unit of the Royal Observatory, Edinburgh. Pictures are copyright© by the Royal Observatory, Edinburgh 1978.

Pictures credited "Hale" were taken by the 200-inch Palomar, 100-inch Mt Wilson and 48-inch Schmidt Telescopes of the Hale Observatories, copyright © California Institute of Technology and the Carnegie Institute of Washington. Reproduced by permission from the Hale Observatories.

Pictures credited KPNO are copyright © Association of Universities for Research in Astronomy Inc, the Kitt Peak National Observatory.

Pictures credited "Lick" are Lick Observatory Photographs.

Pictures credited POSS were prepared from the Palomar Observatory Sky Survey photographed by the 48-inch Schmidt Telescope of the Palomar Observatory © National Geographic Society.

Pictures credited ESO were prepared from the Quick Blue Survey obtained by the 1.0-meter Schmidt Telescope of the European Southern Observatory at La Silla, Chile.

Pictures credited "Lowell" are Lowell Observatory Photographs.

The Lunar Orbiter, Apollo, Mariner, Viking and Pioneer photography of extraterrestrial subjects has been provided by the National Space Science Data Center through the World Data Center A for rockets and satellites.

The Apollo photography of terrestrial subjects is from the National Aeronautics and Space Administration through the EROS Data Center.

Pictures credited CSIRO were obtained by astronomers of the Division of Radiophysics, CSIRO, at their facility at Culgoora NSW.

We are grateful to the following institutions for individual photographs: Lockheed Solar Observatory, American Science and Engineering, Jet Propulsion Lab, Mullard Radio Astronomy Observatory of the University of Cambridge, U.S. Naval Observatory, Westerbork Synthesis Telescope.

We are grateful to individual colleagues who gave us their beautiful pictures, allowed us to reproduce them, or otherwise helped us: Russell Cannon, Bruce Peterson, Ken Elliott, Louise Webster, George Herbig, John Graham, Mark Birkinshaw, R.A. Laing, D.E. Blackwell, Keith Tritton, Sue Tritton, Steve Smerd, Eric Becklin, Dennis Milon, Reed Olsen, Charles T. Kowal, John Brandt, David Carter, Barry Newell, Ed Ney, James L. Anderson, Halton Arp, G. Hunt.

Glossary

absorption lines Missing colors in a spectrum, produced when light passes through a gas. Different atoms in the gas absorb light at particular wavelengths which are a signature of the kind of atom involved. When the spectrum of light is studied by a spectrograph, the absorbed wavelengths appear darker. See *Fraunhofer lines*.

ablation When a meteor or artificial satellite travels very rapidly through air, its leading face may heat to melting point and the molten material is dragged off. This process is called ablation.

angular momentum A measure of rotation. A heavy wheel spinning fast has a large angular momentum, a light wheel spinning slowly has very little angular momentum.

apparent brightness How bright something appears. This is determined both by how much light it emits and by its distance from us (the further the fainter).

astronomical unit The distance between Sun and Earth, 146 million kilometers.

atom Particle of matter consisting of a nucleus orbited by electrons. The number of electrons in a complete atom determines to which element the atom belongs.

billion In U.S. usage (adopted for this book) one thousand millions, 1,000,000,000.

binary star Pair of stars traveling together and orbiting one another.

Bok globule Small spherical opaque cloud of gas and dust named after Dutch-American astronomer Bart Bok.

black hole Very dense object with very large surface gravity. No material object, or even light, can escape from a black hole, so it cannot be seen directly, but its gravitational force can be felt and its effect be observed on matter in the process of falling in to the black hole.

captured satellite A body which now orbits a planet but which formerly traveled freely through space. The planet captured the satellite when the latter passed close by.

common proper motion Pair of stars traveling together through space but not otherwise known to be binary.

constellation Group of stars, usually named after their superficial resemblance in the pattern they make to a mythological creature. The celestial sky is divided into areas named after the constellations, and stars in each area are denoted with a letter or number and the Latin name of the constellation (for example, 36 URSAE MAJORIS, RR CORONAE BOREALIS).

Doppler shift Change of color of light if the source of that light (or the observer) is receding (shift to red) or approaching (shift to blue). This is an analog of the change in pitch of a car engine as it approaches (higher note), passes and recedes (lower note). See also *redshift*.

double star Pair of stars appearing close together in the sky, not necessarily (but often) a binary star.

electromagnetic radiation Light and other radiations of various energies; waves of energy which can travel through a vacuum from their place of origin. The type of radiation is characterized by the wavelength of the waves; for visible light the wavelength is about one 2000th of a millimeter. In order of increasing wavelength the electromagnetic spectrum is: gamma rays, x-rays, ultraviolet radiation, light (violet, blue, green, yellow, orange, red), infrared radiation, radio waves. This is also the order of decreasing energy, so that a much more intense energy source is required to produce x-rays than to produce infrared radiation.

electron Particle of negative electricity, found in orbit around a nucleus in an atom, or freely moving in a hot gas.

element Collections of atoms of one sort (that is with a given number of electrons). Hydrogen (one electron) is the most common element in the Universe, helium (two electrons) the next commonest.

emission lines Colors in a spectrum produced by a hot gas, as when illuminated by very blue stars. The light given off by the gas is a set of particular wavelengths, by which the composition of the gas may be discovered. Thus mercury street lights shine as emission lines, with most of their light given off in the blue and green. Sodium vapor lights on the other hand emit mostly a pair of yellow emission lines.

escape velocity Speed at which a thrown body could leave the surface of a star or planet and not fall back.

field star Stars of our own Galaxy which are unrelated to a celestial object being photographed but which lie along our line of sight to it and therefore appear close to it in the sky.

Fraunhofer lines The most prominent absorption lines in the spectrum of the light from the Sun and other stars.

galactic plane The flat disk in which most of the stars of our Galaxy lie.

gravitational collapse Effect of gravity on a cloud of gas. The cloud becomes smaller and denser and may break up into smaller cloudlets, forming planets, stars, galaxies.

gravity The pull exerted by one object on another. The force of gravity is larger between more massive and closer objects. Most noticeable to us is the pull exerted on us by the Earth because of its size and proximity, but the Moon and Sun also produce clear effects on the height of water in the oceans (tides).
heavy elements To the astronomer these are elements other than hydrogen and helium. Although most of the Earth is made of heavy elements, in the Universe as a whole heavy elements are a trace impurity.
H II region Pronounced "aitch-two region." Gas cloud in which hydrogen has been ionized, that is the hydrogen atoms have lost their electrons (see *ion*).
implosion Sudden collapse of an object when the force supporting it is removed (light bulb or supernova). Contrast *explosion*, a sudden expansion of an object when internal forces become much larger than previously.
infrared radiation Heat waves (see *electromagnetic radiation*).
interstellar dust Tiny, solid grains of dust pervading the Galaxy, especially near the galactic plane.
ion Atom which has lost (or gained) one or more of its orbiting electrons. In astronomical terminology an ion is designated by its chemical symbol and, in Roman numerals, one more than the number of electrons it has lost. Thus H I is hydrogen with no electrons lost, H II is hydrogen with one electron lost.
ionization The process of removing one or more electrons from an atom to make it an ion.
light curve Graphical representation of the changes of brightness of a variable star with time.
light year Distance traveled by light in one year. A little over 9,000 billion (9,000,000,000,000) kilometers.
luminosity Intrinsic brightness. The total amount of light (and/or other radiation) given out by a celestial object (star or galaxy). Does not depend on distance of object (compare *apparent brightness*).
magnetic field A force like gravity, but acting only on certain substances, such as magnetized iron (compass needle) or fast-flowing free electrons (see *synchrotron radiation*).
magnitude Measure of a star's brightness. Brighter stars have smaller magnitudes. A 1st magnitude star is a little over 2.5 times as bright as a 2nd magnitude star.
molecular cloud Collection of gas and dust at very low temperature in space, containing molecules of carbon monoxide, formaldehyde, etc.
molecule Assembly of two or more atoms, only found in cool places such as surface of planets, interstellar space and some of the coolest stars.
nebula Celestial cloud of gas or dust, either glowing brightly (emission nebula), reflecting star light (reflection nebula) or appearing black against a brighter background (dark nebula).
neutral hydrogen Hydrogen gas in which the atoms are intact, having lost no electrons (H I) nor joined together into molecules. Gives rise to radio radiation at a wavelength of 21 centimeters.
neutron star Small, extremely dense star; typically the same mass as the Sun but 20 km in diameter.
nuclear reactions Transformations of one element into another, such as hydrogen to helium inside the Sun.
nucleus Center of an atom. If close enough (for example during a high speed collision in a hot gas) two nuclei can combine in a nuclear reaction.
occultation Hiding of a star by the Moon in its monthly path across the sky.
polarization Property of electromagnetic radiation which describes the direction in which waves vibrate. Light which is polarized in certain directions (for instance, glare polarized horizontally after reflecting from a road) is not transmitted by the vertically polarizing lenses of a pair of sunglasses.
Population I and II stars Two different types of stars distinguished primarily by age but recognized by properties such as color, location in the Galaxy, motion, and chemical composition. Population I stars are blue, confined to the galactic plane, orbit circularly around the Galaxy, and contain heavy elements. Population II stars are red, widespread throughout the Galaxy, have more random motions, and contain fewer heavy elements. Population II stars are older.
proper motion The motion of a nearby star across the sky. In practice the motion is recognized and measured relatively to a pattern of more distant stars whose motions are negligible.
quasar (also **quasi-stellar object, QSO**) Very distant, very luminous object, probably situated within a galaxy and of starlike appearance when photographed.
radial velocity Motion of star or galaxy along the line of sight (radius). Measured by the Doppler effect.
radioactive decay Naturally occurring nuclear process in which the nucleus of one kind of atom spontaneously transforms itself into the nucleus of another kind of atom, releasing energy as it does so.
radio lobes Many radio sources emit their radio radiation in two patches called radio lobes.
radio source Celestial object which emits radio radiation.
reddening Effect of interstellar dust which preferentially removes the blue light from the spectrum of a distant star or galaxy making it redder.
redshift Change in color, due to the Doppler effect, of the spectrum of a receding star or galaxy, in which blue features in its spectrum are moved towards the red, red features towards the infrared and so on.
relativity The Special Theory of Relativity was a theory developed by A. Einstein in which the speed of light figures as the ultimate speed. The theory recognizes the part that the relative position and motion of the observer play in how he sees things. The General Theory of Relativity is Einstein's theory of gravity.
Schmidt telescope Telescope designed to allow large areas of the sky to be photographed.
Seyfert galaxy Galaxy in the center of which a prominent energy source lies.
spectral type Shorthand code to describe a star's surface properties, particularly its temperature.
spectrograph (also **spectroscope**) Instrument which separates light of different colors and spreads them into a spectrum.
spectrum Light from a star or other celestial object spread out so that the different wavelengths are separated along a line. Blue and red lie at opposite ends of a spectrum.
speed of light Speed at which light (and other electromagnetic radiation) travels through space (300,000 km/sec, or 186,000 miles per second).
star Large celestial body like the Sun, shining by electromagnetic radiation and held together by gravity.
supernova Combination explosion and implosion in which a large star generates an enormous amoumt of energy in a very short time, its center collapsing to a neutron star or black hole and its outer layers being blown off into space.
supernova remnant The expanding remains of the ejected part of a star, following its explosion in a supernova.
synchrotron radiation A type of electromagnetic radiation given off mostly at radio wavelengths by electrons spiraling at fast speeds in a magnetic field.
temperature A measure of the energy of the random motion of particles in a gas or of the spread of energy in a spectrum of electromagnetic radiation emitted by the gas. The more energetic the hotter.
ultraviolet radiation Radiation of shorter wavelength than blue or violet light (see electromagnetic spectrum).
variable star Star whose brightness changes with time.

Index

Index